INTRODUCTION TO THERMAL SCIENCES

INTRODUCTION TO THERMAL SCIENCES
THERMODYNAMICS
FLUID DYNAMICS
HEAT TRANSFER

FRANK W. SCHMIDT

ROBERT E. HENDERSON

CARL H. WOLGEMUTH

The Pennsylvania State University

John Wiley & Sons
New York   Chichester   Brisbane   Toronto   Singapore

**To our respective wives
Mary, Annalee, and Lois**

Library of Congress Cataloging in Publication Data:

Schmidt, Frank W., 1929–
    Introduction to thermal sciences.

    Includes index.
    1. Thermodynamics. 2. Fluid dynamics, 3. Heat—
Transmission.   I. Henderson, Robert E. II. Wolgemuth,
Carl H.   III. Title.

TJ265.S39   1984   621.402'1   83-21877
ISBN 0-471-87599-6

Printed in the United States of America

10 9 8 7 6 5 4 3 2 1

PREFACE

This book was written to introduce engineering undergraduates not majoring in mechanical engineering to the thermal sciences—thermodynamics, fluid dynamics, and heat transfer. It is the authors' opinion that the increased emphasis on energy in our society requires that all engineering students have a basic knowledge of the principles of energy: its use, its transfer, and its conversion from one form to another. In this book, an attempt has been made to emphasize the physics describing the fundamental phenomena, while providing a sufficient mathematical description to permit the solution of simple problems in the thermal sciences. This text does not provide in-depth coverage.

The book is intended primarily for use in a one-semester, three-credit course for undergraduates. Prerequisites for this course should include undergraduate physics, chemistry, and mathematics through differential equations and introductory vector calculus. Typical coverage in one semester would include Chapters 1 through 10. The sections on compressible flow and the differential forms of the fluid dynamic equations of motion, Chapter 6, are included to provide a more complete presentation of fluid dynamics. These sections are, however, optional and may be covered at the instructor's discretion.

Chapter 1 serves to identify the thermal sciences and to inform the student of the importance of studying this subject. It also identifies some of the physical concepts that are to be discussed in the later chapters.

Chapters 2 and 3 define the concepts and properties that are encountered in thermodynamics. This includes a discussion of heat, work, equilibrium, and reversible processes. The first and second laws of thermodynamics are presented in Chapter 4 and their application to the analysis of thermodynamic systems is discussed. This includes their application to the analysis of thermodynamic cycles, the definition of entropy and reversibility, and the use of temperature–entropy (T–s) diagrams.

The principles associated with the analysis of a system, a fixed quantity of mass, presented in Chapter 4 are extended in Chapter 5 to the analysis of a control volume that has mass flow across its boundary. A general relationship, the Reynolds transport theorem, is developed, and relates the characteristics of a system and a control volume. The relationships describing the conservation of mass, linear momentum, and energy for a control volume are then developed in a one-dimensional form. Special cases of the linear momentum equation are presented in Chapter 6; these cases lead to Bernoulli's equation and describe fluids at rest or experiencing a constant acceleration.

Chapter 6 also presents a brief, but optional, introduction to compressible flows that emphasizes the significant influence of density changes in flow systems. One-dimensional, isentropic duct flows are presented and serve to emphasize the inter-relationship of fluid dynamics and thermodynamics.

Chapters 7 and 8 deal with the effects of viscosity of fluid flow and heat transfer by convection. Chapter 7 is concerned with external flows. The concepts of a fluid and a thermal boundary layer are introduced. The fluid boundary layer is examined and relationships are developed for estimating the friction drag and pressure drag experienced by a surface or object. The thermal boundary layer is investigated for both forced and natural convection heat transfer alone and the two in combination. Both laminar and turbulent flows are considered. Chapter 8 presents a similar treatment of internal flows and examines the flow and heat transfer in piping systems and heat exchangers.

Chapter 9 discusses heat transfer by the conduction or diffusion process. This includes the equations describing conduction for both one- and two-dimensional problems. The analogy between conduction heat transfer and the flow of electrical current is discussed. Finally, transient heat transfer is presented for several surface geometries; namely the sphere, infinite cylinder, and infinite slab.

The principles of thermal radiation and procedure for the calculation of heat transfer by radiation are presented in Chapter 10. This includes a discussion of black, gray, and real body radiation. The effect of the radiation characteristics and the geometrical orientation of the surfaces on the rate of heat transfer are discussed.

The material in this text uses SI units exclusively. Example problems are presented to emphasize the practical application of the various principles presented. A similar concept has been applied to the homework problems.

We are sincerely grateful to Professor C. Birnie, Jr. for providing many constructive comments and for teaching from the draft notes of this book. Thanks also to all of the students who prepared detailed critiques of these notes. Finally, we wish to express our appreciation to Kathy Ishler, who very ably typed the numerous draft copies of this manuscript.

We would also like to thank the following people, whose comments while reviewing the text were most helpful: Robert D. Fox, Purdue University; D. J. Helmers, Texas Tech University; Frank Incropera, Purdue University; Michael A. Paolino, U. S. Military Academy; Jerald D. Parker, Oklahoma State University; Sriram Somasundaram, Texas A. & M. University; and James Zaiser, Bucknell University.

Frank W. Schmidt
Robert E. Henderson
Carl H. Wolgemuth

CONTENTS

*Optional Section

SYMBOLS

	TITLE	UNITS
A	Area	m²
A_c	Cross-sectional area	m²
a	Acceleration	m/s²
b	Width	m
C	Heat capacity of fluid stream	W/K
C_D	Total drag coefficient	
C_P	Pressure coefficient	
$\overline{C_f}$	Average friction drag coefficient	
C_{fx}	Local friction drag coefficient	
c	Acoustic velocity	m/s
c_p	Constant-pressure specific heat	J/kg·K
c_v	Constant-volume specific heat	J/kg·K
D_F	Viscous (friction) drag force	N
D_h	Hydraulic diameter	m
D_P	Pressure drag force	N
D_T	Total drag force	N
d	Diameter	m
E, e	Energy, specific energy	J, J/kg
E_b	Rate of blackbody radiation energy	W/m²
E_λ	Rate of monochromatic radiation energy	W/m²·μm
$E_{\lambda,b}$	Rate of monochromatic radiation energy of a blackbody	W/m²·μm
F, f	Force, force per unit volume	N, N/m³
$F_{i,j}$	Radiation shape factor	
$F_{[0-\lambda]}$	Fraction of blackbody radiation in a wavelength interval	
f	Friction factor	
G	Irradiation	W/m²
g	Gravitational acceleration	m/s²
H, h	Enthalpy, specific enthalpy	J, J/kg
H_T	Total fluid head	m
$\overline{h}$	Average convection heat transfer coefficient	W/m²·K
h_f	Head loss due to friction	m
h_L	Total head loss	m
h_m	Minor head loss	m
h_o	Stagnation enthalpy	J/kg
h_r	Mean roughness height	m
h_x	Local convection heat transfer coefficient	W/m²·K
I	Irreversibility	J/K
$\dot{I}$	Rate of irreversibility	W/K
i, j, k	Unit vectors in x, y, and w directions	
J	Radiosity	W/m²

	TITLE	UNITS
K	Minor head loss coefficient	
KE	Kinetic energy	J
k	Thermal conductivity	W/m·K
L	Length	m
L_c	Characteristic length	m
M	Mass	kg
$\dot{m}$	Mass flow rate	kg/s
n	Unit normal vector	
n	Polytropic exponent	
P	Pressure	N/m², Pa
P_{CR}	Critical pressure	N/m², Pa
P_o	Stagnation pressure	N/m², Pa
P_T	Total pressure	N/m², Pa
PE	Potential energy	J
Q, q	Heat transfer, heat transfer per unit mass	J, J/kg
$\dot{Q}, \dot{q}$	Rate of heat transfer, rate of heat transfer per unit mass	W, W/kg
$\dot{Q}^*, \dot{q}^*$	Internal heat generation, internal heat generation per unit volume	W, W/m³
$\dot{q}''$	Rate of heat transfer per unit area	W/m²
$\dot{q}_w''$	Uniform rate of heat transfer per unit area at wall	W/m²
$\dot{q}_x''$	Local rate of heat transfer per unit area	W/m²
R, r	Radius	m
R	Specific gas constant	J/kg·K
R	Reaction force	N
R	Equivalent radiation resistance	m⁻²
R	Dimensionless radius	
R_0	Universal gas constant	J/mol·K
R_c	Radius of curvature	m
R_t	Thermal resistance	K/W
S	Shape factor	
S, s	Entropy, specific entropy	J/K, J/kg·K
s, n	Streamline coordinates	
T	Temperature	°C, K
T	Dimensionless temperature	
T_b	Bulk fluid temperature	K
T_{cR}	Critical temperature	K
T_o	Stagnation temperature	K
t	Time	s
U, u	Internal energy, specific internal energy	J, J/kg
U	Free-stream velocity	m/s
U	Overall heat transfer coefficient	W/m²·K
u, v, w	Velocity components in x, y, and z directions	m/s
V, v	Volume, specific volume	m³, m³/kg
$\dot{V}$	Volumetric flow rate	m³/s
V	Velocity	m/s
V	Average or uniform velocity	m/s
W, w	Work, work per unit mass	J, J/kg
$\dot{W}, \dot{w}$	Rate of energy transfer as work, rate of energy transfer as work per unit mass	W, W/kg
W_s	Shaft work	J
X	Dimensionless distance	
x	Quality	

	TITLE	**UNITS**
x, y, z	Spatial coordinates	m
Z	Compressibility factor	
z	Elevation	m

GREEK SYMBOLS

α	Absorptivity	
α	Thermal diffusivity	m²/s
β	Volume coefficient of expansion	K^{-1}
β_R	Coefficient of performance—refrigeration unit	
β_{HP}	Coefficient of performance—heat pump	
γ	Ratio of specific heats	
δ	Fluid boundary layer thickness	m
δ_T	Thermal boundary layer thickness	m
ε	Emissivity	
ε	Heat exchanger effectiveness	
η	Efficiency	
η	Coordinate	
η_{th}	Thermal efficiency	
θ	Angular coordinate	rad
κ	Isothermal compressibility	m²/N
λ	Wavelength	μm
μ	Dynamic viscosity	N·s/m², kg/m·s
ν	Kinematic viscosity	m²/s
ρ	Density	kg/m³
ρ	Reflectivity	
ρ_o	Stagnation density	kg/m³
σ	Total stress	N/m²
τ	Transmissivity	
τ	Shear stress	N/m²
ϕ, Φ, φ	Arbitrary fluid property, specific arbitrary fluid property	

SCRIPT

$\mathscr{F}$	Fouling factor	m²·K/W
$\mathscr{M}$	Molecular weight	
$\mathscr{P}$	Perimeter	m

SUBSCRIPTS

a	Actual
atm	Atmospheric
b	Blackbody
c	Cold, cold fluid
CR	Critical
CS	Control surface
CV	Control volume
℄	Centerline
d	Discharge, exit
f	Saturated liquid, final
G	Gauge
g	Saturated vapor
grav	Gravitational

	TITLE	UNITS
H	High temperature	
h	Hot, hot fluid	
Irr	Irreversible	
i	Inlet	
L	Low temperature	
lam	Laminar	
max	Maximum	
min	Minimum	
n	Normal	
0	Initial	
pres	Pressure	
Rev	Reversible	
rad	Radiation	
sur	Surface, surroundings	
sys	System	
tur	Turbulent	
vis	Viscous	
w	Wall	
x, y, z	Spatial coordinates	
λ	Wavelength	
∞	Surrounding fluid	

DIMENSIONLESS GROUPS

	Name	Definition
Bi	Biot	hL_c/k
Fo	Fourier	$\alpha t/L_c^2$
Gr	Grashof	$g\beta(T_w - T_\infty)L_c^3/\gamma^2$
M	Mach	V/c
Nu	Nusselt	hL_c/k
Pe	Peclet	$\dfrac{L_c \rho c_p \mathbf{U}}{k}$, RePr
Pr	Prandtl	$c_p\mu/k$
Ra	Rayleigh	$\dfrac{g\rho^2 c_p \beta(T_w - T_\infty)L_c^3}{k\mu}$
Ra*	Modified Rayleigh	$\dfrac{g\rho^2 c_p \beta \dot{q}''L_c^4}{\mu k^2}$
Re	Reynolds	$L_c\rho\mathbf{U}/\mu$
St	Stanton	$h/\rho c_p \mathbf{U}$, Nu/RePr

INTRODUCTION TO
THERMAL SCIENCES

1

Introduction

1.1 INTRODUCTION

There are a number of words in our language which are universally used but whose meanings are not always clearly understood by the users. The word energy is one such word and daily reference to its cost, availability, type, utilization, and conservation can be found in our personal conversations and in the written and visual communication media. The word energy is commonly misused and misunderstood. The scientific and engineering communities, however, must be quite exact in their usage of the word and distinctions must be clearly made when discussing the various forms of energy.

Acceptable definitions for energy are numerous and one would not be surprised to find the number of definitions of such a basic quantity to be nearly equal to the number of definers. We will not attempt to define energy at this time. The inability to agree upon a universally accepted definition of energy should not deter the student from a study of this very important area. The discussion in this book will be based upon and consistent with all the accepted definitions of energy and the concepts presented are universal and founded on well-documented physical phenomena.

Since energy takes a number of different forms it is usually preceded by an adjective. Electrical energy, nuclear energy, chemical energy, and solar energy are examples of terms with which we are all familiar. There is a certain awareness that we use energy to do work but the direct relationship between work and energy is not clearly appreciated, particularly in the nontechnical community. This causes further confusion in energy related discussions.

A description of the various forms of energy and a study of processes that convert energy from one form to another are presented in this book. Particular attention will be given to limits imposed on the conversion processes. Energy transport processes will also be discussed. These involve a number of different processes ranging from the movement of an energy transporting fluid to the transfer of thermal energy due to a temperature difference. After a clear understanding of the phenomena associated with these processes is obtained, the engineer will be more adequately prepared to practice his or her profession in an energy-concious society.

1.2 THERMAL ENERGY SCIENCES

This book will be restricted to a discussion of the basic principles of the thermal energy sciences. These are usually referred to as thermodynamics, fluid dynamics, and heat transfer. We can define these three sciences more specifically as

Thermodynamics. The science that encompasses the study of energy transformation and the relationship among the various physical quantities of a substance which are affected by or cause these transformations.

Fluid dynamics. The science that deals with the transportation of energy and the resistance to motion associated with flowing fluids.

Heat transfer. The science that describes the transfer of a specific form of energy as a result of the existence of a temperature difference.

The three thermal sciences are closely interrelated. The most basic science is thermodynamics, which together with the laws of dynamics, provides the knowledge upon which we develop the relationships used in the study of fluid dynamics and heat transfer. In many respects thermodynamics is more conceptual than the other two thermal sciences. In a thermodynamics analysis a minimum amount of thought is given to the actual mechanism used to transport the fluid from one location to another or to the design of a device to transform energy from one form to another by a specific thermodynamic process. As an example, the performance of a refrigeration cycle used in a domestic freezer depends not only upon the operating conditions selected through a thermodynamic analysis of the cycle but upon one's ability to design the components of the cycle needed to achieve the specific operating conditions. The design of the condenser, evaporator, compressor, and control valves are based on the principles of heat transfer and fluid dynamics.

The engineer who has a knowledge of the basic fundamentals of thermodynamics, fluid dynamics, and heat transfer is thus in a unique position to handle energy-related problems. The electrical, industrial, civil, and architectural engineer frequently is forced to make design decisions based upon factors associated with the thermal sciences. In most practical situations, all aspects of thermal sciences are involved. It is a rare occasion when a decision involving the thermal design aspect of a device can be made on the basis of only one of the thermal sciences. For this reason the basic fundamentals of thermodynamics, fluid dynamics, and heat transfer are presented in a unified fashion.

1.3 BASIC PRINCIPLES

Before a detailed study is made of the individual thermal sciences a brief summary of the basic principles associated with each of the sciences will be presented to allow the reader to form an overview of these areas.

1.3.1 Thermodynamics

The science of thermodynamics involves the study of the energy associated with a certain amount of material or a clearly defined volume in space. The fixed amount of material is called the thermodynamic system while the clearly defined volume in space is called the control volume. Initially we will confine our attention only to the thermodynamic system.

The study of the energy of a thermodynamic system is really quite simple in concept. Energy can enter or leave, be transferred to or from, a system in only two ways, either as heat or as work. If the energy transfer is due to a difference in temperature between the system and its surroundings, the energy transfers as heat, otherwise it transfers as work. Heat transfer will be studied in the latter portion of this text. The word transfer is redundant and is used only to emphasize that heat is the energy being transferred. In the early chapters we will generally assume that the rate of heat transfer is given information. An exception to this would be the computation of a rate of heat transfer required to produce a given change in a system. For example, we may wish to know at what rate heat must be removed from a system to produce a given rate of temperature decrease. This calculation involves only energy considerations and we would not determine precisely how this rate of transfer would physically take place. The chapters on heat transfer address that problem.

Energy transfer as work will be studied in more detail than heat transfer in the early chapters. There are no subsequent chapters on "work transfer." Since work is energy transferred across a system boundary due to some driving potential other than temperature, there are many types or modes of work. There is mechanical work where a force acts through a displacement of the boundary of the system such as the piston of an internal combustion engine. There is electrical work where an electrical potential acts over an electrical charge at the boundary of the system. Although other forms of work exist, we will confine our attention in the text to these two, with the greater emphasis on mechanical work. Many systems involve mechanical work since whenever a force acts on a moving system boundary, mechanical work is involved.

The first law of thermodynamics is really a statement of the conservation of energy. Intuitively, you would expect the algebraic sum of all energy crossing the boundary of the system to equal the net change of energy stored within the system. Since heat and work are the only forms of energy crossing the boundary, the algebraic sum of heat and work must equal the net change of energy stored within, or possessed by, the system. The energy possessed by the system may be kinetic energy, potential energy and internal energy. From a study of basic physics you will recall that the kinetic energy is computed as

$$KE = \frac{M\mathbf{V}^2}{2} \tag{1-1}$$

where M is the system mass and $\mathbf{V}$ is the velocity of the system. In the earth's gravitational field the potential energy is given by

$$PE = M\mathbf{g}z \tag{1-2}$$

where $\mathbf{g}$ is the acceleration due to gravity and z is the elevation of the system above some datum level. At sea level on the earth the standard value for $\mathbf{g}$ is 9.807 m/s². The acceleration $\mathbf{g}$ is a vector, having both a magnitude and a direction that is always toward the center of the earth; however, we will more frequently be concerned only with the magnitude, since the direction is understood, and the symbol g will be used.

To evaluate the energy stored within the system we must know something about the behavior of the material and the relationship between certain properties of the material. These property relationships are sometimes in the form of algebraic equations and sometimes in the form of tabulated data. In general the system will undergo changes with time, thus its properties will change with time. The change in properties over a specified period of time must be determined so that the change in energy stored in the system can be computed. For this reason, properties of pure substances are described in Chapter 3 and provide the methods required to evaluate changes in system energy.

Certain changes in properties of a substance proceed in one direction only. This natural direction is given by the second law of thermodynamics. If a block slides with uniform velocity down an inclined plane in a gravitational field, the decrease in potential energy is dissipated by friction between the block and the plane. Even if we assume this frictional energy is somehow stored in the block, or the plane, there is no way we can use it to return the block to its original position on the plane. Thus, there is a natural direction for this process and the second law will tell us that the reverse direction is impossible. Perhaps more significantly, the second law tells us that work can be completely and continuously converted into heat, but that the reverse process is impossible. Whenever a system involves the continuous conversion of heat into work only a portion of the heat supplied to the system can be converted into work; the remainder must be rejected. We will see that there is a theoretical limit to the fraction of the heat supplied which can be converted into work for a continuous process. This limit is independent of the properties of the substance or the type of process or machinery being used. Solid-state devices, reciprocating machinery, rotating machinery, and the like, all have the same theoretical limit.

The second law is also useful in providing a measure of the departure of an actual process from an ideal process, one that is reversible. This measure enables comparisons between actual processes and is useful in selecting the most efficient process.

The concepts developed for the thermodynamic system are extended to a control volume in Chapter 5. This broadens the applicability of the laws of thermodynamics to additional problems of interest. The science of fluid dynamics is introduced since we can now consider the transportation of material through a volume or region in space. When devices operate continuously over a period of time, we frequently analyze the process using a control volume where the conditions do not change with time. Such a process is called a steady-state, steady-flow process (abbreviated SSSF) and requires that

(a) The properties of the mass at any location in the control volume do not change with time.

(b) The properties and flow rate of the mass entering and leaving the control volume do not change with time.

Throughout the study of thermodynamics these concepts and definitions play an important role in successfully understanding and applying the basic principles. The student is urged to master these concepts and definitions early so that confusion will be minimized in subsequent sections of the course.

1.3.2 Fluid Dynamics

Once a source of energy has been identified, the useful employment of this energy usually requires that it be transported from one spatial location to another. For example, a hot water or hot air heating system creates a source of thermal energy by the combustion of oil or gas in, say, the basement of a building. To provide the remainder of the building with heat, the energy must be transported from the basement throughout the building. This is accomplished by transferring the energy to a working fluid, water, or air, and then moving, or pumping, the fluid throughout the building where the energy is then removed from the fluid. The study of the movement of the fluid is called fluid dynamics.

A fluid is a substance that deforms continuously when it experiences a shearing or tangential stress, that is, it flows. A solid on the other hand resists a shearing stress by undergoing an initial deformation, but does not continue to deform. The differences between these two substances can be observed by considering the effect of rubbing one's hand over the surface of a table and the surface of water in a sink. Fluids exist as a liquid (water, gasoline, crude oil); as a gas (air, hydrogen, natural gas); or as a combination of liquid and gas (wet steam, nasal spray).

While a fluid provides a means for the transportation of energy, this transportation process itself requires an expenditure of energy. For example, electrical energy is required to drive the electric motor/pump system in the heating system discussed above. This electrical energy is required to overcome the forces that act on the fluid and oppose its motion. It is very important that an engineer understand the generation of the forces that oppose the fluid motion and how to predict their magnitudes and directions in order to (1) design the solid structural surfaces containing the fluid and (2) minimize the amount of energy required to transport the fluid between two locations.

This text will emphasize these two points in the chapters dealing with the science of fluid dynamics. In order to do this the concept of a control volume, a clearly defined boundary in space through which the fluid passes, will be used rather than the concept of a system of a fixed amount of fluid. Mathematical relationships developed in Chapter 5 which describe the flow of a fluid through a control volume will allow the determination of

- The drag (resistance to motion) of a car, ship, airplane, or train.
- The wind loads on a building.
- The power required to pump fluids between different locations.
- The force on the nozzle of a fire hose.
- The effect of dimples on the flight of a golf ball.

The development of these relationships will rely on the use of mathematical models of the fluid dynamics. In cases where these models are too complex or have not yet been developed, empirical descriptions based on well-controlled experiments will be presented. It should be noted that the study of fluid dynamics relies very heavily on experimentation.

The special case when the fluid velocity is zero is also encountered. Such cases are termed fluid statics.

1.3.2.A Types of Flows

Since the science of fluid dynamics is concerned with the spatial movement of fluids, the properties of a fluid are, in general, a function of three spatial dimensions and time. This functional dependence on four independent variables makes the general study of fluid dynamics very complex. Less complex flows must therefore be considered to clearly illustrate the basic principles involved.

The different types of flows considered are classified by the characteristics of a number of flow and fluid properties. To define these properties the fluid will be assumed to be a continuum. This assumption means that all fluid properties have a definite value at a given point in space and time and are identified using a macroscopic rather than a microscopic approach. Therefore, molecular variations are not important and the molecular spacing is small compared to the volume of fluid being considered. These properties can be grouped as flow field, transport, and fluid properties. The individual properties in each of these groups will be discussed below.

Flow Field Properties.   The flow field is a representation of the motion of the fluid through space at different instances of time. The property that represents the flow field is the velocity $\mathbf{V}\{x, y, z, t\}$. Note that the velocity is a vector quantity and has components in the x, y, and z directions.

A visual representation of a flow field is obtained by introducing a tracer material into the flow and photographing the resulting flow field. Examples of such tracers are colored dyes in water and smoke in air. Such photographs, Fig. 1–1, depict the flow *streamlines*, defined as a continuous line which is tangent to the velocity vectors throughout the flow field at an instant of time. By this definition, there is no flow crossing a streamline. Therefore the solid boundary or wall that encloses a flow is also a streamline.

When one observes the path of a given fluid particle as a function of time the particle *pathline* can be determined. In steady flows, the streamline and pathlines are coincident. If the flow is a function of time, unsteady, the streamlines and pathlines are different.

If the velocity of a flow is known, $\mathbf{V} = \mathbf{i}u + \mathbf{j}v + \mathbf{k}w$, the acceleration of the fluid particles, $\mathbf{a}$, can be determined as the total change of velocity with respect to time.

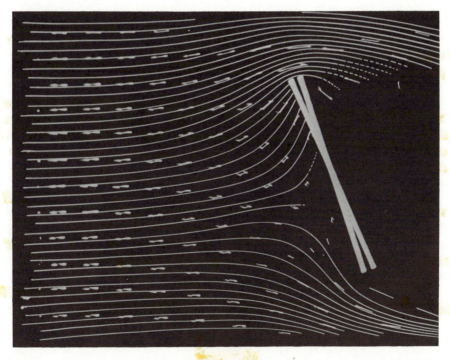

Figure 1-1   Instantaneous streamlines past an oscillating flat plate. (From *Illustrated Experiments in Fluid Mechanics*, National Committee for Fluid Mechanics Films, Educational Development Center, Inc.) Used with permission.

$$\mathbf{a} = \frac{D\mathbf{V}}{Dt} = \frac{\partial \mathbf{V}}{\partial t} + \frac{\partial \mathbf{V}}{\partial x}\frac{dx}{dt} + \frac{\partial \mathbf{V}}{\partial y}\frac{dy}{dt} + \frac{\partial \mathbf{V}}{\partial z}\frac{dz}{dt}$$

$$= \underbrace{\frac{\partial \mathbf{V}}{\partial t}}_{\substack{\text{local}\\\text{acceleration}}} + \underbrace{\mathbf{u}\frac{\partial \mathbf{V}}{\partial x} + \mathbf{v}\frac{\partial \mathbf{V}}{\partial y} + \mathbf{w}\frac{\partial \mathbf{V}}{\partial z}}_{\substack{\text{convective}\\\text{acceleration}}} \tag{1-3}$$

The notation

$$\frac{D}{Dt} \equiv \frac{\partial}{\partial t} + \mathbf{u}\frac{\partial}{\partial x} + \mathbf{v}\frac{\partial}{\partial y} + \mathbf{w}\frac{\partial}{\partial z} \tag{1-4}$$

is called the substantial or particle derivative. The total acceleration involves both the change in velocity with time, the local acceleration, and the change in velocity due to the fluid's spatial motion, the convective acceleration. If a steady-state, steady-flow (SSSF) is considered the acceleration of the fluid is only due to convective acceleration. An example is the SSSF of a fluid through a duct whose cross-sectional area is decreasing. Although the flow is steady it will accelerate since the duct area and, hence, the fluid velocity is changing.

Transport Properties.   While all fluids deform continuously when acted upon by a shearing stress, the rate of deformation is different for different fluids. The thermophysical property that relates the shear stress and the strain rate associated with the movement of a fluid is the dynamic or absolute viscosity, μ. The shear stress within the fluid will determine the local fluid velocity which is directly related to the momentum of the fluid. The dynamic viscosity can thus be directly associated with the transport of momentum and is classified as a transport property. Other transport properties are the thermal conductivity k, which is associated with the transport of thermal energy, and the diffusivity D, which is associated with the transport of mass. While the dynamic viscosity directly effects the flow of a fluid, the other two have an indirect effect.

A Newtonian fluid (air, water, gasoline) exhibits a linear relationship between the applied shear stress τ and the strain rate in the fluid. It will be shown in Chapter 7 that the fluid strain rate is equal to the change in fluid velocity in the direction normal to the flow. The dynamic viscosity μ of a Newtonian fluid is the constant of proportionality in this linear relationship. Fluids that do not exhibit this linear relationship are called non-Newtonian fluids (silly putty, tar).

If a Newtonian fluid experiences a velocity u in the x direction, the shear stress on the fluid in the x direction at any location in the fluid is (note that τ is a vector)

$$\tau_x = \mu\frac{\partial \mathbf{u}}{\partial y} \tag{1-5}$$

The fluid adjacent to a solid surface or wall experiences what is called the wall shear stress τ_w. The movement of a fluid past a solid surface imposes a special condition at the surface. At the solid surface, the relative velocity between the fluid and the surface must be zero, the no-slip condition. As we move away from the surface, the fluid velocity relative to the surface will increase from zero to a finite value, Fig. 1–2. This results in the velocity gradient $\partial u/\partial y$ in eq. 1–5. The shear stress τ_x acts to resist the motion of the fluid, and is a maximum at the surface where there is no fluid motion relative to the surface. Conversely, if the surface is moved through the fluid, as the movement of a car, train, submarine, or airplane, there is a force, or drag, on the surface which resists the motion. The viscosity of a fluid is thus associated with energy that is not recoverable when a fluid is transported, or with the energy required to move an object through a fluid.

The dynamic coefficients of viscosity of a number of Newtonian fluids are presented in Fig. A–13. Non-Newtonian fluids, ones that do not satisfy eq. 1–5, will not be discussed in this text.

Fluid Properties.   There are several properties or characteristics of a fluid which distinguish it from other fluids, but which are not dependent upon the motion of the fluid. These include:

density, the mass of the fluid per unit volume;

vapor pressure, the pressure at which a liquid boils and is in equilibrium with its own vapor;

surface tension, the molecular attraction in a liquid near a surface or another fluid;

acoustic velocity, the speed at which a pressure or acoustic wave moves through the fluid.

For the discussions in this book the most important of these is the density. The specific volume, which is the reciprocal of the density g is also used.

If the density is constant through the flow field, the flow is said to be incom-

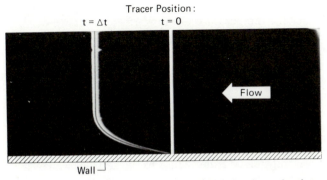

Figure 1-2 Displacement of a vertical marker showing boundary layer development and no slip condition. (From *Illustrated Experiments in Fluid Mechanics*, National Committee for Fluid Mechanics Films, Educational Development Center, Inc.) Used with permission.

pressible. The existence of an incompressible flow greatly simplifies the analysis of the flow field. To determine if a flow is incompressible (ρ = constant) or compressible ($\rho \neq$ constant), the ratio of the magnitude of the velocity, V, to the acoustic velocity, c, in the fluid is examined. If this ratio, called the Mach number, M, is less than 0.3 the flow can be considered to be incompressible. When $M = 1.0$ a critical flow regime is experienced which is referred to the "sound barrier" in the flight of aircraft. If $M > 1.0$ the flow is supersonic. Usually, the flow of a liquid will be incompressible since the acoustic velocities of liquids are large, for example, $c_{WATER} = 1500$ m/s.

1.3.2.B Classification of Fluid Dynamics

A classification of the science of fluid dynamics can be made as a function of the dependency of the flow on time, the velocity of the flow, the fluid viscosity, and the fluid density. Such a classification is made in order to simplify the analysis of fluid dynamic problems since it is possible to study the individual flows separately, for example, compressible and incompressible flows.

In this text only time-independent or steady-state, steady-flows will be considered. This means that we will be considering only time-mean flow properties and characteristics. In most cases we will consider only one- and two-dimensional flows. Such assumptions allow us to obtain very good solutions to many complex flow problems. They also allow us to obtain an understanding of the physical phenomena that is occurring. A great deal of this text is devoted to the consideration of the effects of viscosity. Viscous flows are classified as being either laminar or turbulent depending upon the value of the ratio of the inertia force on the fluid to the fluid viscous force. This ratio is called the Reynolds number and is discussed in Chapters 7 and 8. We will see that the majority of flows are turbulent. Many flows of practical interest can also be considered to be incompressible.

1.3.3 Heat Transfer

We have all observed that when two substances at different temperatures are brought together the temperature of the warmer substance decreases while the temperature of the cooler substance increases. As an illustration of this phenomena recall what happens when a can of soda is removed from the refrigerator and placed on a table. The temperature of the soda starts to increase because of the flow of energy to it from the warmer air surrounding the can. After a period of time, enough energy has been transferred to the soda to enable one to sense by touch that the temperature of the soda has increased. Because of the large amount of air surrounding the can we will probably not be able to sense a decrease in the temperature of the air although intuition tells us that energy has been transferred from the air to the soda. If we continue to observe the soda we will note that its temperature will increase until the temperature of the soda and air are equal. We may thus conclude that if a temperature difference is present energy will be transferred. The physical phenomena involved and the parameters, other than the temperature difference, that govern the rate and amount of energy transfer are not obvious. When the energy transfer is the result of only a temperature difference, and no work is done on or by the substance, the energy transfer is referred to as heat transfer.

The science of heat transfer identifies the factors that influence the rate of energy transfer between solids and fluids or combinations thereof. This information is then used for the prediction of the temperature distribution and the rate of heat transfer in thermodynamic systems.

There are three general categories used to describe the manner in which heat is transferred. These are referred to as conduction, convection, and radiation.

1.3.3.A Conduction

Heat transfer by conduction is the transfer of energy through a substance, a solid or a fluid, as a result of the presence of a temperature gradient within the substance. This process is also referred to as the diffusion of energy or heat. Although the transfer process occurs on the microscopic level, the engineer uses a macroscopic approach to perform engineering calculations of the transfer process.

The basic relationships used for calculating the conduction or diffusion of energy in a substance is Fourier's law. The rate of energy transfer per unit area is called the heat flux and is a vector quantity, $\dot{\mathbf{q}}''$. Fourier's law states that the heat flux is directly proportional to the magnitude of the component of the temperature gradient in the direction of the flux. In a three-dimensional cartesian coordinate system, the temperature in a substance will be a function of position and time, $T(x, y, z, t)$. The mathematical expression for the heat flux vector is

$$\dot{\mathbf{q}}'' \equiv \mathbf{i}\dot{q}_x'' + \mathbf{j}\dot{q}_y'' + \mathbf{k}\dot{q}_z'' \qquad (1\text{–}6)$$

where

$$\dot{q}_x'' \propto \frac{\partial T}{\partial x}, \quad \dot{q}_y'' \propto \frac{\partial T}{\partial y}, \quad \text{and} \quad \dot{q}_z'' \propto \frac{\partial T}{\partial z}$$

Each of the heat flux components can be time dependent. These expressions can be transformed into equalities by introducing the thermal conductivity of the substance, k. For an isotropic material, $k_x = k_y = k_z$, the expressions become

$$\dot{q}_x'' = -k\frac{\partial T}{\partial x}, \quad \dot{q}_y'' = -k\frac{\partial T}{\partial y}, \quad \text{and} \quad \dot{q}_z'' = -k\frac{\partial T}{\partial z} \qquad (1\text{–}7)$$

The sign convention used in writing these relationships assumes the heat flux to be positive if it is in the direction of the coordinate axis. Since heat or energy flows in the direction of decreasing temperature, a negative temperature gradient, we are required to insert a negative sign on the right side of the equation to be consistent with our sign convention.

The thermal conductivity is a thermophysical property of the substance. Good heat conductors, such as most metals, possess large thermal conductivities, while insulating materials possess low thermal conductivities. Values of the thermal conductivity for a number of different materials are given in the appendix.

Frequently the transfer of heat to a solid will cause a portion of it to undergo a change in phase. The solid material may sublimate, change directly to a vapor, or it may melt, change to the liquid state. The reverse situation is also experienced when a liquid or vapor solidifies because of the removal of heat. In both of these processes the rate of heat transfer is governed by conduction.

1.3.3.B   Convection

Heat transfer by convection is the transfer of energy between a fluid and a solid surface by two different phenomena which are the result of the presence of a temperature difference. The first phenomena is the diffusion or conduction of energy through the fluid because of the presence of a temperature gradient within the fluid. The second is the transfer of energy within the fluid which is associated with the movement of the fluid from one thermal environment, temperature field, to another. As we have already noted, diffusion or conduction is a molecular transport phenomena whose rate is controlled by the thermophysical properties of the substance as well as the thermal environment. The second phenomena is associated with the macroscopic characteristics, the movement or flow of the fluid, as well as the thermophysical properties of the fluid and the thermal characteristics of the solid surface.

In convection heat transfer, the temperature difference that causes the energy to flow is that which exists between the surface and a representative location in the fluid. If the surface is immersed in a large body of fluid, the effect of the energy transfer process on the temperature of the fluid will be negligible. This situation is classified as external flow and the driving temperature difference for the transfer of energy is the difference between the free stream temperature of the fluid and the temperature of the surface.

If the fluid is flowing in a duct, the flow is classified as internal flow. In this case, energy will be transferred if a temperature difference exists between the walls of the duct and the bulk temperature of the fluid. The transfer of energy to or from the fluid will cause the temperature of the fluid to change as it moves through the duct. The temperature difference used in the calculation of the heat transfer is that which exists between the surface of the duct and the bulk temperature of the fluid. Both are measured at the same position in the direction of the flow.

The magnitude of the rate of energy transfer by convection, in a direction perpendicular to the surface–fluid interface, $\dot{Q}$, is obtained by use of an expression referred to as Newton's law of cooling,

$$\dot{Q} = hA\Delta T \qquad (1-8)$$

where A is the surface area of the body which is in contact with the fluid, ΔT is the appropriate temperature difference, and h is the convection heat transfer coefficient. The rate of heat transfer is related to the heat flux by $\dot{q}'' = \dot{Q}/A$.

One of the most important tasks of the heat transfer engineer is to accurately predict the magnitude of the convection heat transfer coefficient. Since this quantity is a composite of both microscopic and macroscopic phenomena, many factors must be taken into consideration. A detailed discussion of these factors will be given in Chapters 7 and 8. Some of the items to be considered will, however, be listed now.

Thermophysical properties of the fluid.   The density, thermal conductivity, dynamic viscosity, and specific heat are the major fluid properties affecting the heat transfer coefficient. When the fluid experiences a change in phase during the energy transfer process, boiling or condensation, the latent heat is also important. Generally speaking, the largest heat transfer coefficients are encountered in a flow that experiences a change in phase. In single-phase flow the smallest heat transfer coefficients are present when the fluid is a gas.

Method of fluid movement.   There are two phenomena that will result in the flow of a fluid. One is called forced convection and results when a pump, blower, or other piece of equipment is used to move the fluid. The velocity of the fluid can be controlled to a certain degree by the design of the ducting and the selection of the piece of equipment which is used to force the fluid over the surface or through the duct. The second phenomena is called natural or free convection. The movement of the fluid is entirely governed by the buoyant forces created in the fluid because of density gradients created by the nonuniform temperature field. The magnitudes of these gradients are determined by the rate of energy transfer to the fluid. An example of such a flow is the movement of air and the heat loss from the sides of an electric toaster. When the fluid is flowing at a very low velocity, both forced and free convection effects must be considered. The heat transfer coefficients are much larger in forced convective flows.

Flow characteristics.   As discussed earlier, the value of the ratio of the inertia force on the fluid to the fluid viscous force will determine if the flow is laminar or turbulent. The mixing action present in turbulent flows will enhance the macroscopic transfer of energy and the heat transfer coefficient is much larger.

This list is certainly not complete but it does contain the main factors that must be given consideration when predicting the convection heat transfer coefficient for a given application.

1.3.3.C   Radiation

The transfer of energy by electromagnetic waves is called radiation heat transfer. All matter at temperatures greater than absolute zero will radiate energy. Energy can be transferred by thermal radiation between a gas and solid surface or between two or more surfaces. The transfer of heat from a wood stove to a person standing a meter away is an example of surface-to-surface radiation heat transfer. The heat transfer from the flames of a fire in a room to the walls of the room illustrate a gas-to-surface radiation exchange.

Since radiation heat transfer is a wave phenomena it is possible for it to occur simultaneously with either conduction, if the solid is transparent to a portion of the thermal radiation, or convection. If a body is transferring heat by natural convection, radiation heat transfer must be taken into consideration when calculating the total heat lost or gained by the surface since it can represent a significant amount of the total heat transfer. In most forced convection applications the heat transfer coefficient is of such a magnitude that radiation effects will be comparatively small, unless the temperature of the surface or gas is very large.

Chapter 10 is devoted to a discussion of radiation heat transfer. There are several important factors that must be considered when calculating the rate of heat transfer by radiation. As previously noted any surface will radiate energy if its absolute temperature is greater than zero. The rate of energy emitted by an ideal surface, frequently called a blackbody, is given by the Stefan–Boltzmann law

$$E_b = \sigma T^4 \tag{1–9}$$

T is the absolute temperature and σ is the Stefan–Boltzmann constant. The radiation leaving a real surface will depend upon the surface's radiation characteristic: polished, oxidized, and other factors, and will be less than that leaving a blackbody which is at the same temperature.

The exchange of radiant energy between two or more surfaces or a gas and several surfaces is a very complicated process. Since electromagnetic waves travel in straight lines, the geometrical orientation of the surfaces exchanging radiant energy must be considered. Techniques for taking into consideration the geometrical orientation of surfaces and radiation surface characteristics in the calculation of the rate of heat transfer for simple configurations are presented in Chapter 10.

1.4 UNITS

In this text we will use the International System of Units (SI). We will use five fundamental dimensions or base units. They are length, mass, time, temperature, and electric current. All other units will be derived from this set of five. Table 1–1 lists the fundamental quantity, unit name, and unit symbol. We will consistently use these units and symbols. The only exception will be the frequent use of the Celsius temperature scale. The conversion between Celsius and Kelvin is given by

$$K = {}^{\circ}C + 273.15$$

Table 1–1 SI Units for Fundamental Quantities

Fundamental quantity	Unit name	Unit symbol
Length	Meter	m
Mass	Kilogram	kg
Time	Second	s
Temperature	Kelvin	K
Electric current	Ampere	A

The degree symbol is used with the Celsius scale but it is not used with the Kelvin scale. Note that a temperature change or a temperature difference is the same in both scales; thus a $\Delta T = 1\,°C = 1\,K$. The Kelvin scale corresponds to a thermodynamic temperature scale and is necessary for thermodynamic analysis. The Celsius scale is used here only because it is a commonly used scale.

The derived quantities that will be frequently used in this text are listed in Table 1–2 along with the appropriate symbol and units. From this table it can be seen that 1 newton is a force that accelerates a 1 kilogram mass with an acceleration of 1 meter per second squared. Similarly, 1 pascal is a pressure of 1 newton per meter squared, a very low pressure. Normal atmospheric pressure is 101,325 pascals. It is often convenient to use multiples of the SI Units. Table 1–3 lists the multiplication factor along with the prefix and prefix symbol used.

1.5 REFERENCE BOOKS

As noted in the Preface, it is not the intent of the authors to present an indepth coverage of thermodynamics, fluid dynamics, and heat transfer. It is recognized, however, that the reader may wish to consult other books that discuss specific subjects in more detail. In order to assist the reader in a quest for more detailed information, the following list of books is given. In compiling this list, no attempt was made to include every book published on these topics and thus the list should be considered to be only a partial or representative listing of reference books.

Table 1–2 SI Units for Derived Quantities

Derived quantity	Unit name	Unit symbol	Relationship to other units
Force	Newton	N	$m \cdot kg/s^2$
Pressure or stress	Pascal	Pa	N/m^2
Energy	Joule	J	$N \cdot m$
Power	Watt	W	J/s
Electric charge	Coulomb	C	$A \cdot s$
Electric potential	Volt	V	W/A
Electric resistance	Ohm	Ω	V/A

Table 1–3   SI Unit Prefixes

Multiplication factor	Prefix name	Prefix symbol
10^{-12}	Pico	p
10^{-9}	Nano	n
10^{-6}	Micro	μ
10^{-3}	Milli	m
10^{3}	Kilo	k
10^{6}	Mega	M
10^{9}	Giga	G
10^{12}	Tera	T

1.5.1 Thermodynamics

Hatsopoulos, G. N., and Keenan, J. H., *Principles of General Thermodynamics*, Wiley, New York, 1965.

Holman, J. P., *Thermodynamics*, 3rd ed., McGraw-Hill, New York, 1980.

Morse, P. M., *Thermal Physics*, 2nd ed., Benjamin/Cummings Publishing, Reading, Mass., 1969.

Obert, E. F., and Gaggioli, R. A., *Thermodynamics*, McGraw-Hill, New York, 1963.

Reynolds, W. C., *Thermodynamics*, McGraw-Hill, New York, 1965.

Van Wylen, G. J., and Sonntag, R. E., *Fundamentals of Classical Thermodynamics*, 2nd ed., Revised Printing, SI Version, Wiley, New York, 1978.

Wark, K., *Thermodynamics*, McGraw-Hill, New York, 1983.

Zemansky, M. W., *Heat and Thermodynamics*, 4th ed., McGraw-Hill, New York, 1957.

Zemansky, M. W., and VanNess, H. C. *Basic Engineering Thermodynamics*, McGraw-Hill, New York, 1966.

1.5.2 Fluid Dynamics

General

Fox, R. W., and McDonald, A. T., *Introduction to Fluid Mechanics*, Wiley, New York, 1978.

Rouse, H., and Ince, S. *History of Hydraulics*, Dover, New York, 1963.

White, F. M., *Fluid Mechanics*, McGraw-Hill, New York, 1979.

Viscous Flows

Hoerner, S. F., *Fluid Dynamic Drag*, published by the author, Midland Park, NJ, 1965.

Schlichting, H., *Boundary Layer Theory*, 7th ed., McGraw-Hill, New York, 1979.

White, F. M., *Viscous Fluid Flow*, McGraw-Hill, New York, 1974.

Inviscid Flows

Milne-Thomson, L. M., *Theoretical Hydrodynamics*, 4th ed., Macmillian, New York, 1968.

Robertson, J. M., *Hydrodynamics in Theory and Application*, Prentice-Hall, Englewood Cliffs, NJ, 1965.

Compressible Flows

Leipmann, H. W., and Roshko, A., *Elements of Gasdynamics*, Wiley, New York, 1957.

Shapiro, A. H., *The Dynamics and Thermodynamics of Compressible Fluid Flow*, 2 Vols., Roland, New York, 1953.

Thompson, P. A., *Compressible Fluid Dynamics*, McGraw-Hill, New York, 1972.

Fluid Mechanics Films

National Committee for Fluid Mechanics Films, *Illustrated Experiments in Fluid Mechanics*, M.I.T. Press, Cambridge, Mass., 1972.

1.5.3 Heat Transfer

General

Bird, R. B., Stewart, W. E., and Lightfoot, E. N., *Transport Phenomena*, Wiley, New York, 1966.

Eckert, E. R. G., and Drake, R. M., Jr., *Analysis of Heat and Mass Transfer*, McGraw-Hill, New York, 1972.

Gebhart, B., *Heat Transfer*, 2nd ed., McGraw-Hill, New York, 1971.

Holman, J. P., *Heat Transfer*, 5th ed., McGraw-Hill, New York, 1981.

Incropera, F. P., and DeWitt, D. P., *Fundamentals of Heat Transfer*, Wiley, New York, 1981.

Karlekar, B. V., and Desmond, R. M., *Engineering Heat Transfer*, West, St. Paul, 1977.

Kreith, F., and Black, W. Z., *Basic Heat Transfer*, 4th ed., Harper & Row, New York, 1980.

Rohsenow, W. M., and Hartnett, J. P., eds., *Handbook of Heat Transfer*, McGraw-Hill, New York, 1973.

Conduction

Arpaci, V. S., *Conduction Heat Transfer*, Addison-Wesley, Reading, Mass., 1966.

Carslaw, H. S., and Jagger, J. C., *Conduction of Heat in Solids*, 2nd ed., Oxford University Press, London, 1959.

Kern, D. Q., and Kraus, A. D., *Extended Surface Heat Transfer*, McGraw-Hill, New York, 1972.

Myers, G. E., *Analytical Methods in Conduction Heat Transfer*, McGraw-Hill, New York, 1971.

Ozisik, M. N., *Boundary Value Problems of Heat Conduction*, International Textbook, Scranton, PA, 1968.

Convection

Kays, W. M., and Crawford, M. E., *Convective Heat and Mass Transfer*, 2nd ed., McGraw-Hill, New York, 1980.

Schlichting, H., *Boundary Layer Theory*, 7th ed., McGraw-Hill, New York, 1979.

Shah, R. K., and London, A. L., *Laminar Flow Forced Convection in Ducts*, Academic Press, New York, 1978.

Radiation

Hottel, H. C., and Sarofin, A. F., *Radiative Transfer*, McGraw-Hill, New York, 1967.

Planck, M., *The Theory of Heat Radiation*, Dover, New York, 1959.

Siegel, R., and Howell, J. R., *Thermal Radiation Heat Transfer*, 2nd ed., McGraw-Hill, New York, 1981.

Sparrow, E. M., and Cess, R. D., *Radiation Heat Transfer*, Brooks/Cole, Belmont, CA, 1966.

Numerical

Dusinberre, G. M., *Heat Transfer Calculations by Finite-Difference*, International Textbook, Scranton, PA, 1961.

Patankar, S. V., *Numerical Heat Transfer and Fluid Flow*, McGraw-Hill, New York, 1980.

Heat Exchangers

Fraas, A. P., and Ozisik, M. N., *Heat Exchanger Design*, Wiley, New York, 1965.

Kays, W. M., and London, A. L., *Compact Heat Exchangers*, 2nd ed., McGraw-Hill, New York, 1964.

Schlunder, E. U., et al., *Heat Exchanger Design Handbook*, Hemisphere, Washington, 1982.

2

Thermodynamic Concepts and Definitions

2.1 CLASSICAL THERMODYNAMICS

Classical thermodynamics uses the macroscopic approach to the study of energy transformations as opposed to statistical thermodynamics, which uses a microscopic approach. As the name implies, the *macroscopic* approach utilizes a large number of molecules for study so that average values of properties can be defined and used to describe the behavior of a substance. A *microscopic* approach uses a very small quantity of a substance for study and attempts to describe the behavior of each molecule. The microscopic analysis is used extensively in texts on kinetic theory and statistical mechanics. The approach to be followed in this text will be macroscopic, although microscopic concepts may be introduced occasionally to assist in the understanding of a particular phenomenon. In the macroscopic approach, fluids are treated as continuous rather than made up of a number of individual particles. The macroscopic approach has wide application and can be used to answer many questions in the study of thermodynamics; however, it is not valid in situations where very few molecules are involved or where the behavior of individual molecules is sought.

2.2 THERMODYNAMIC SYSTEM

A thermodynamic *system* is defined as that portion of the mass of the universe chosen for a thermodynamic analysis. The system is separated from the surroundings by a *boundary*. Thus, a system is a fixed mass and all matter is either in the system or in the surroundings. A very important word in the definition of a system is "chosen." The individual doing the analysis can place the system boundary wherever he or she wishes. The only constraint is that the boundary must enclose a mass large enough for a macroscopic analysis. In many cases it is fairly obvious where the boundary should be placed, but in other cases experience is useful in making a wise choice. Frequently the analysis can be made easier by proper location of the system boundary. In this regard, the system in thermodynamics is like the free-body diagram in solid mechanics, where the boundary is chosen to include (or exclude) certain forces. In thermodynamics the system boundary is chosen to include (or exclude) certain energy transfers. In mechanics, the main objective of the analysis is to determine the forces acting on the free body and hence its motion. In thermodynamics the objective is to determine energy transfers and hence the change in the state of the system. It is important to note that only energy crosses the system boundary, and this boundary may be fixed or it may move in space. As you will see later the energy crossing a system boundary is either one of two forms, heat or work. A schematic diagram of a system is shown in Fig. 2–1.

Frequently it is necessary to analyze a device where the flow of a fluid occurs and the fixed mass concept of the thermodynamic system can present some difficulties in its application. To analyze devices where mass flow occurs, it is advantageous to perform the thermodynamic analysis on some region in space. This region, shown schematically in Fig. 2–2, is called a *control volume*. The control volume is separated from the surroundings by a *control surface*, which is analogous to the boundary of a system; however, mass transfer may occur across the control surface. The control volume may move in space and may have its volume change with time. In general, there is no requirement that the volume of a control volume be fixed, although in many cases a stationary, constant volume, control volume can be used.

Thus a system is a fixed quantity of matter and has no mass transfer across its boundary; only energy may flow across a system boundary. On the other hand, both energy and mass may flow across the control surface of a control volume. In Chapters 2, 3, and 4 we will deal only with thermodynamic systems. A control volume analysis will be introduced in Chapter 5, where the laws of

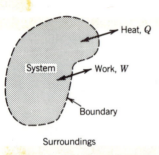

Figure 2-1   The thermodynamic system.

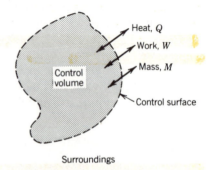

Figure 2-2   The control volume.

classical thermodynamics will be extended to apply to a control volume. A discussion of the location of the control surface is contained in Chapter 5.

Sometimes a system boundary can be chosen where energy does not flow across it. Such a boundary encloses an *isolated* system because there is no communication between the system and its surroundings.

A *homogeneous system* is one in which the mass is uniformly distributed throughout the system volume. If the mass is not uniformly distributed, the system is *heterogeneous*.

2.3 THERMODYNAMIC PROPERTIES

A thermodynamic *property* can be any observable characteristic of a substance. Properties are generally divided into two categories. Extensive properties are those that depend upon the amount of mass in the system and intensive properties are independent of the amount of mass in the system. Pressure and temperature are examples of intensive properties while volume is an example of an extensive property. If 5 g of helium in a balloon constitute system A, and system B is made up of 1 mg of helium in the center of the balloon, the pressure and temperature of systems A and B are identical, while the volume of systems A and B differ.

The thermodynamic *state* of a system is its condition as described by its physical characteristics, that is, its properties. The state can be specified or fixed when the independent properties for that system have been specified. Later we shall learn how to determine the number of independent properties of a system. Not all properties are independent. Specifying the length of one side of a cube fixes all other geometric properties of that cube. Specifying the pressure, temperature, and mass of helium in a balloon may fix the volume of that balloon.

2.4 THE THERMODYNAMIC PROPERTIES: PRESSURE, VOLUME, AND TEMPERATURE

Any system when subjected to external forces will exert a force on its boundary. In fluid systems the force per unit area exerted normal to the boundary area is called the *pressure*, P, and is defined as being positive if directed outward from the boundary. (In solid systems this is usually called a stress.) The pressure at any point in a fluid system in equilibrium is the same in any direction. The thermodynamic property pressure is an absolute pressure. Most pressure measuring devices or gauges, however, measure the difference between the system pressure and the ambient pressure. The measured pressure is called gauge pressure. The ambient pressure must be added to the gauge pressure to convert it to absolute pressure. Thus,

$$P = P_G + P_{ambient} \tag{2–1}$$

In this text the word pressure will always mean absolute pressure and if a gauge pressure is used it will be labeled gauge pressure. When a gauge measures a vacuum (a pressure less than ambient), the gauge pressure is negative.

The *volume* of a system, V, is an extensive thermodynamic property. The volume per unit mass $V/M = v$, the specific volume, is a very useful intensive thermodynamic property. The reciprocal of specific volume is density, designated by the symbol ρ.

$$\rho = \frac{1}{v} \tag{2-2}$$

In thermodynamics the specific volume is the more commonly used property while in fluid mechanics and heat transfer the density is more commonly used.

An absolute thermodynamic temperature scale will be established in Chapter 4 with the aid of the second law of thermodynamics; however, at this time we need a working definition of temperature. The measurement of this property can be achieved with the use of a device we call a thermometer and the zeroth law of thermodynamics. The zeroth law of thermodynamics states: *when any two bodies are in thermal equilibrium with a third they are also in thermal equilibrium with each other*. Thus, if you insert a thermometer (body A) into another system (body B) and wait until some property of body A becomes constant, for example, length (the property could be also volume, pressure, electrical resistance, etc.), you then conclude that bodies A and B are in thermal equilibrium. If you then insert the thermometer (body A) into another system (body C) and measure the same value for the length of body A, you may conclude that body B and body C are in thermal equilibrium, hence are at the same temperature.

While this zeroth law of thermodynamics may seem obvious and unnecessary, it cannot be derived from other laws and for certain kinds of equilibrium (other than thermal) an equivalent relationship does not apply. For example "when a zinc rod and a copper rod are dipped in a solution of zinc sulfate, both rods come to electrical equilibrium with the *solution*. If they are connected by a wire however, it is found that they are not in electrical equilibrium with *each other*, as evidenced by an electric current in the wire."[1]

The thermometer used in the illustration of the zeroth law can then be scribed with any convenient temperature scale. The Fahrenheit and Celsius scales are two commonly used scales. The absolute thermodynamic temperature scale in the SI system of units is the *Kelvin* scale, designated by a K (without the degree symbol). The practical temperature scale is the *Celsius* scale (°C) which is related to the Kelvin scale by

$$K = °C + 273.15° \tag{2-3}$$

On the Celsius scale the triple point (where the solid, liquid, and vapor phases exist together in equilibrium) of water is 0.01 °C and the normal boiling point of water is 100 °C. Note that the size of the Celsius degree is the same as the Kelvin, *thus a change (or difference) in temperature is numerically the same in both scales.* For a more detailed discussion of the Celsius temperature scale see a more complete text on thermodynamics such as *Fundamentals of Classical Thermodynamics* by Gordon J. Van Wylen and Richard E. Sonntag.[2]

2.5 CHANGES IN STATE

A system undergoes a *process* when it changes from one state to another. The series of states through which the system passes in going from its initial state to its final state constitute the *path* for that process. Suppose the air in a bicycle tire leaks very slowly from an initial pressure of 0.4 to 0.1 MPa while the temperature of the air remains constant. Choosing the mass of air inside the tire at the end of the process as the system (See Fig. 2–3), we can say the system experienced a constant-temperature (isothermal) process. The path of the process is shown in Fig. 2–3 on a pressure versus specific volume plot.

A series of processes can be put together such that the system returns to its initial state. This series of processes is called a *thermodynamic cycle*. We will look at cycles in more detail later, but two examples are shown in Fig. 2–4.

2.6 THERMODYNAMIC EQUILIBRIUM

A system is in *thermodynamic equilibrium* when it is incapable of spontaneous change, even if it is subjected to a catalyst or minor disturbance. A stoichiometric mixture of hydrogen and oxygen may constitute a system, but this system is not in thermodynamic equilibrium because it is capable of a spontaneous change when subjected to a catalyst or a small spark.

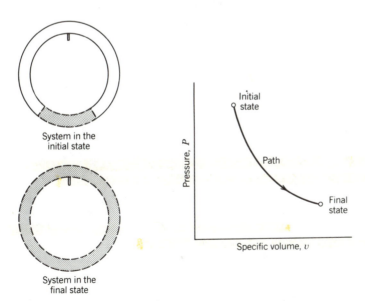

Figure 2-3 Illustration of path.

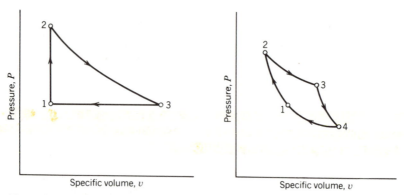

Figure 2-4 Illustration of cycles.

Thermodynamic equilibrium requires that the system be in thermal equilibrium, mechanical equilibrium, and chemical equilibrium. Thermal equilibrium requires that the temperature of the system be uniform, mechanical equilibrium requires that the pressure be uniform throughout, and chemical equilibrium requires that the system be incapable of a spontaneous change in composition.

In order for a system to undergo a process it may be necessary to disturb the system to drive it in the desired direction. Thus the temperature may not be completely uniform if energy is to be transferred across the boundary as heat. The pressure may not be completely uniform if a boundary moves to create a volume change process. When these deviations from equilibrium are very very small, we call the process a *quasiequilibrium* process and assume that the system is in complete thermodynamic equilibrium. Thus the path of the process can be shown on any thermodynamic property diagram since the system is assumed to be in equilibrium and the properties are defined. Processes that occur relatively slowly can be imagined to be quasiequilibrium processes, while those that occur relatively rapidly may deviate significantly from equilibrium. In this latter case the system may be in equilibrium in the initial and final states, but the exact path of the process could not be shown on a property diagram because the properties may not be defined for the system when it is not in equilibrium.

2.7 REVERSIBLE PROCESSES

If a process can be made to reverse itself completely in all details and follows the exact same path it originally followed, then it is said to be *reversible*. If a reversible process occurs and is followed by the reverse process back to its original state, there is no evidence in the system or the surroundings of any process(es) having occurred. A necessary, but not sufficient, condition that a process be reversible is that it be a quasiequilibrium process. If a process occurs very rapidly the system may not be in (or infinitesimally removed from) equilibrium states along the path, hence the process would not be reversible. A simple example illustrating this is a frictionless piston moving in a cylinder that contains a gas. If the piston moves outward (away from the gas) at a velocity equal to the average velocity of the gas molecules, the force exerted on the piston by the gas will be fairly low since few of the molecules will contact the piston and those that do contact the piston will have a low velocity relative to the piston. Much less work is obtained on this very rapid expansion than on a relatively slow expansion. If the piston moves inward (toward the gas) at a velocity equal to the average velocity of the gas molecules as in a compression process, more molecules impact on the piston and the velocity relative to the piston of those hitting the piston is much higher, creating a larger force on the piston and more work of compression than on a relatively slow process. It is clear that the work of compression is much larger than the work of expansion, hence the process is not reversible because of the rapid rate at which the process took place.

Another factor that precludes reversibility is friction. Friction present in any way will render a process irreversible. To be reversible the process must be both a quasiequilibrium process and frictionless. It should be noted that in the real world there is no such thing as a reversible process, but it is a useful concept in thermodynamics for presenting a limiting case. In some cases actual processes can quite closely approach reversibility and a direct comparison between the actual and ideal (reversible) provides a measure of the efficiency of the device. There are other processes that are inherently irreversible, that is, there is no conceivable idealization whereby the process can be thought of as reversible (examples are electrical current flow through a resistor, and fluid flow through a porous medium). A generalization of this concept is that any process where there is a flow across a finite difference in the driving potential, and where less than the maximum theoretical work is produced, is irreversible.

There are other factors besides the two mentioned here (rapid rate and friction) which can cause a process to be irreversible and more will be said about some of these when the second law of thermodynamics is discussed.

2.8 HEAT

Heat is defined as energy in transition not associated with mass transfer and due to a difference in temperature. This is a very specific and precise definition and may be somewhat different from previous definitions where it was related to calorimetry; thus the main features of this definition will be discussed.

The first point in this definition is that heat is energy in transition. It is energy crossing the system boundary or the control surface of a control volume. It is not possessed by the system or control volume, but is energy identified as heat only when it is crossing the boundary or control surface. Heat cannot be stored and must be converted to some other form of energy after crossing the system boundary or control surface. Note that the term energy is not defined. It is a derived quantity such as force and the term is more fully understood when its many forms and transformations are understood, typically from a study of thermodynamics.

The second point in the definition of heat is that it is not accompanied by mass transfer. If an energy transfer occurs across a control surface due to mass transfer, (the mass flowing into or out of the control volume carries some energy with it), that energy transfer is not heat. In Chapter 5 we will account for energy transport due to the mass transport, but at this time we only wish to emphasize that energy transported by mass is not heat.

The third point is that the driving force for the transfer of heat must be a temperature difference. If the flow of energy across the boundary is caused by any driving force other than the temperature difference between the system and surroundings, it is not called heat. As your experience would indicate, and later you will see that the second law of thermodynamics demands, heat flows from the high-temperature region to the low-temperature region.

The requirement that heat is not energy stored or possessed by a system or control volume, means that it is not a property (recall that a property was defined as any observable characteristic of a system). Thus, we do not speak of the heat in a system or the heat of a system; this makes absolutely no sense in view of this definition. The heat transfer to or from a system does require a change in state of that system and the magnitude of the heat transfer is a function of the path the system follows during the process that causes the change of state. Thus we say that heat is a path function (in mathematical terms an inexact differential) and use the symbol δQ to indicate an infinitesimal quantity of heat transfer. This helps distinguish it from a property that is a point function (in mathematical terms an exact differential) and an infinitesimal change is indicated by dY where Y is any property.

When the inexact differential is integrated to get the heat transfer to or from a system undergoing a process from state 1 to state 2, we write

$$\int_1^2 \delta Q = {}_1Q_2 \qquad (2\text{–}4)$$

where the ${}_1Q_2$ represents the total amount of heat transferred when the system changes from state 1 to state 2. Note that we do *not* write the value of this integral as $Q_2 - Q_1$ since Q_1 and Q_2 have no meaning; the system does not possess a value of Q at state 2 or at state 1.

The concept of a path function is not unique to thermodynamics. A simple two-dimensional geometric example may serve to illustrate this fact. The coordinates x and y in Fig. 2–5 are spatial coordinates. The distance traveled in going from point 1 to point 2 on the surface is clearly a function of the path taken, and is different for paths A, B, and C in the figure. The distance traveled, S, is found by integrating δS, that is,

$$\int_1^2 \delta S = {}_1S_2 \qquad (2\text{–}5)$$

$$\int_1^2 \delta S \neq S_2 - S_1 \qquad (2\text{–}6)$$

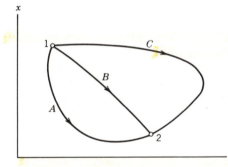

Figure 2-5 Geometric example illustrating the distance traveled between points 1 and 2 is a path function.

We do not use S_2 and S_1 since the distance traveled at point 1 or point 2 has no meaning. On the other hand, the change in the coordinate x between the two locations is independent of the path and

$$\int_1^2 dx = x_2 - x_1 \qquad (2\text{--}7)$$

Thus we see that some functions are path functions (inexact differentials) and some functions are point functions (exact differentials).

The sign convention for heat transfer is positive for heat added to a system or control volume and negative for heat removed from a system or control volume. Thus, $_1Q_2 = -100$ kJ indicates that as the system changed from state 1 to state 2, 100 kJ of energy was removed as heat. A process in which absolutely no heat transfer occurs ($_1Q_2 = 0$) is called an *adiabatic* process.

Very often it is desirable to compute the rate of heat transfer and this rate is represented by $\dot{Q}$. Thus,

$$\dot{Q} = \frac{\delta Q}{dt} \qquad (2\text{--}8)$$

The units for the rate of heat transfer are joules per second or watts. It is also common to compute the heat transferred per unit mass which would be represented by

$$_1q_2 = \frac{_1Q_2}{M} \qquad (2\text{--}9)$$

Details on the variables governing the rate of heat transfer will be presented in Chapters 7 through 10. At this point we should simply note that heat transfer can take place by conduction, by convection, and by radiation. While the convection heat transfer does involve the flow of a fluid, the fluid flowing does not cross the control surface where heat transfer occurs but usually flows parallel to it. In any case the heat transfer by convection across a control surface is independent of any mass transfer across that control surface.

2.9 WORK

Work is defined as energy in transition not associated with mass transfer and due to a difference in a potential other than temperature. The similarity between this definition and that for heat is obvious. There are only two ways a system can exchange energy with its surroundings, either as heat or as work. If the driving force for the energy transfer is temperature, then the energy transfer is called heat; if the driving force is anything other than temperature, the energy transfer is called work.

The same major points that applied to the definition of heat apply here also. Work is energy crossing a boundary; this energy cannot be stored as work. Energy transfer as work is not associated with mass transfer. The fact that work is not energy stored or possessed by a system or control volume means that it is not a property. Thus the work is a function of the path the system follows when it changes state, and we use the symbol δW to indicate an infinitesimal quantity of work. When this inexact differential is integrated to get the work for a system undergoing a process from state 1 to state 2, we obtain

$$\int_1^2 \delta W = {}_1W_2 \tag{2–10}$$

where the quantity of ${}_1W_2$ represents the amount of work done on or by the system when it changes from state 1 to state 2. Again note that we do *not* write the value of this integral as $W_2 - W_1$, since W_1 and W_2 have no meaning. The system does not possess a value of W at state 2 or at state 1.

The sign convention for work is opposite to that for heat. Work is *positive* when it is produced *by* a system (energy leaving the system) and *negative* when done *on* the system (energy added to the system). This unusual sign convention originated a long time ago and is based on the fact that a system producing work (energy out) is desirable and useful and thus was given a positive sign. The units for work are units of energy and are the same as the units for heat.

The rate of energy transfer as work is defined as power, $\dot{W}$. Thus

$$\dot{W} = \frac{\delta W}{dt} \tag{2–11}$$

The unit of power is the watt, which is 1 joule per second. Note that the units for power and the rate of heat transfer are identical (watts, kilowatts, etc.) but it is not thermodynamically correct to refer to a heat transfer rate as power.

2.9.1 Mechanical Work

In previous courses in physics and mechanics the mechanical work was computed as the product of the force and the displacement of that force in the same direction as the force, or as the dot product of the vectors $\mathbf{F}$ and $d\mathbf{S}$,

$$\delta W = \mathbf{F} \cdot d\mathbf{S} \tag{2–12}$$

If we consider a fluid system where the force on any part of the boundary is the product of the pressure and the area of that boundary, then

$$\delta W = \mathbf{F} \cdot d\mathbf{S} = P\mathbf{A} \cdot d\mathbf{S} \tag{2-13}$$

However, the product $\mathbf{A} \cdot d\mathbf{S}$ represents the volume change of the system, dV, so that

$$\delta W = P\,dV \tag{2-14}$$

To evaluate the work done at the moving boundary during a process, the above equation must be integrated. Before we integrate we must note that the system pressure must be the same as the pressure at the boundary: equilibrium conditions must prevail. We can, however, get by with the quasiequilibrium assumption and conclude that for a quasiequilibrium process the work done by a boundary moving normal to the force is given by

$$_1W_2 = \int_1^2 \delta W = \int_1^2 P\,dV \tag{2-15}$$

If the pressure remains constant during the process 1–2, then the equation is readily integrated; however, in general it can be seen that some relationship between P and V is required. This should be apparent when you recall that work is a path function.

Property diagrams are useful tools in thermodynamics. When calculating the mechanical work during a reversible process a pressure–volume diagram showing the path of the process is particularly useful. The shaded area under the path in Fig. 2–6 is proportional to the magnitude of the work, $\int P\,dV$. Note that the work is positive for a process in which the volume increases. If a different path were used between states 1 and 2, the magnitude of the work would be different, again illustrating that work is a path function. When comparing processes, dia-

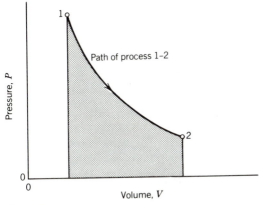

Figure 2-6 A pressure–volume diagram showing $P\,dV$ work for the quasiequilibrium process from state 1 to state 2.

grams like this are useful to indicate at a glance which process produces more work.

EXAMPLE 2–1

Gas is contained in a cylinder behind a movable piston as shown in Fig. E2–1. The initial volume is 0.0001 m³, the initial pressure is 1.0 MPa, the initial temperature is 25 °C, and the specific gas constant, R, is equal to 0.297 kJ/kg · K. The gas obeys the ideal gas equation of state, $PV = MRT$. The gas expands to a volume of 0.001 m³ on a quasiequilibrium process such that the product $PV = \text{const}$ (an isothermal process).

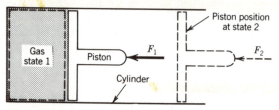

Figure E2-1 Expansion of a gas in a piston–cylinder device.

(a) What is the mass of the gas contained in the cylinder behind the piston?

(b) How much work is done as the gas expands?

SOLUTION

Choose the gas in the cylinder behind the piston as the system. Let subscripts 1 and 2 represent the initial and final states, respectively. Recall that 1 Pa = 1 N/m² so that the product of P and V, with P in units of kilopascals and V in units of cubic meters, will have units of kilonewton-meter or kilojoule.

(a) The ideal gas equation of state can be used to determine the mass

$$M = \frac{P_1 V_1}{RT_1} = \frac{(1000)\,(0.0001)}{(0.297)\,(298.15)} = 1.129 \times 10^{-3}\ \text{kg}$$

(b) Since the boundary adjacent to the piston moves, and there is a force in the direction of motion acting on that boundary, we know that some work is involved in this process. The work can be evaluated by integrating $P\,dV$ because this is a quasiequilibrium process and the relationship between P and V is known. Thus

$$_1W_2 = \int_1^2 P\,dV$$

and $PV = \text{const} = P_1V_1 = P_2V_2$. Therefore

$$P = \frac{P_1V_1}{V}$$

$${}_1W_2 = \int_1^2 \frac{P_1V_1}{V}\, dV = P_1V_1 \int_1^2 \frac{dV}{V} = P_1V_1(\ln V_2 - \ln V_1)$$

$${}_1W_2 = P_1V_1 \ln\left(\frac{V_2}{V_1}\right)$$

$${}_1W_2 = \underset{\text{kPa}}{(1000)}\underset{\text{m}^3}{(0.0001)}\ln\left(\frac{0.001}{0.0001}\right) = 0.2303\ \text{kJ}$$

COMMENT

The work is positive in sign; therefore it is work done by the system (on the surroundings) and represents energy transferred out of the system.

2.9.2 Other Work Modes

There are many other work modes, depending upon the nature of the system being studied. These work terms are all of the general form of a dot product between a "force" and a "displacement" and are valid for quasiequilibrium processes. A few examples will be briefly stated here.

The $P\, dV$ work can be regarded as due to boundary motion normal to the force acting on the boundary. There can also be work done due to tangential boundary motion. This work would be computed as

$$_1W_2 = -\int_1^2 \mathbf{F}_t \cdot d\mathbf{S} \tag{2-16}$$

where $\mathbf{F}_t$ is the shear force acting on the moving boundary and $\mathbf{S}$ is the boundary displacement. The work in stretching a wire or elastic band can be computed as

$$_1W_2 = -\int_1^2 \mathfrak{S} \cdot d\mathbf{L} \tag{2-17}$$

where $\mathfrak{S}$ is the tension in the wire and $\mathbf{L}$ is the length of the wire. The work done in boundary stretching as in a soap film or a balloon is given by

$$_1W_2 = -\int_1^2 \boldsymbol{\sigma} \cdot d\mathbf{A} \tag{2-18}$$

where σ is the surface tension (force/length) and A is the surface area. The work done in boundary twisting as in a shaft is given by

$$_1W_2 = -\int_1^2 \tau \cdot d\theta \tag{2-19}$$

where τ is the torque and θ is the angular displacement. Electrical work is computed by

$$_1W_2 = -\int_1^2 \mathfrak{E} \cdot d\mathbf{Z} \tag{2-20}$$

where $\mathfrak{E}$ is the electrical potential and $\mathbf{Z}$ is the electrical charge. Magnetic work is given by

$$_1W_2 = -\int_1^2 \mu_0 \mathfrak{H} \cdot d(V\,\mathfrak{M}) \tag{2-21}$$

where μ_0 is the permeability of free space, $\mathfrak{H}$ is the strength of the applied magnetic field, and $\mathfrak{M}$ is the magnetization vector (magnetic dipole moment per unit volume).

For further details on these various modes of work a more complete thermodynamics textbook [2,3,4,5] or a basic physics text should be consulted. Because of the definition of work, only forces acting at the boundary where motion occurs can cause work. Body forces acting at the interior of the system which result from the presence of some field do not produce work but the potential energy associated with these conservative force fields must be included in the first law of thermodynamics equation.

To determine whether any form of mechanical work is present for a system or control volume the system boundary must be examined. If there is a force acting on the boundary and that boundary has a component of motion in the direction of the force, mechanical work is present. If no forces are present on the boundary or if forces are present but no motion occurs in the direction of the forces, then there is no mechanical work done. To determine if other forms of work are present, the boundary must be examined for evidence of this work such as electrical conductors crossing the boundary.

2.9.3 Irreversible Work

The equations in this chapter for computing the magnitude of the work are valid for reversible processes. For example, the work done on a gas compression process in a piston-cylinder device may be computed correctly by $\int P\,dV$ if the process is reversible, but if the process is irreversible the work will be greater than for the reversible case and probably cannot be easily computed. There are certain forms of mechanical work which are inherently irreversible, for example, work due to frictional forces. Quite frequently the frictional force can be computed and the mechanical work can then be calculated.

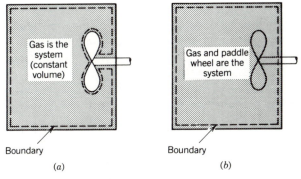

Figure 2-7 Paddle wheel work.

Another form of irreversible work is "paddle wheel" work illustrated in Fig. 2–7a. Here the system consists of a gas in a constant volume tank. The paddle wheel is not part of the system. Work is done by the paddle wheel on the gas because the gas (system) exerts a force on the boundary (on the paddle wheel) and that boundary is moving. The process is irreversible. Suppose that after adding 100 kJ of energy as work to the gas by turning the paddle wheel you stopped it. It is inconceivable that the paddle wheel would then begin to turn and deliver 100 kJ of energy as work from the gas.

The magnitude of the work done by the paddle wheel on the gas cannot be computed by integrating $P\,dV$. In fact, the value of this integral is zero, indicating that no reversible work was done, thus all the work is irreversible. To compute the magnitude of the work, the system boundary can be changed so that it cuts the propeller shaft, as shown in Fig. 2–7b. At this point there is a force (torque) acting on a moving boundary and if the torque and the angular displacement are known, the magnitude of the work can be determined. The system now consists of the gas and the paddle wheel but under steady-state conditions the work done by the shaft on the paddle wheel is equal to the work done by the paddle wheel on the gas. Further discussion of irreversible processes takes place in Chapter 4 after the second law of thermodynamics has been presented.

EXAMPLE 2–2

A rigid tank is divided into two volumes by a thin diaphragm as shown in Fig. E2–2. In the initial state the section to the left of the diaphragm contains a gas,

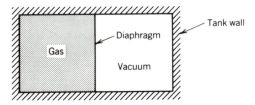

Figure E2-2 Rigid tank in the initial state.

while the section on the right is completely evacuated. If the diaphragm is ruptured so that in the final state the gas occupies the total volume, how much work is done on or by the gas in going from the initial state to the final state?

SOLUTION

If we assume our system is the complete inside volume of the tank, we could argue that the only mass contained inside the boundary is the mass of the gas. In this case the boundary does not move and we would conclude that no work is done on or by the system. Thus, $_1W_2 = 0$.

If on the other hand we choose the initial system boundary to include only the left volume (containing the gas initially), we see that the boundary of that system does move and we may be tempted to integrate $P\ dV$ to evaluate the work. The use of $P\ dV$ is not valid in this case because it is not a quasiequilibrium process; we would have difficulty defining the pressure on the moving boundary since it moves very fast. In fact there is no opposing force whatsoever on that moving boundary so we again conclude that there is no work done by the gas. Thus, $_1W_2 = 0$.

REFERENCES

1. Lee, J. F., and Sears, F. W., *Thermodynamics*, Addison–Wesley, Cambridge, Mass., 1955.

2. Van Wylen, G. J., and Sonntag, R. E., *Fundamentals of Classical Thermodynamics*, 2nd ed., Revised Printing, SI Version, Wiley, New York, 1978.

3. Zemansky, M. W., *Heat and Thermodynamics*, 4th ed., McGraw–Hill, New York, 1957.

4. Wark, K., *Thermodynamics*, McGraw–Hill, New York, 1966.

5. Hatsopoulos, G. N., and Keenan, J. H., *Principles of General Thermodynamics*, Wiley, New York, 1965.

PROBLEMS

2–1   A pressure gauge on a cylinder of gas reads 1.05 MPa when the barometer reads 95.3 kPa. What is the pressure of the gas?

2–2   A block of ice at $-10\ °C$ is placed in water at $25\ °C$. What is the absolute temperature of the ice and of the water? What is the initial temperature difference between the ice and the water in degrees Celcius and Kelvin?

2–3   The properties of a certain gas may be related using the ideal gas equation of state, $PV = MRT$. The specific gas constant for this gas, R, has a value of 0.297 kJ/kg $\cdot$ K.

(a) How many independent properties are needed to specify the state of a fixed mass of this gas?

(b) Which properties in the equation of state are intensive and which are extensive?

(c) Rewrite the equation of state in terms of intensive properties.

(d) Sketch the path taken by several isothermal (constant temperature) processes on a pressure (ordinate) versus specific volume (abscissa) graph.

(e) What is the density of the gas when it is at 20 °C and a pressure gauge indicates 1.0 MPa while the ambient pressure is 0.1 MPa?

2–4   A piston that has a mass of 2.5 kg is closely fitted into a cylinder with a diameter of 0.080 m. The local acceleration of gravity is 9.80 m/s² and the local barometric pressure is 0.100 MPa. A certain mass, M, is placed on top of the piston as shown in Fig. P2–4, and the pressure gauge indicates 12.0 kPa. Calculate the value of the mass M and the absolute pressure of the gas.

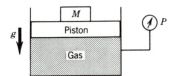

Figure P2-4   Piston–cylinder device.

2–5   A method for measuring small pressure differences employs a device called a manometer. In its simplest form it is a hollow tube in a U shape (see Fig. P2–5) filled with a suitable fluid. Mercury, water, glycerine, and light oils are some of the fluids used. Ignoring the density of the atmospheric air and the gas in the tank, a hydrostatic analysis shows that the pressure difference between the fluid in the tank and the atmosphere is given by $P = \rho g z$, where ρ = the manometer fluid density, g = the local acceleration of gravity, and z = the difference in height between the two columns of manometer fluid. Determine the gauge pressure in the tank (in kilopascals) if the manometer fluid is mercury, $z = 10$ cm, and $g = 9.80$ m/s².

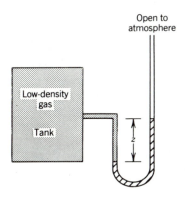

Open to
atmosphere

Figure P2-5   Sketch of a manometer.

2–6   Figure P2–6 shows a piston contained in a cylinder. The mass of the piston is 15 kg. The absolute pressures at A and B are 100 kPa and 125 kPa, respectively. Determine the magnitude and the direction of $\mathbf{F_R}$ necessary to maintain the piston

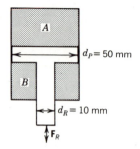

$d_P = 50$ mm

$d_R = 10$ mm

F_R

Figure P2-6   Piston–cylinder device.

in static equilibrium. Assume atmospheric pressure acts on the exposed piston rod.

2–7   An electrical heating element is installed in an insulated tank containing water. When the electrical current flows to the heater and the temperature of the heater exceeds that of the water, is there any heat transfer to or from the system if the system is:
(a)   The water only?
(b)   The tank (including the water, heating element, and the tank walls)?

2–8   The *italicized* word or words identifies the system for the process indicated. State whether the net work is positive, negative, or zero for each process.
(a)   A *spring* is stretched.
(b)   A *capacitor* is slowly discharged through a resistor.
(c)   A capacitor is slowly discharged through an *electric motor*. Assume the motor has an efficiency of 100% (electrical work in = mechanical work out).
(d)   An *electric motor* driven by a battery slowly lifts a weight. The motor efficiency, defined as the ratio of mechanical work out to electrical work in, is 95%.
(e)   An inflated *elastic balloon* is deflated. The rubber membrane was stretched significantly when inflated. Only the rubber material is the system; the air in the balloon is not part of the system.
(f)   The *air* contained in a tire and connected tire pump. The plunger of the pump is pushed down, forcing the air into the tire.
(g)   *Liquid and vapor water* in a rigid closed container is put on a stove where heat is added causing the pressure and temperature of the water to increase.

2–9   The pressure of a gas in a piston–cylinder device remains constant at a value of 0.1 MPa while heat is added and the volume increases reversibly from 0.1 to 0.2 m^3. For the gas as the system, determine the magnitude and sign of the work.

2–10   A piston-cylinder device contains 2.5 kg of air initially at 150 kPa and 30 °C. The air, assumed to be an ideal gas, $(PV = MRT)$, is compressed reversibly and isothermally. The work done during the compression process is equal to 150 kJ. Determine:
(a)   The final volume of the cylinder.
(b)   The final pressure of the air in the cylinder.

2–11   A battery energizes a glow plug electric heater in a diesel engine for 1 min during which the average current is 10 A and the voltage is 12 V.
(a)   How much work is done on or by the battery?
(b)   How much power is produced by the battery?

(c)   How much work is done on or by the glow plug?

(d)   Is there any heat transfer to or from the glow plug?

2–12   A large spherical balloon is initially flat. When it is filled, it has a diameter of 15 m. How much work is required to slowly fill it when the atmospheric pressure is 0.101 MPa? Assume the fabric of the balloon walls does not stretch.

2–13   A 20-kg mass is placed on a horizontal frictionless surface. The mass is accelerated at a constant value of 5.0 m/s² by a horizontal force. Taking the mass as the system, determine the work done in a 3s interval of time.

2–14   Air contained in a piston–cylinder device has an initial volume of 0.10 m³ and an initial pressure of 150 kPa. The air, which may be assumed to be an ideal gas, undergoes a quasiequilibrium process such that PV^{-1} = const. If the final volume is 0.25 m³, determine:

(a)   The final pressure of the air in the cylinder.

(b)   The work done if the air is the system.

2–15   A piston–cylinder device contains 1.0 kg of air which undergoes a reversible process where $PV^{1.33}$ = const. The initial pressure and temperature of the air are 400 kPa and 200 °C, respectively, while the final temperature is 100 °C.

(a)   Determine the work for the air as the system.

(b)   Sketch the process on a P–V diagram.

2–16   Figure P2–16 shows a piston–cylinder device containing air. The area of the piston is 0.10 m² and the initial volume of the cylinder is 0.01 m³. At the initial state the pressure inside the cylinder is 150 kPa which just balances the atmospheric pressure on the outside of the piston plus the weight of the piston. A spring, which has a linear spring constant of 200 kN/m, is just touching the piston but exerts no force on it. Heat is then added to the air in the cylinder causing it to expand reversibly against the spring until it reaches a final volume of 0.03 m³.

(a)   Sketch the process on P–V coordinates.

(b)   What is the final pressure of the air in the cylinder?

(c)   Calculate the work done by the air in the cylinder.

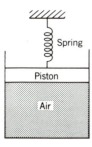

Figure P2-16   Piston–cylinder device with a spring.

2–17   A copper wire, 1 m long and 0.01 cm² in cross section, has a tension of 0.1 kN applied. If the tension is increased during a reversible isothermal process to 0.2 kN, compute the work for this process. Assume Young's modulus for copper is 110×10^6 kPa.

3

Properties of Pure Substances

3.1 DEFINITIONS

The systems we will consider contain substances for which certain property relationships must be known. If we wish to describe the state or change of state of a system we must know something about the characteristics of the substance that makes up the system. The simplest system is one consisting of a pure substance. A *pure substance* is one that is uniform and invariant in chemical composition. While a pure substance may exist in the solid, liquid, or vapor phase, the chemical composition in each phase is identical. Elements and stable compounds are examples of pure substances, while mixtures such as alloys, liquid solutions, and mixtures of gases are not pure substances. In this text a mixture of gases will be assumed to be a pure substance provided that no component of that mixture changes from the gas phase. Thus air will be considered to be a pure substance if the temperatures are high enough so that no condensation of any of the components takes place.

At this stage we are interested in describing the system by specifying the minimum number of properties necessary. Obviously there are a large number of properties that may change when the system undergoes a change of state but many of these may not be relevant to describing the behavior we wish to study. For example, we may be interested in determining the pressure change of a gas when the volume is changed. Properties of the gas such as electrical conductivity, thermal conductivity, and viscosity are not relevant to this process. To help narrow our scope we define a *simple compressible substance* as any pure substance where surface tension, magnetic, electrical, gravitational, and motion

39

effects are not significant. The net result of this definition is that the only reversible work mode is volume change ($P \, dV$) work; no other reversible work modes are possible for a simple compressible substance.

The state principle is given as: *The number of independent properties required to specify the thermodynamic state of a system is equal to the number of possible reversible work modes plus one.* The state principle tells us that specifying two independent properties specifies the state of a simple compressible substance. An important point concerning this principle is that the two properties must be independent, which means the one cannot be determined solely from a knowledge of the other. In certain phase change processes for example, pressure and temperature are not independent. Since a system is defined as a fixed mass, specifying the mass does not constitute specifying one of the two independent properties since the mass cannot be independently changed.

3.2 PHASE EQUILIBRIUM

To illustrate the phase changes for a simple compressible substance let us consider a solid in a vertical cylinder, Fig. 3–1. The solid, which is defined as the system, has a weighted floating piston placed on it so that a constant pressure is imposed on the system. Heat is added to the solid so that the system undergoes a constant-pressure heat addition process.

We will follow the path of this process on the temperature–volume diagram shown in Fig. 3–2. As heat is added both the volume and the temperature of the solid increase as shown on Fig. 3–2 path 1–2. This path is very short since we are assuming the expansion is very small so that the volume can be considered essentially constant. At state 3, the solid is at its melting temperature for the given pressure and the heat addition causes some of the solid to change into the liquid phase as shown in Fig. 3–1. Note that the liquid is floating on the solid. It is less dense than the solid; hence, this is a substance that expands on melting (contracts on freezing). While most substances behave in this manner, *water*

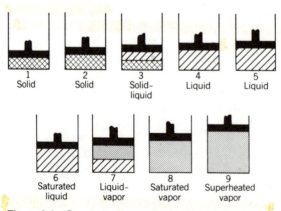

Figure 3-1 Constant-pressure heat addition.

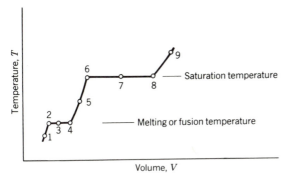

Figure 3-2 Temperature–volume diagram corresponding to the heating process in Fig. 3-1.

does not; it expands on freezing. We can conclude that the pure substance considered in this example is not water.

As more heat is added the remaining solid continues to melt until all the solid has been changed into the liquid phase, at state 4. In the region where the solid and liquid phases coexist in equilibrium (2–4) both the pressure and the temperature remain constant. Thus P and T are not independent properties in this region; if one is known the other is known.

Continued heat addition to the liquid causes its volume and temperature to increase through state 5 until a temperature is reached at which the substance starts to change from a liquid to a vapor, state 6. The liquid at this state is called a *saturated liquid* because further addition of heat converts some of the liquid into a vapor. The temperature and pressure at which this liquid–vapor phase change occurs (boiling or condensation) are called *saturation temperature* and *saturation pressure* respectively. At state 7 both the liquid and vapor exist in equilibrium, and along this line (6–7–8), the pressure and temperature are again dependent properties. When enough heat has been added to vaporize all the liquid, but the system is still at the saturation temperature for the given pressure, the system is at state 8 where the vapor is called *saturated vapor* (removal of a small amount of heat from the vapor would cause condensation, hence the vapor is saturated).

If further heat addition occurs, the temperature and volume of the vapor increase as shown in path 8–9 and the vapor is called a *superheated vapor*, because it is at a temperature greater than the saturation temperature for that pressure.

3.2.1 Temperature–Volume Diagrams

The temperature–volume diagram (Fig. 3–2) used to illustrate the path of the heating process just described is a valuable tool to visualize the process as well as to assist in determining the state of the system. A more complete diagram can be even more helpful. Let us confine our attention to the liquid, liquid–vapor, and vapor regions and repeat the previously described heating process for several different system pressures. A number of paths similar to path 5–6–

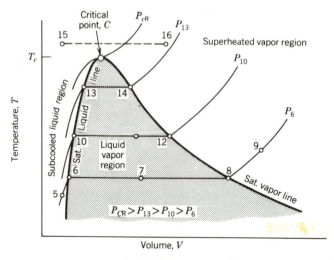

Figure 3-3   Temperature–volume diagram for the liquid and vapor regions.

7–8–9 would be obtained as shown in Fig. 3–3. When the pressure reaches a value of P_{cR}, a distinct phase change between the liquid and the vapor no longer occurs. The state, labeled c in Fig. 3–3, is the *critical point* for this substance and the critical pressure and critical temperature are labeled P_{cR} and T_{cR}, respectively. The heavy line 6–10–13–c is the locus of saturated liquid states, called the *saturated liquid line* and the heavy line 8–12–14–c is the *saturated vapor line*. These two lines meet at the critical point. The region to the right of the saturated vapor line is the superheated vapor region, and the region to the left of the saturated liquid line is the subcooled or compressed liquid region. The region is called *subcooled* because the temperature is below the saturation temperature for any given pressure and it is called *compressed* because the pressure is above the saturation pressure for any given temperature.

At temperatures above the critical temperature, the fluid can pass from a region where its properties are like those of a liquid (point 15 in Fig. 3–3) to a region where its properties are like those of a vapor (point 16 in Fig. 3–3) without going through a sudden and distinct phase change. The change is rather gradual and continuous. The problem of deciding whether to call the substance a liquid or a vapor in the region above but near to the critical point can be solved by simply referring to it as a fluid.

The abscissa shown in Figs. 3–2 and 3–3 is volume, but with a system (fixed mass) specific volume could also have been used. Since no numerical values are shown, the abscissa can also be regarded as specific volume.

3.2.2   Quality of a Saturated Liquid–Vapor Mixture

A property that will be widely used for equilibrium mixtures of saturated liquid and saturated vapor is the *quality*. Quality is the mass fraction of the liquid–vapor mixture which is saturated vapor. A value of zero indicates only saturated liquid is present and a value of 1 indicates only saturated vapor is present. The quality, represented by the letter x, can be written as

$$x = \frac{\text{mass of sat. vapor}}{\text{mass of sat. vapor} + \text{mass of sat. liquid}}. \qquad (3\text{--}1)$$

always > 0 but < 1

The quality is a property useful in computing other properties in the liquid–vapor region. Consider a system containing a total mass, M, of liquid–vapor mixture and occupying a volume V. The specific volume of the saturated liquid is represented by v_f and the specific volume of the saturated vapor is represented by v_g (the subscripts f and g are used in most tables of thermodynamic properties to represent saturated liquid and saturated vapor, respectively). From the definition of quality we have

$$M_g = xM \qquad (3\text{--}2)$$

and

$$M_f = (1 - x)M \qquad (3\text{--}3)$$

Using eqs. 3–2 and 3–3 in the definition of specific volume we obtain

$$v = \frac{V}{M} = \frac{V_f + V_g}{M} = \frac{M_f v_f + M_g v_g}{M} = \frac{(1 - x)v_f M + xMv_g}{M}$$

$$v = (1 - x)v_f + xv_g \qquad (3\text{--}4)$$

The specific volume of the mixture, v, can be readily computed from eq. 3–4 if the quality and the specific volumes of the saturated liquid and saturated vapor are known. Typically the values of the properties along the saturated liquid and saturated vapor lines (v_f and v_g in this case) are obtained from tables of saturation properties such as Tables A–1.1 and A–1.2 in the appendix.

EXAMPLE 3–1

Find the quality of a liquid–vapor mixture where $v_f = 0.00101$ m³/kg, $v_g = 0.00526$ m³/kg, and the total mass of 2.0 kg occupies a volume of 0.01 m³.

SOLUTION

Using eq. 3–4 and solving for x we have

$$x = \frac{v - v_f}{v_g - v_f}$$

where

$$v = \frac{V}{M} = \frac{0.01}{2.0} = 0.005 \text{ m}^3/\text{kg}$$

$$x = \frac{0.005 - 0.00101}{0.00526 - 0.00101} = 0.9388$$

COMMENT

At the particular state given, the mixture consists of 93.88% saturated vapor and 6.12% saturated liquid by mass.

___

3.2.3 Pressure–Temperature Diagram

Another useful property diagram is the pressure–temperature diagram. Figure 3–4 shows such a diagram for a substance that contracts on freezing. A diagram for a substance that expands on freezing is similar except that the fusion line slopes slightly to the left of vertical as the pressure increases instead of to the right as shown in Fig. 3–4. The states for the constant-pressure heating process shown in Figs. 3–1 and 3–2 are also shown in this figure. The two-phase (liquid–vapor) region shown on Fig. 3–3 appears as the vaporization line.

The *triple point* is that pressure and temperature at which all three phases, solid, liquid and vapor, coexist in equilibrium. It is a point only on a *P–T* diagram. On temperature–volume and pressure–volume diagrams the three phases can coexist in equilibrium at a number of states represented by a line (called the triple line). In some other three-dimensional property diagrams the triple states lie on a bounded surface. In any case, the pressure and temperature at any triple state are the same unique values as the temperature and pressure at the triple point. Obviously, pressure and temperature are not independent properties when the three phases coexist in equilibrium.

It is possible to go directly from the solid region into the vapor region by crossing the sublimation line as illustrated by path 17–18 in Fig. 3–4. Solid carbon dioxide (dry ice) at atmospheric pressure illustrates this process. Carbon dioxide has a triple-point pressure of approximately 0.52 MPa, which is well above atmospheric pressure, so that at 1 atm pressure carbon dioxide sublimates at about −77 °C. Figure 3–4 is schematic, while Fig. 3–5 shows actual vaporization lines for some pure substances.

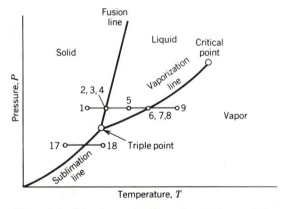

Figure 3-4   Pressure–temperature diagram for a substance that contracts on freezing.

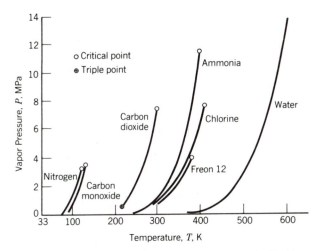

Figure 3-5   Pressure–temperature relationships in the liquid–vapor region for real fluids.

3.2.4   Pressure–Specific Volume Diagram

The information contained in Fig. 3–3 can be plotted as pressure versus specific volume so that a diagram such as Fig. 3–6 is obtained. The saturated liquid and saturated vapor lines are shown as heavy lines meeting at the critical point. The critical isotherm, $T = T_{CR}$, has a point of inflection at the critical point. Other lines of constant temperature are also shown.

3.2.5   Pressure–Specific Volume–Temperature Surfaces

The state principle tells us that for a simple compressible substance, two independent properties are needed to specify the state. We have been discussing three properties, pressure, specific volume, and temperature and in general only

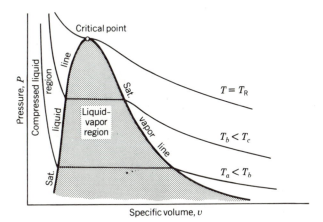

Figure 3-6   Pressure–specific volume diagram for the liquid and vapor regions.

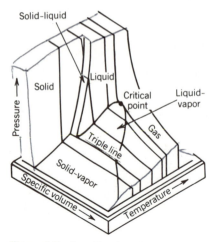

Figure 3-7 *P–v–T* surface for a substance that contracts on freezing.

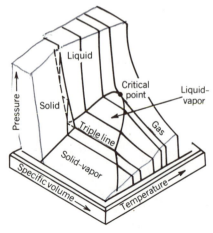

Figure 3-8 *P–v–T* surface for a substance that expands on freezing.

two of these are independent so that the third can be written as a function of the other two, for example, $T = f(P, v)$. Since this defines a surface in space we would expect to obtain a surface if a three-dimensional plot of P, and v, and T were constructed. That is, all equilibrium states for a simple compressible substance lie on this *P–v–T* surface. The *P–v–T* surface is extremely helpful in visualizing property changes for certain processes, particularly phase change processes.

Figure 3–7 shows such a representation for a substance that contracts on freezing. Views of this surface from the directions parallel to the volume axis and the temperature axis are shown in Fig. 3–9. The *P–v–T* surface for a substance that expands on freezing is shown on Fig. 3–8. This surface is a bit more

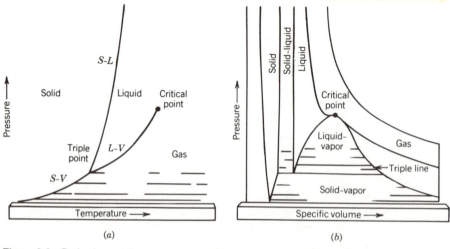

Figure 3-9 Projections of the surface in Fig. 3-7 on the *P–T* and *P–v* planes.

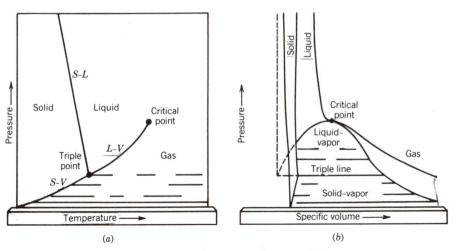

Figure 3-10 Projections of the surface in Fig. 3-8 on the *P–T* and *P–v* planes.

difficult to show since the freezing solid increases in volume and masks some of the liquid and liquid–vapor regions when viewed in the *P–v* plane (see Fig. 3–10*b*).

3.3 TABULAR PROPERTIES

The thermodynamic properties for many common substances have been measured and tabulated in books on thermodynamic properties[1-3]. This text includes tabulated values of the properties of water and Freon 12 in the appendix. Table A–1.1 contains saturation data for water with temperature as the argument (independent variable) and Table A–1.2 contains the same basic information with pressure as the argument. Water in this text refers to the substance H_2O regardless of the phase. The vapor phase is usually referred to as steam, the solid phase as ice, and the liquid phase is frequently called water; however, to be certain that we are referring to the liquid phase, we shall refer to it in this phase as liquid water.

The tabular values in the appendix will be used to solve problems in the text to illustrate the theory and the behavior of the substance. Note that only intensive properties are tabulated. When frequent use of these properties is required, the data are stored in a computer, either in equation or tabular form, and fairly simple routines are coded to solve for the desired properties.

The subscripts *f* and *g* used in these tables refer to saturated liquid and saturated vapor, respectively. From eq. 3–4, the general expression for *v* as a function of the quality and the saturation data is given as

$$v = (1 - x)v_f + xv_g \tag{3–5}$$

where *v* is the specific volume of any state in the liquid–vapor region. Several

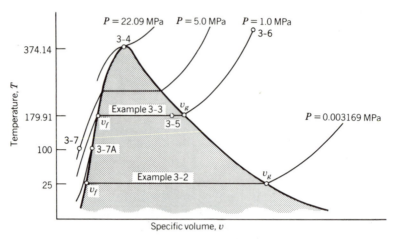

Figure 3-11   Temperature–specific volume diagram for Examples 3-2 to 3-7 (not to scale).

examples are given to illustrate the use of these tables and the states given in these examples are shown on a T–v diagram in Fig. 3–11.

EXAMPLE 3–2

Find the pressure, v_f and v_g for saturated water at 25 °C.

SOLUTION

> From Table A–1.1 read (at T = 25 °C)
> P = 3.169 kPa, v_f = 0.001003 m³/kg, v_g = 43.36 m³/kg

COMMENT

Both v_f and v_g are illustrated in Fig. 3–11.

EXAMPLE 3–3

Find the temperature, v_f and v_g for saturated water at P = 1.0 MPa.

SOLUTION

> From Table A–1.2 read (at P = 1.0 MPa)
> T = 179.91 °C, v_f = 0.001127 m³/kg, v_g = 0.19444 m³/kg

COMMENT

Both v_f and v_g are shown in Fig. 3–11.

EXAMPLE 3–4

Find the temperature, v_f and v_g for saturated water at P = 22.09 MPa.

SOLUTION

> From Table A–1.2 read (at P = 22.09 MPa)
> T = 374.14 °C, v_f = 0.003155 m³/kg = v_g

COMMENT

This is the critical point for water and is shown as state 3–4 in Fig. 3–11.

EXAMPLE 3–5

Find the temperature and specific volume for water at P = 1.0 MPa and x = 0.95.

SOLUTION

> From Table A–1.2 read (at P = 1.0 MPa)
> T = 179.91 °C, v_f = 0.001127 m³/kg, v_g = 0.19444 m³/kg
> Compute v from eq. 3–4
> $v = (1 - x)v_f + xv_g$
> $v = (0.05)(0.001127) + (0.95)(0.19444) = 0.18477$ m³/kg

COMMENT

This state is shown as 3–5 in Fig. 3–11. The fact that the quality was given in the statement of the problem is a clear indication that the state is in the liquid–vapor region.

EXAMPLE 3–6

Find the specific volume for water at 400 °C and P = 1.0 MPa.

SOLUTION

> From Table A–1.3 read (at $P = 1.0$ MPa and $T = 400\,°C$)
> $v = 0.3066$ m³/kg

COMMENT

This state is in the superheated vapor region where T and P are independent properties. The state is labeled 3–6 in Fig. 3–11.

Examples 3–2 through 3–6 do not include any states in the compressed liquid region. Tabular data for the compressed liquid region do exist for a few common substances; however, there are no such tables in this text. Whenever a property is needed for a substance in a compressed liquid state, an approximate value for that property can be obtained by using the value for the saturated liquid at the *same temperature* as the compressed liquid. This approximation is reasonably accurate provided the state is not close to the critical point.

EXAMPLE 3–7

Find the specific volume of water at $T = 100\,°C$ and $P = 5.0$ MPa (a state in the compressed liquid region).

SOLUTION

> $v_{100\,°C,\ 5\ MPa} \approx v_{f,100\,°C}$
> From Table A–1.1 read (at 100 °C) $v_f = 0.001044$ m³/kg, $v \approx 0.001044$ m³/kg

COMMENT

The correct value for v in this case is 0.001041 m³/kg. This approximation is within 0.3% of the correct value. The approximate value is shown in Fig. 3–11 as state 3–7A, while the true state is shown as state 3–7. Figure 3–11 indicates that these states are significantly different. As we have already noted, states 3–7 and 3–7A are actually extremely close to each other, but this diagram (Fig. 3–11) has been distorted to show the lines distinctly. If the diagram were drawn to scale, the saturated liquid line in the vicinity of 100 °C would appear to be vertical and the constant pressure lines of 5 and 1 MPa would lie in the compressed liquid region less than one line width away from the saturated liquid line. The distortions are made to clearly distinguish lines and states in this region.

The reader should also verify that this state is in the compressed liquid region. Using the temperature of 100 °C the saturation pressure can be found in Table A–1.1 to be 0.10135 MPa. Since the actual pressure of 1.0 MPa is greater than the saturation pressure of 0.10135 MPa, the liquid is in a compressed liquid state. Alternately the pressure of 5.0 MPa can be used to find from Table A–1.2 that the saturation temperature is 263.99 °C. Since the actual temperature (100 °C) is less than the saturation temperature, we say the liquid is in a sub-cooled liquid state. Recall that the terms subcooled liquid and compressed liquid refer to the same region.

3.4   THE IDEAL GAS EQUATION OF STATE

The ideal gas equation of state is valid for substances in the superheated vapor phase at very low density. The common form of this equation is

$$\bigstar\bigstar \qquad PV = MRT \qquad\qquad (3\text{--}6)$$

where R is the specific gas constant given by

$$R = \frac{R_0}{\mathfrak{M}} \qquad\qquad (3\text{--}7)$$

$\mathfrak{M}$ is the molecular weight of the substance and R_0 is the universal gas constant, 8.31434 J/mol · K (or 8.31434 kJ/kmol · K). Both the pressure and the temperature must be *absolute values* in this equation. The equation can also be written in terms of either specific volume or density as

$$P = \frac{RT}{v} = \rho RT \qquad\qquad (3\text{--}8)$$

Since this is a very simple equation of state, it is very convenient to use and the question usually arises as to its validity for various substances in the vapor phase. As noted, it is valid at low density, but what constitutes low density? Certainly near the critical point the density would not be considered low. The low density would occur at high temperatures and low pressures but what constitutes a high temperature or a low pressure and could these values be different for different substances? An equally appropriate question is what kind of precision is sought in the value of a property computed by this equation? Certain rules can be given but they do not speak to the question of precision. For purposes of this text we will adopt the following rule. If either $T/T_{\text{CR}} > 2.0$ or $P/P_{\text{CR}} < 0.1$ *and* the substance is in the vapor phase we consider the ideal gas equation of state valid. T_{CR} and P_{CR} represent the critical temperature and pressure, respectively, for the substance in question. Values of T_{CR} and P_{CR} for some common substances are given in Table A–7. This rule does allow use of the ideal gas equation of state up to the saturated vapor line if the pressure is below one-tenth of the critical pressure. At the saturated vapor line there can be considerable error involved. For example, water vapor at a pressure of 2.0 MPa has a value of P/P_{CR} less than 0.1 and the temperature of the saturated vapor at this pressure is 212.42 °C. The ideal gas equation, solved for v yields

$$v = \frac{RT}{P} = \frac{8.31434(485.67)}{18(2000)} = 0.11217 \text{ m}^3/\text{kg}$$

while Table A–1.2 indicates the correct value is 0.09963 m³/kg. Thus an error of about 12.6% results in the use of this equation at this point. Of course, at lower pressures the accuracy of the ideal gas equation improves.

3.5 OTHER EQUATIONS OF STATE

Any relationship giving the P–v–T behavior of a substance is called an equation of state. In regions where the ideal gas equation of state is not sufficiently accurate, a number of other equations of state are available and may be applicable for that substance over the range of interest. The generalized compressibility factor, Z, defined by

$$Z = \frac{Pv}{RT} \tag{3–9}$$

may be found from a generalized compressibility chart such as shown in Fig. A–3. The reduced coordinates $P_r = P/P_{CR}$ and $T_r = T/T_{CR}$ are used to find values for Z, and these values will provide a fairly accurate equation of state, provided the value of Z for the substance at the critical point, Z_{CR}, is close to 0.27.

Other equations of state such as the van der Waals equation, the Beattie–Bridgman equation, the virial equation, and the Redlich–Kwong equation are available and have been used with success in certain situations. These equations, listed in Table A–4, are discussed in more detail in the text by Wark[4].

Tables of property data can be used instead of algebraic equations. Tables listing values for the thermodynamic properties of the superheated vapor provide precise data for the P–v–T relationship (equation of state) for some common substances such as steam, refrigerants R–12 and R–22, carbon dioxide, ammonia, hydrogen, oxygen, and mercury. While these tabulated values may prove useful they are not as convenient to work with as a simple analytical expression. If precise property calculations are to be made for a particular substance, the digital computer is used to perform the calculations using either tabulated data or other fairly complex equations of state.

In this text properties are tabulated for water in the superheated vapor region and problems requiring the use of this table, Table A–1.3, are included. In the superheated vapor region two independent properties are needed to define the state and most tables are set up with pressure and temperature as the arguments.

Since entries in the table are discreet points, interpolations in one or two directions is often required. To avoid these time-consuming operations with the numbers, most of the problems in this text have been chosen so that a minimum amount of interpolation is required.

EXAMPLE 3-8

Find the mass of water in a tank of 0.1 m³ volume for the three systems given below. Each system is at a temperature of 200 °C.

(a) System A: $P = 0.5$ MPa
(b) System B: Saturated vapor
(c) System C: $x = 0.9$

SOLUTION

(a) From Table A–1.3 obtain the specific volume at 200 °C and 0.5 MPa as $v_A = 0.4249$ m³/kg. Then

$$M_A = \frac{V}{v_A} = \frac{0.1}{0.4249} = 0.2353 \text{ kg}$$

(b) From Table A–1.2, obtain the specific volume of the saturated vapor at 200 °C as $v_B = v_g = 0.12736$ m³/kg. Then

$$M_B = \frac{V}{v_B} = \frac{0.1}{0.12736} = 0.7852 \text{ kg}$$

(c) From Table A–1.2 also obtain that at 200 °C, $v_f = 0.001157$ m³/kg. Then

$$v_C = (1 - x)v_f + xv_g$$

$$v_C = 0.1(0.001157) + 0.9(0.12736) = 0.11474 \text{ m}^3/\text{kg}$$

$$M_C = \frac{V}{v_C} = \frac{0.1}{0.11474} = 0.8715 \text{ kg}$$

3.6 THERMODYNAMIC PROPERTIES OF A SIMPLE COMPRESSIBLE SUBSTANCE OTHER THAN P–V–T

There are a large number of properties of a pure substance which will be used in this text and an even larger number which may be of interest in other situations where magnetic, dielectric, or optic effects are present. A manufacturer of the compound sulfur hexafluoride lists values for over 17 properties in a data bulletin where some of these properties are tabulated as a function of temperature and pressure. Some of the other more commonly used properties will be described briefly in this section.

The *volume coefficient of expansion*, β, also called the isobaric (constant pressure) compressibility, is defined as

$$\beta \equiv \frac{1}{v}\left(\frac{\partial v}{\partial T}\right)_P = -\frac{1}{\rho}\left(\frac{\partial \rho}{\partial T}\right)_P \tag{3-10}$$

This property represents the fractional volume change per unit change in temperature for a constant-pressure process. The *isothermal compressibility*, κ, is defined as

$$\kappa \equiv -\frac{1}{v}\left(\frac{\partial v}{\partial P}\right)_T = \frac{1}{\rho}\left(\frac{\partial \rho}{\partial P}\right)_T \tag{3-11}$$

This property represents the fractional volume change per unit change in pressure when the temperature is held constant. As seen from the defining equations, both β and κ are functions of the P–v–T relationship for the substance, that is, given the equation of state, numerical values for β and κ can be obtained at any defined state. These properties are particularly useful when dealing with substances in the liquid or solid phase.

• The *internal energy, U*, cannot be precisely defined until the first law of thermodynamics is introduced; however, at this point we can say that it is a measure of energy stored in, or possessed by, the system due to the microscopic kinetic and potential energy of the molecules of the substance in the system.
• The *specific internal energy u* will also be widely used, where

$$u = \frac{U}{M} \tag{3-12}$$

• The *enthalpy, H*, is defined by

$$H \equiv U + PV \tag{3-13}$$

This property is introduced because it is an extremely useful property when dealing with control volumes. The *specific enthalpy, h* is given by

$$h = \frac{H}{M} = u + Pv \tag{3-14}$$

• The *constant-volume specific heat, c_v*, is defined by

$$c_v \equiv \left(\frac{\partial u}{\partial T}\right)_v \qquad KJ/kg \cdot K \tag{3-15}$$

and the *constant-pressure specific* heat is defined by

$$c_p \equiv \left(\frac{\partial h}{\partial T}\right)_P \qquad \text{kJ/kg·K} \qquad (3\text{--}16)$$

These are very useful properties although historically they were somewhat incorrectly named. They are intensive properties of a system and are in general unrelated to any heat transfer process. Figure 3–12 shows that c_v is the slope of a constant-volume line on a u versus T plot, while c_p is the slope of a constant-pressure line on an h versus T plot. The above definitions indicate they are useful in determining changes in internal energy and enthalpy. The ratio of specific heats, γ, is defined as

$$\gamma \equiv \frac{c_p}{c_v} \qquad (3\text{--}17)$$

The property related to the transport of energy in the substance is the thermal conductivity k. This property is defined as the heat flux per unit driving potential. Thus the units for this property are watt/meter · Kelvin. More will be said about this property in Chapter 9.

The property related to the transport of momentum in a fluid is the dynamic viscosity or viscosity coefficient μ. It is the proportionality constant between the shear stress and the velocity gradient normal to a flow, and for Newtonian fluids is a thermodynamic property varying with temperature and pressure. More will be said about this property in Chapter 7.

3.7 PROPERTY RELATIONSHIPS FOR IDEAL GASES

An ideal gas is one for which the ideal gas equation of state is valid. Experimental measurements show that the specific internal energy of an ideal gas is a function of temperature only. This fact can also be proved analytically after the second law of thermodynamics has been introduced. If the temperature of the ideal gas is known, the internal energy can be determined independent of the pressure or specific volume.

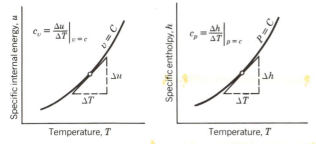

Figure 3-12 c_v and c_p as slopes of constant specific volume and constant-pressure lines.

In general u is a function of two independent properties $u = f(T, v)$; hence

$$du = \left(\frac{\partial u}{\partial T}\right)_v dT + \left(\frac{\partial u}{\partial v}\right)_T dv \qquad (3\text{--}18)$$

For an ideal gas u is a function of T only; hence

$$\left(\frac{\partial u}{\partial v}\right)_T = 0 \qquad (3\text{--}19)$$

and

$$du = \left(\frac{\partial u}{\partial T}\right)_v dT = c_v\, dT \qquad (3\text{--}20)$$

Changes in specific internal energy between any two states can be evaluated by integrating this equation to obtain

$$\int_1^2 du = u_2 - u_1 = \int_1^2 c_v\, dT \qquad (3\text{--}21)$$

CONSTANT

From the fact that u is a function of temperature only, this last equation tells us that c_v is a function of temperature only and if this functional relationship is known the integral can be evaluated.

The specific enthalpy for an ideal gas can be written as

$$h = u + Pv = u + RT \qquad (3\text{--}22)$$

Since u is a function of T only and R is a constant, it is obvious that h is a function of temperature only. Repeating the procedure used for u we can write

$$h = f(T, P) \qquad (3\text{--}23)$$

$$dh = \left(\frac{\partial h}{\partial T}\right)_P dT + \left(\frac{\partial h}{\partial P}\right)_T dP \qquad (3\text{--}24)$$

but

$$\left(\frac{\partial h}{\partial P}\right)_T = 0 \qquad (3\text{--}25)$$

so that

$$dh = \left(\frac{\partial h}{\partial T}\right)_P dT = c_p\, dT \qquad (3\text{--}26)$$

and

$$\int_1^2 dh = h_2 - h_1 = \int_1^2 c_p\, dT \qquad (3\text{--}27)$$

CONSTANT

Again we know that c_p is a function of temperature only (because h is a function of temperature only) and the integral can be evaluated if this relationship is known. Table A–5 lists some equations for c_p as a function of temperature for several substances. Caution must be exercised that these relationships are not used where the behavior of the substance is far removed from that of an ideal gas.

From the equation $h = u + RT$ we obtain

$$dh = du + R \, dT \qquad (3\text{–}28)$$

or

$$c_p \, dT = c_v \, dT + R \, dT \qquad (3\text{–}29)$$

Thus

$$c_p - c_v = R \qquad (3\text{–}30)$$

for ideal gases.

3.7.1 The Special Case of a Reversible Adiabatic Process for a Stationary System

The relationship between P and V can be determined for an ideal gas with constant specific heats undergoing a *reversible adiabatic process* in a stationary system. A stationary system is a system where negligible motion takes place; therefore, no changes in potential or kinetic energy occur.

Reversible means that

$$\delta W = P \, dV$$

Adiabatic means that

$$\delta Q = 0$$

Ideal gas means that

$$PV = MRT$$

and

$$du = c_v \, dT, \quad dh = c_p \, dT, \quad c_p - c_v = R$$

In the next chapter we will show that the first law of thermodynamics for a stationary system is

$$\delta Q - \delta W = dU$$

which for this special case reduces to

$$0 - P\,dV = dU \tag{3-31}$$

$$-P\,dV = dU = M\,du = Mc_v\,dT = \frac{PV}{RT}c_v\,dT$$

or

$$-\frac{dV}{V} = \frac{c_v\,dT}{RT} \tag{3-32}$$

Since $c_p - c_v = R$ and $\gamma = c_p/c_v$, it can be shown that $c_v/R = 1/(\gamma - 1)$. Thus

$$-\frac{dV}{V} = \left(\frac{1}{\gamma - 1}\right)\frac{dT}{T} \tag{3-33}$$

Equation 3–33 can be integrated, since γ is assumed constant, to determine the relationship between T and V for this process. However, if the P–V relationship is sought, the temperature must be eliminated from this equation and replaced by the pressure. The ideal gas equation of state provides this relationship. By taking the logarithm of both sides of the equation, then differentiating, the desired form is obtained,

$$\ln M + \ln R + \ln T = \ln P + \ln V \tag{3-34}$$

$$\frac{dT}{T} = \frac{dP}{P} + \frac{dV}{V}$$

This may be substituted into eq. 3–33 to yield

$$-\frac{dV}{V} = \frac{1}{\gamma - 1}\frac{dP}{P} + \frac{1}{\gamma - 1}\frac{dV}{V}$$

$$-\frac{dV}{V}\left(1 + \frac{1}{\gamma - 1}\right) = \left(\frac{-\gamma}{\gamma - 1}\right)\frac{dV}{V} = \frac{1}{\gamma - 1}\frac{dP}{P}$$

$$\gamma\frac{dV}{V} + \frac{dP}{P} = 0 \tag{3-35}$$

Integration of this equation yields

$$\gamma \ln V + \ln P = \ln(\text{const}) = \ln(PV^\gamma) \tag{3-36}$$

hence

$$PV^\gamma = \text{const} = P_1 V_1^\gamma = P_2 V_2^\gamma \tag{3-37}$$

or

$$\frac{P_1}{P_2} = \left(\frac{V_2}{V_1}\right)^\gamma \tag{3-38}$$

Equation 3–37 is an important and useful property relationship for an ideal gas with constant specific heats undergoing a reversible adiabatic process where the macroscopic kinetic and potential energy effects are negligible.

3.7.2 Polytropic Processes

In some actual processes involving ideal gases, the heat transfer and irreversibilities are very small and can be assumed negligible, so that eq. 3–38 can be used. However, in many actual processes the heat transfer is not negligible, and the equation PV^n = const more accurately describes the pressure–volume relationship. Such a process is called a *polytropic process*, and n, called the polytropic exponent, is frequently determined empirically. In the case of a reversible polytropic process, n can be related to the ratio of the heat transfer to reversible work (see Ref. 4). The expression for reversible work for a polytropic process can be obtained by substituting PV^n = const and integrating $P\,dV$ to obtain

$$_1W_2 = \int_1^2 P\,dV = \text{const}\int_1^2 \frac{dV}{V^n} = \frac{P_2V_2 - P_1V_1}{1 - n} \qquad (3\text{–}39)$$

This expression is not valid for $n = 1$, but when $n = 1$ we have an isothermal process and the integral of $P\,dV$ in that case is given by

$$_1W_2 = \int_1^2 P\,dV = MRT\int_1^2 \frac{dV}{V} = MRT\ln\left(\frac{V_2}{V_1}\right) = P_1V_1\ln\left(\frac{V_2}{V_1}\right) \quad (3\text{–}40)$$

Using PV^n = const and the ideal gas equation of state it can be shown that for ideal gases

$$\frac{P_2}{P_1} = \left(\frac{V_1}{V_2}\right)^n = \left(\frac{T_2}{T_1}\right)^{n/(n-1)} \qquad (3\text{–}41)$$

These relationships are valid for any value of n (except $n = 1$) and are used quite frequently for $n = \gamma$. Paths for processes where PV^n = const are shown in Fig. 3–13 for several values of n.

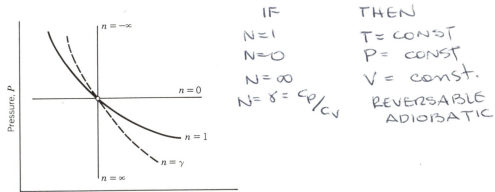

Figure 3-13 Polytropic processes for an ideal gas.

EXAMPLE 3–9

A bicycle tire containing air at 0.5 MPa and 25 °C (the ambient temperature) leaks rapidly such that in one second the pressure is reduced to ambient, 0.1 MPa. The volume of the tire is 0.001 m³. Assuming air to be an ideal gas with constant specific heats, find the temperature of the air remaining in the tire after 1 s and find the work done on or by this air remaining in the tire.

SOLUTION

Choose as the system the air that *remains* in the tire. Initially (state 1) it will occupy only a portion of the total tire volume. Figure E3–9 shows a sketch of the system along with a P–V diagram for the process. We may assume that this system undergoes a reversible adiabatic process. For a reversible adiabatic process (from eq. 3–41)

$$\frac{T_2}{T_1} = \left(\frac{P_2}{P_1}\right)^{(\gamma-1)/\gamma}$$

Using $\gamma = 1.4$ from Table A–6 we have

$$T_2 = 298.2 \left(\frac{0.1}{0.5}\right)^{0.4/1.4} = 188.3 \text{ K}$$

Since the boundary of the system moves, and there is a force on that moving boundary, there will be mechanical work involved in this process.

The work can be computed from either eq. 3–39 (with $n = \gamma$) or from the first law, eq. 3–31. Choosing eq. 3–39 we have

$$_1W_2 = \frac{P_2V_2 - P_1V_1}{1 - \gamma} = \frac{MR(T_2 - T_1)}{1 - \gamma}$$

$$_1W_2 = \frac{P_2V_2(T_2 - T_1)}{T_2(1 - \gamma)}$$

$$_1W_2 = \frac{100(0.001)(188.3 - 298.2)}{188.3(1 - 1.4)} = 0.1459 \text{kJ}$$

COMMENT

The work is positive, indicating that this system does work in expelling air from the tire. The boundary of the system moves and a force is acting on that boundary. Because work is done by the system (energy leaves the system)

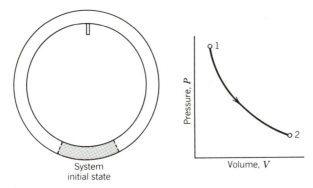

Figure E3-9 Bicycle tire.

and the system is adiabatic, the internal energy, hence temperature, of the system decreases. The temperature decreased to a very low value but a check on the critical temperature and pressure for oxygen and nitrogen indicates that by our rule the ideal gas equation of state is valid. The value of γ does change slightly over this temperature range but we will consider this calculated value of temperature satisfactory.

Justification for the reversible adiabatic assumption is as follows. The process is slow enough that pressure equilibrium can be maintained (pressure waves travel at the speed of sound in the air and pressure equilibrium could be maintained on the order of milliseconds) and the quasiequilibrium assumption is valid. There are no appreciable friction forces acting on the system. Thus the *assumption* of reversibility is valid. The adiabatic assumption is also valid since the time available for heat transfer is very short. (If the leak had occurred over a period of 1 hr, then an isothermal assumption would have been more realistic.)

REFERENCES

1. *Properties of Commonly-Used Refrigerants*, published by Air Conditioning & Refrigerating Machinery Association, Washington, DC.
2. *Steam and Air Tables in SI Units*, edited by Thomas F. Irvine, Jr. and James P. Hartnett, Hemisphere Publishing Corporation, Washington, 1976.
3. Keenan, J. H., and Keyes, F. G., *Thermodynamic Properties of Steam*, Wiley, New York, 1967.
4. Wark, K., *Thermodynamics*, McGraw-Hill, New York, 1966.

PROBLEMS

3–1 Using Table A–1.1 or A–1.2, determine whether the following states of water are in the compressed liquid, liquid–vapor, or superheated vapor regions or on the saturated liquid or saturated vapor line.

(a) $P = 1.0$ MPa $T = 207$ °C
(b) $P = 1.0$ MPa $T = 107.5$ °C
(c) $P = 1.0$ MPa $T = 179.91$ °C, $x = 0.0$
(d) $P = 1.0$ MPa $T = 179.91$ °C, $x = 0.45$
(e) $T = 340$ °C $P = 21.0$ MPa
(f) $T = 340$ °C $P = 2.1$ MPa
(g) $T = 340$ °C $P = 14.586$ MPa, $x = 1.0$
(h) $T = 500$ °C $P = 25$ MPa
(i) $P = 50$ MPa $T = 25$ °C

3–2 Find the specific volume at states b, d, and h in Prob. 3–1.

3–3 Water is a pure substance. As such, any two independent properties will determine its state. In each of the following cases, determine its phase and fill in the missing property.
(a) $P = 900$ kPa; $v = 0.035^3$/kg
Phase_____; $T = $ _____°C
(b) $P = 10$ MPa; $x = 0.33$
Phase_____; $v = $ _____m³/kg
(c) $P = 200$ kPa; $T = 50$ °C
Phase_____; $v = $ _____m³/kg
(d) $T = 250$ °C; $v = 1.00$ m³/kg
Phase_____; $P = $ _____MPa
(e) $P = 6.50$ MPa; $T = 515$ °C
Phase_____; $v = $ _____m³/kg

3–4 Ammonia at $P = 150$ kPa, $T = 0$ °C is in the superheated vapor region and has a specific volume and enthalpy of 0.8697 m³/kg and 1469.8 kJ/kg, respectively. Determine its specific internal energy at this state.

3–5 A rigid tank contains saturated steam at a pressure of 0.1 MPa. Heat is added to the steam to increase the pressure of the steam to 0.3 MPa. What is the final temperature of the steam?

3–6 A rigid tank with a volume of 1.00 m³ contains a mixture of H_2O in the liquid–vapor region at 20 °C. The mass of this mixture is 943.4 kg.
(a) If heat is added to the mixture in this tank, at what temperature will the tank contain only liquid?
(b) What would happen if the heating process were to continue beyond this point?
(c) If the tank contained only 100 kg of mixture (instead of 943.4 kg), would the tank reach the "all liquid" state?

3–7 A rigid tank with a volume of 0.002 m³ contains an equilibrium mixture of liquid and vapor water at a temperature of 150 °C. The mass of the mixture is 0.5 kg.
(a) What is the quality of the mixture?
(b) What fraction of the tank volume is occupied by the liquid?

3–8 May the ideal gas equation of state be considered valid for the following substances at the indicated states? See Table A–7 for data.
(a) Carbon dioxide; $P = 0.1$ MPa, $T = 20$ °C
(b) Carbon dioxide; $P = 8.0$ MPa, $T = 40$ °C
(c) Butane; $P = 1.0$ MPa, $T = 20$ °C
(d) Nitrogen; $P = 3$ MPa, $T = 20$ °C
(e) Hydrogen; $P = 10$ MPa, $T = 0$ °C

3–9   Tank A, with a volume of 0.1 m³, contains air at 1.0 MPa and 20 °C, the ambient temperature. Tank B with a volume of 0.2 m³ is initially evacuated. A very small valve connecting the two tanks is opened and air flows very slowly from tank A into tank B. The process is slow enough that you may assume heat transfer can take place to keep the temperature of the air at ambient in each tank. What is the pressure in tank B when pressure equilibrium between the tanks is reached?

3–10   From the data in Table A–1.3, calculate c_p for steam at 10 MPa and 500 °C.

3–11   Using the generalized compressibility chart in Fig. A–3 and the data of Table A–7, find the specific volume of carbon dioxide at 92 °C and 11.085 MPa.

3–12   Air is compressed reversibly and adiabatically from a pressure of 0.1 MPa and a temperature of 20 °C to a pressure of 1.0 MPa.
(a)   What is the temperature of the air after compression?
(b)   What is the density ratio (after compression to before compression)?
(c)   How much work is done in compressing 2 kg of air?
(d)   How much power is required to compress 2 kg/s?

3–13   Compute the specific heat at constant volume for carbon monoxide gas at $T = 800$ °C using the equations in Table A–5 of the appendix. Compare your value with that given in Table A–7.

3–14   Nitrogen gas is heated at constant pressure from 100 to 1500 °C. Determine the change in specific enthalpy in kilojoules per kilogram as a result of this process, taking into account the variation in specific heat of the nitrogen with temperature.

3–15   Air is compressed polytropically from an initial pressure, temperature, and volume of 100 kPa, 200 °C, and 10 m³ to a final volume of 1.5 m³. Determine the final temperature and pressure if the polytropic exponent is:
(a) $n = 0$
(b) $n = 1.0$
(c) $n = 1.33$
Sketch the three processes on P–V coordinates.

3–16   Compute the specific volume of carbon dioxide at a pressure of 15 MPa and a temperature of 93 °C using the generalized compressibility chart of Fig. A–3 and the critical constants in Table A–7.

4

Systems Analysis—First and Second Laws

4.1 THE FIRST LAW OF THERMODYNAMICS

The first law of thermodynamics is a statement of the conservation of energy applied to a system. This conservation principle states that the algebraic sum of all the energy transfers across the system boundary must be equal to the change in energy of the system. Since heat and work are the only forms of energy that may cross a system boundary, we can write the first law in differential form as

$$\delta Q - \delta W = dE \tag{4-1}$$

The minus sign appears with the work term because of the sign convention adopted for work (discussed in Chapter 2). Equation 4–1 defines the property E which is called the energy of the system. In the absence of electric, magnetic, and surface tension this energy quantity consists of three terms:

1. The internal energy, U, represents the energy possessed by the molecules of the substance by virtue of their microscopic kinetic and potential energy.
2. The macroscopic kinetic energy, KE, represents the kinetic energy of the system due to its motion.
3. The macroscopic potential energy, PE, represents the potential energy of the system due to its position in a gravitational field.

Thus

$$E = Me = U + KE + PE \tag{4-2}$$

and

$$dE = M\, de = dU + d(KE) + d(PE) \tag{4-3}$$

The kinetic energy of the system is given by

$$KE = \frac{MV^2}{2} \tag{4-4}$$

where V is the velocity of the system, and the potential energy of the system is given by

$$PE = Mgz \tag{4-5}$$

where g is the local gravitational acceleration and z is the elevation of the system above some datum.

Equation 4–1 can be integrated for any process to obtain

$$_1Q_2 - {}_1W_2 = E_2 - E_1 = U_2 - U_1 + KE_2 - KE_1 + PE_2 - PE_1 \tag{4-6}$$

$$= M(u_2 - u_1) + M\left(\frac{v_2^2}{2} - \frac{v_1^2}{2}\right) + Mg(z_2 - z_1)$$

There are many real processes that are time dependent. If the time rate at which the properties change is relatively small, the quasiequilibrium assumption is valid. The *rate form* of the first law is useful in solving many such problems. The rate form of the first law states

$$\dot{Q} - \dot{W} = \frac{dE}{dt} \tag{4-7}$$

where $\dot{Q}$ is the rate of heat transfer, the time rate at which energy is being transferred across the boundary as heat, $\dot{W}$ is the power, the rate at which energy is crossing the boundary as work, and dE/dt is the time rate of change of system energy.

If a system is *stationary* there is no change in kinetic or potential energy so that the first law may be expressed as

Differential form:

$$\delta Q - \delta W = dU \tag{4-8}$$

Rate form:

$$\dot{Q} - \dot{W} = \frac{dU}{dt} \tag{4-9}$$

Integrated form:

$$_1Q_2 - {}_1W_2 = U_2 - U_1 \tag{4-10}$$

EXAMPLE 4–1

One hundredth of a kilogram (0.01 kg) of air is compressed in a piston–cylinder device. Find the rate of temperature rise at an instant of time when $T = 400$ K, the rate at which work is being done on the air is 8.165 kW, and heat is being removed at a rate of 1.0 kW.

SOLUTION

IDEAL GAS: $Mc_v(T_2 - T_1) + M\left(\frac{V_2^2}{2} - \frac{V_1^2}{2}\right) + Mg(z_2 - z_1)$

Choose the air as the system. Assume air to be an ideal gas ($Pv = RT$, $du = c_v dT$) and assume $E = U$ (the system is assumed stationary). Using the rate form of the first law we have

$$\dot{Q} - \dot{W} = \frac{dE}{dt} = \frac{dU}{dt} = \frac{d(Mu)}{dt} = \frac{d}{dt}(Mc_v T)$$

$$\dot{Q} - \dot{W} = Mc_v \frac{dT}{dt}$$

$$-1 - (-8.165) = 0.01(0.7165)\frac{dT}{dt}$$

$$\frac{dT}{dt} = \frac{7.165}{0.007165} = 1000 \text{ K/s}$$

COMMENT

This may appear to be a very high rate of temperature increase; however, it can be shown to be realistic. Consider a reciprocating compressor running at 360 rpm. A compression stroke occurs in $\frac{1}{12}$ of a second so that $\frac{1}{12}(1000) = 83.3$ K is the temperature change per stroke if this rate of change is assumed constant throughout the stroke.

When a system undergoes a series of processes which returns it to its original state we say the system has executed a thermodynamic cycle. The value of any property of the system at the end of the cycle is identical to its value at the beginning of the cycle; thus

$$\oint dY = 0 \tag{4–11}$$

where Y represents any property and $\oint$ represents the integral over the cycle. Since E is a property we have

$$\oint \delta Q - \oint \delta W = \oint dE = 0 \tag{4–12}$$

Therefore

$$\oint \delta Q = \oint \delta W \tag{4–13}$$

or

$$Q_{\text{cycle}} = W_{\text{cycle}} \tag{4–14}$$

The *net* work done on or by the system when executing a cycle is identically equal to the *net* heat transferred during the cycle.

EXAMPLE 4–2

An ideal gas is compressed reversibly and isothermally from a volume of 0.01 m³ and a pressure of 0.1 MPa to a pressure of 1.0 MPa. How much heat is transferred during this process?

SOLUTION

Choose the gas as the system and assume $E = U$ (stationary). Since the temperature is constant, u is constant and the first law, $\delta Q - \delta W = dE = dU$ reduces to

$$\int_1^2 \delta Q - \int_1^2 \delta W = \int_1^2 dU = \int_1^2 M\, du = 0$$

or

$$_1Q_2 = {_1W_2}$$

The work can be evaluated by integrating $P\, dV$ for the process

$$\int_1^2 P\, dV = MRT \int_1^2 \frac{dV}{V} = MRT \ln\left(\frac{V_2}{V_1}\right) = {_1W_2}$$

but $MRT = P_1V_1 = P_2V_2$; thus $P_1/P_2 = V_2/V_1$. Therefore

$$_1Q_2 = {_1W_2} = P_1V_1 \ln\left(\frac{P_1}{P_2}\right) = 100(0.01) \ln\left(\frac{0.1}{1.0}\right)$$

$$_1Q_2 = -2.303 \text{ kJ}$$

COMMENT

The negative sign indicates that this heat is removed from the system. Note that for a reversible isothermal process the heat transfer is independent of the gas, provided the gas exhibits ideal gas behavior.

EXAMPLE 4–3

The volume below the weighted piston in the cylinder shown in Fig. E4–3 contains 0.01 kg of water. The piston area is 0.01 m² and the mass on the piston is 102 kg. The top of the piston is exposed to the atmosphere which is at a pressure of 0.1 MPa. Initially the water is at 25 °C and the final state of the water is saturated vapor. How much heat is added to the water and how much work is done on or by the water in going from the initial to the final state?

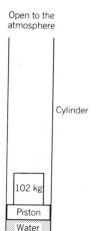

Open to the
atmosphere

Cylinder

102 kg

Piston

Water

Figure E4-3 Piston–cylinder device containing water.

SOLUTION

The process is a quasiequilibrium process at constant pressure so that the work can be calculated by

$$_1W_2 = \int_1^2 P \, dV = P(V_2 - V_1)$$

The pressure must first be calculated as

$$P = \frac{F}{A} = \frac{P_{atm}A + Mg}{A} = P_{atm} + \frac{Mg}{A}$$

$$P = 100 + \frac{102(9.807)}{0.01(1000)} = 200 \text{ kPa}$$

The initial specific volume, v_1, is found from Table A–1.1 at 25 °C. Note that the pressure of 200 kPa is well above the saturation pressure for this temperature so the state of the liquid is in the compressed liquid region, but the approximate specific volume may be obtained by using the saturated liquid specific volume at the same temperature.

$$V_1 = Mv_1 = 0.01(0.001003) = 10.03 \times 10^{-6} \text{ m}^3$$

The final specific volume, v_2, is obtained from Table A–1.2 at a pressure of 0.2 MPa ($v_2 = v_g$ at 0.2 MPa) and V_2 is calculated as

$$V_2 = Mv_2 = 0.01(0.8857) = 8.857 \times 10^{-3} \text{ m}^3$$

Then

$$_1W_2 = 200(8.857 \times 10^{-3} - 10.03 \times 10^{-6}) = 1.769 \text{ kJ}$$

The first law states

$$_1Q_2 - {_1W_2} = E_2 - E_1 = U_2 - U_1 = M(u_2 - u_1)$$

Note that we assumed $E_2 - E_1 = U_2 - U_1$. We have neglected the fact that the center of mass of the system (the water) shifts to a higher elevation in the final state.

$$_1Q_2 = {_1W_2} + U_2 - U_1 = 1.769 + (2529.5 - 104.88)0.01$$

where u_2 and u_1 were obtained from Tables A–1.2 and A–1.1, respectively.

$$_1Q_2 = 26.01 \text{ kJ}$$

The sign is positive, indicating that this heat was added to the system.

COMMENT

Although the solution to the problem is completed, a more detailed examination of several aspects may further your understanding of the solution. We determined that the system did 1.769 kJ of work on the surroundings. This work was done on the atmosphere and in increasing the potential energy of the mass on the piston. The work done on the atmosphere is

$$-W_{atm} = \int P \, dV = P_{atm}(V_2 - V_1)$$

$$= 100(8.857 \times 10^{-3} - 10.03 \times 10^{-6}) = 0.8847 \text{ kJ}$$

The work done on the mass on the piston is the increase in potential energy of this mass. Applying the first law to the piston as a system yields

$$PE_{P2} - PE_{P1} = M_P g(z_2 - z_1)_p = -W_p$$

but $V = Az$; therefore,

$$z_2 - z_1 = \frac{V_2 - V_1}{A} = \frac{8.857 \times 10^{-3} - 10.03 \times 10^{-6}}{0.01}$$

$$= 0.8847 \text{ m}$$

$$PE_{P2} - PE_{P1} = 102 \frac{9.807(0.8847)}{1000} = 0.8850 \text{ kJ} = -W_p$$

The sum of the work done on the piston (-0.8850 kJ) and the work done on the atmosphere (-0.8847 kJ) is equal to the work done by the water. The increase in elevation of the center of mass of the water is $(z_2 - z_1)/2 = 0.4424$ m. The increase in potential energy of the system due to this increase in elevation is thus

$$PE_2 - PE_1 = 0.01(9.807)\frac{(0.4424)}{1000} = 43.39 \times 10^{-6} \text{ kJ}$$

This change is indeed negligible on a process where over 20,000 times this amount of energy represented the smallest energy transfer considered and over 500,000 times this amount of energy was added as heat on the process. Whether or not you should neglect changes in kinetic and potential energy in problems such as this can be more easily determined as you gain experience in working these problems. When in doubt, an estimate of the change can be made to determine the amount of error if the change is neglected. There are some problems, such as a small element of water moving from the surface of a large reservoir down to the exit of a hydraulic turbine at a much lower elevation, where the change in potential energy is not negligible. Similarly, the change in kinetic energy of a small element of water as it moves through the nozzle on the end of a fire hose would not be negligible.

4.2   THE SECOND LAW OF THERMODYNAMICS

The first law of thermodynamics was neither proved by some simple experiment nor was it derived from some other fundamental considerations. It was simply stated and its proof lies in the fact that violations of that law have not been observed. The second law of thermodynamics is similar in this regard; its proof lies in the fact that violations have not been observed. It too is somewhat negative in its approach. The first law pointed out that energy *cannot* be created or destroyed, while the second law dictates that certain processes *cannot* occur. The first law does not recognize a distinction between heat and work, but the second law makes a very clear distinction between heat and work.

We will begin the discussion of the second law by observing that our experience indicates that certain processes do not occur naturally. There seems to be a natural direction for certain processes. Examples are

1.   A gas can undergo a free expansion but a "free compression" has not been observed to occur naturally.
2.   Fuel oil and air react to form carbon dioxide and water but carbon dioxide and water do not naturally react to form fuel oil.
3.   A hot cup of coffee will cool to ambient temperature but a cup of coffee at ambient temperature will not become a hotter cup naturally.

Thus processes of a mechanical, chemical, and thermal nature seem to have directions in which they will naturally proceed and directions in which they will not proceed unless assisted by external forces. The second law of thermodynamics provides a formal means for determining the natural direction of such processes.

A microscopic approach to the second law shows that this natural direction for the process is toward the system's state of maximum probability, its most random state. This natural direction for processes is from order to disorder. We will not pursue this microscopic approach since our main interest is in a macroscopic study of thermodynamics. It is mentioned only to aid in understanding the significance of the second law. More details on this microscopic approach can be found in Refs. 1–4.

4.2.1. Classical Statements of the Second Law

We will examine two different statements of the second law.

The Clausius statement: *It is impossible to construct a device that operates in a cycle and produces no effect other than the transfer of heat from a cooler body to a hotter body*.

The Kelvin–Planck statement: *It is impossible to construct a device that operates in a cycle and produces no effect other than the production of work and exchange of heat with a single reservoir*.

Both of these statements are negative statements; they note that something is impossible. They both are concerned with devices that operate in a cycle. This cyclic requirement is necessary for continuously operating devices. Operation on a process would require an end to the process; it could not continue indefinitely in time. The Kelvin–Planck statement uses the term reservoir which means a large body from which heat may be removed, or to which heat may be added, without changing the temperature of the reservoir. A reservoir thus has a constant temperature. When heat is removed from the reservoir it is frequently called a source and when heat is added to the reservoir it is called a sink.

The device the Clausius statement prohibits is shown schematically in Fig. 4–1 while a device that directly violates the Kelvin–Planck statement is shown in Fig. 4–2. The first law, when applied to cycle A in Fig. 4–1, requires that $Q_H = Q_L$, and when applied to cycle B in Fig. 4–2 requires that $Q_H = W$.

It should be clear that these statements are giving the natural direction of certain processes. The Clausius statement is telling us that heat cannot flow from a low temperature to a high temperature. The natural direction is the opposite, from a high temperature to a low temperature. The Kelvin–Planck statement is telling us that heat cannot continuously and completely be converted into work. Experience shows us that the reverse process is the natural process; work can be completely and continuously converted into heat.

It is important to keep the cyclic or continuous requirement in mind. In Example 4–2 the work output for an isothermal expansion *process* for an ideal gas was shown to be equal to the heat added. But that was a process, not a cycle. The volume increased on that process; however, at some point the process would have to stop because it could not continue indefinitely.

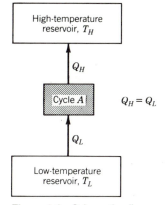

$Q_H = Q_L$

Figure 4-1 Schematic diagram of a Clausius statement violator.

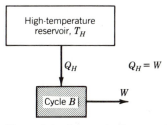

Figure 4-2 Schematic diagram of a Kelvin–Planck statement violator.

4.2.2. Heat Engines and Heat Pumps

A *heat engine* is a device that operates in a thermodynamic cycle and produces a net positive work while receiving heat from a high-temperature source and rejecting heat to a low-temperature sink. Figure 4–3 shows a schematic diagram of a heat engine. The heat engine operates in a cycle so that it can operate continuously. The working substance in the heat engine may be a solid, a liquid, a vapor, or it may undergo changes in phase. A steel wire, a rubber band, or a thermoelectric converter are examples of solid-phase working substances. Most of the working substances considered in this text will deal with fluids, either ideal gases or substances undergoing phase changes.

The large electrical generating stations used by utilities to produce electrical power use heat engines that have water as the working fluid. The boiler produces a vapor from the liquid and the condenser returns the vapor to the liquid phase. Figure 4–4 shows a very simplified schematic diagram of such a system. The heat added to the water in the boiler could come from a high-temperature zone (the source) created by the combustion of coal or from the nuclear fission process. The heat rejected from the condenser ultimately finds its way into the environment (the sink), typically through the use of a cooling tower. The similarity

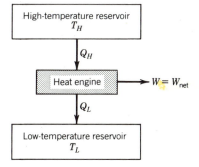

Figure 4-3 Schematic diagram of a heat engine.

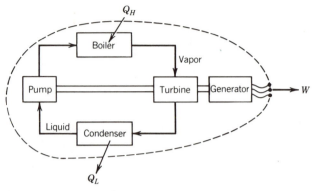

Figure 4-4 Simplified schematic diagram of an electrical power generation station.

between Figs. 4–3 and 4–4 should be evident. Figure 4–4 simply provides a few more details on the heat engine.

A first law analysis on the heat engines of Figs. 4–3 and 4–4 yields

$$\oint \delta Q = \oint \delta W \tag{4–15}$$

$$Q_H - Q_L = W \tag{4–16}$$

Note that a minus sign has been inserted before Q_L in eq. 4–16 so that Q_L must be considered a positive quantity. The proper sign for Q_L is negative if the heat engine is the system and is positive if the sink is the system. To avoid confusion on this point, the subscript L is used to identify the direction and Q_L is always considered a positive quantity. Thus, the appropriate sign must be included in an equation where Q_L appears.

Equation 4–16 expresses the concept that all of the heat supplied to the heat engine is not converted into work; some energy is rejected as heat to a low-temperature sink as required by the second law. A measure of how much heat is converted into work is given by the *thermal efficiency* (sometimes called the cycle efficiency or cycle thermal efficiency) defined as

$$\eta_{th} \equiv \frac{W}{Q_H} = \frac{Q_H - Q_L}{Q_H} \tag{4–17}$$

In this equation W represents the net cycle work but Q_H represents only that energy that is *supplied* as heat, and is *not* the net cycle heat transfer. The thermal efficiency can also be expressed as a ratio of rates as

$$\eta_{th} = \frac{\dot{W}}{\dot{Q}_H} = \frac{\dot{Q}_H - \dot{Q}_L}{\dot{Q}_H}$$

The efficiency is a ratio of the useful work output (what a power station sells) to the input (what a power station buys in coal or nuclear fuel). The Kelvin–Planck statement of the second law tells us this efficiency cannot be 100%, but how high can it go? Does it have a limit? A simple heat engine such as shown in Fig. 4–4 might have a thermal efficiency of about 30%, while a modern coal-fired power station (which is a lot more complex than Fig. 4–4) might have a thermal efficiency of slightly over 40%. But this does not answer the question regarding the existence of a limit, and if a limit exists, what is it? We shall address this issue in Section 4.2.4.

If the heat engine of Fig. 4–3 were reversed we would reverse the three arrows shown in that figure and have a device that absorbed work. It would take heat in from a low-temperature source and reject heat to a high-temperature sink. Such a device is called a heat pump, because it moves energy as heat from a low-temperature region to a higher temperature region, somewhat like a pump would move water from a low-pressure (or elevation) region to a high-pressure (or elevation) region. A schematic diagram of such a cycle is shown in Fig. 4–5. This device does not violate the second law since work is required to transfer the heat.

Refrigerators are a special class of heat pumps where the low temperature, T_L, is below ambient and the "useful" energy is Q_L. A measure of performance of heat pumps is called the *coefficient of performance*, β, defined as

$$\beta \equiv \frac{\text{magnitude of useful output}}{\text{magnitude of required input as work}}$$

The magnitude of the energy quantities is used to obtain a positive value of β. Thus, for a refrigerator we have

$$\beta_R = \frac{Q_L}{W} = \frac{Q_L}{Q_H - Q_L} = \frac{1}{(Q_H/Q_L) - 1} \tag{4-18}$$

and for a heat pump we have

$$\beta_{HP} = \frac{Q_H}{W} = \frac{Q_H}{Q_H - Q_L} = \frac{1}{1 - (Q_L/Q_H)} \tag{4-19}$$

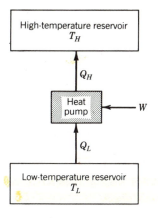

Figure 4-5 Schematic diagram of a heat pump.

EXAMPLE 4–4

A heat pump delivers heat at a rate of 10 kW to a house while using 4 kW of power. What is the coefficient of performance of this heat pump?

SOLUTION

In the statement of the problem we are given rates of energy transfer rather than energy transfers, but the ratio of the rates is equal to the ratio of energy transfers so that

$$\beta_{HP} = \frac{Q_H}{W} = \frac{\dot{Q}_H}{\dot{W}} = \frac{10}{4} = 2.5$$

EXAMPLE 4–5

If a refrigeration cycle used in a house air conditioning system delivers heat to the outside air at a rate of 10 kW while using 4 kW of power, what is its coefficient of performance?

SOLUTION

$$\beta_R = \frac{Q_L}{W} = \frac{\dot{Q}_L}{\dot{W}} = \frac{\dot{Q}_H - \dot{W}}{\dot{W}} = \frac{\dot{Q}_H}{\dot{W}} - 1 = \frac{10}{4} - 1 = 1.5$$

COMMENT

Note that for the same values of $\dot{Q}_H$ and $\dot{W}$ the heat pump coefficient of performance is equal to the refrigerator coefficient of performance plus 1.0.

4.2.3 Externally Reversible Cycles; the Carnot Cycle

The concept of reversibility was introduced in Chapter 2. If all irreversible effects within the system are ignored, the process is said to be reversible. At this point we wish to extend this concept to a cycle and include effects at the boundary of the system. If no irreversibilities occur within the system during a process, the process is said to be *internally reversible*. If, *in addition*, no irreversibilities occur at the boundary, the process is said to be *externally reversible.* Thus externally reversible processes are a subset of internally reversible processes. The type of irreversibility that most commonly occurs at a boundary is the transfer of heat through a finite temperature difference. To be externally reversible the process must first be internally reversible and the system and surroundings must be at the same temperature when direct heat transfer takes place. An externally reversible cycle is one in which all the processes are externally reversible.

There are many externally reversible cycles and the conclusions reached from the study of any one externally reversible cycle are valid for the others. The cycle commonly used to represent these cycles is the Carnot cycle, named after Sadi Carnot (1796–1832), a French engineer who in 1824 published a small book *Reflections on the Motive Power of Heat and on Machines Fitted to Develop that Power.* This work, although based on the caloric theory, is the first work to contain the basic concept of a cycle and the limitations imposed by the second law of thermodynamics.

Regardless of the working substance, the Carnot cycle heat engine consists of four externally reversible processes. They are

1–2	Reversible addition of heat, Q_H, at constant temperature, T_H
2–3	Reversible adiabatic decrease in temperature from T_H to T_L
3–4	Reversible rejection of heat, Q_L, at constant temperature, T_L
4–1	Reversible adiabatic increase in temperature from T_L to T_H

The Carnot cycle using an ideal gas as the working fluid is shown on P–v coordinates in Fig. 4–6. The isothermal lines ($n = 1$) are not as steep as the reversible adiabatic lines ($n = \gamma$) as noted in Chapter 3. Work is involved in each of the four processes as evidenced by the areas under the paths. The area enclosed by the cycle on this figure is proportional to the net cycle work.

4.2.4 An Absolute Temperature Scale and the Carnot Efficiency

All externally reversible heat engines operating between the same two reservoirs have the same thermal efficiency. The validity of this statement can be shown by *assuming* that it is not true and showing that this assumption yields a direct

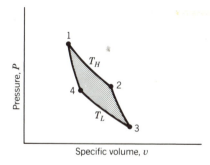

Figure 4-6 The Carnot cycle for an ideal gas working substance.

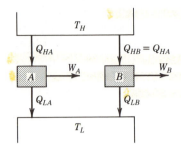

Figure 4-7 Externally reversible heat engines A and B operating between the same two reservoirs.

violation of the Kelvin–Planck statement of the second law. Figure 4–7 shows two externally reversible heat engines, A and B, operating between two reservoirs. Let us first *assume* that engine A is more efficient than engine B and then set $Q_{HA} = Q_{HB}$, thus supplying the same amount of heat to each engine. The work output of A will be greater than the output of B since A has the higher efficiency. Let us now reverse engine B to make it operate as a heat pump and drive it with engine A. We may reverse engine B since it is an externally reversible device. Engine A produces more work than required to drive the heat pump; thus we get a net work output from the combination as shown in Fig. 4–8. The high temperature reservoir can be eliminated since Q_{HA} and Q_{HB} are equal in magnitude and opposite in direction relative to this reservoir. The net result (shown in Fig. 4–8) is a device operating in a cycle and producing work while exchanging heat with a single reservoir. This is a direct violation of the Kelvin–Planck statement and is clearly impossible. Everything that we did in arriving at this configuration was legitimate except the initial assumption that one externally reversible heat engine could have a higher thermal efficiency than another. This assumption was clearly incorrect so that we can conclude that all externally reversible heat engines operating between the same two reservoirs (temperatures) have the same thermal efficiency.

The thermal efficiency of an externally reversible heat engine must therefore be a function only of the temperatures of these reservoirs since temperature is the only relevant property of a reservoir. Thus

$$\eta_{th} = \frac{W}{Q_H} = \frac{Q_H - Q_L}{Q_H} = 1 - \frac{Q_L}{Q_H} = f_1\left(T_H, T_L\right) \qquad (4\text{-}20)$$

or

$$\frac{Q_L}{Q_H} = f_2\left(T_H, T_L\right) \qquad (4\text{-}21)$$

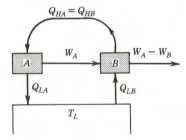

Figure 4-8 Heat engine B reversed and driven by engine A. (B and A are the heat engines shown in Fig. 4-7).

We can apply eq. 4–21 to the Carnot heat engines shown in Fig. 4–9 to obtain

$$\frac{Q_1}{Q_2} = f(T_1, T_2)$$

$$\frac{Q_2}{Q_3} = f(T_2, T_3)$$

$$\frac{Q_1}{Q_3} = f(T_1, T_3)$$

but

$$\frac{Q_1}{Q_3} = \left(\frac{Q_1}{Q_2}\right)\left(\frac{Q_2}{Q_3}\right)$$

or

$$f(T_1, T_3) = f(T_1, T_2)f(T_2, T_3) \qquad (4\text{–}22)$$

Since the left side of eq. 4–22 is not a function of T_2, the right side cannot be a function of T_2. This requires that the functional relationship, f, be of such a form that the T_2 dependence cancels in the right side. The form

$$f(T_1, T_2) = \frac{g(T_1)}{g(T_2)} \qquad (4\text{–}23)$$

satisfies this requirement. Substituting eq. 4–23 into eq. 4–22 yields

$$\frac{g(T_1)}{g(T_3)} = \frac{g(T_1)g(T_2)}{g(T_2)g(T_3)} = f(T_1, T_3) = \frac{Q_1}{Q_3} \qquad (4\text{–}24)$$

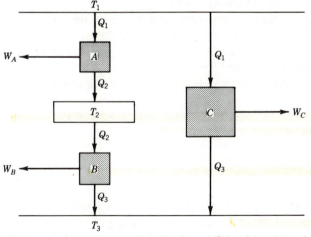

Figure 4-9 Carnot heat engines *A*, *B*, and *C* for determining the thermodynamic temperature scale.

or

$$\frac{Q_1}{Q_3} = \frac{g(T_1)}{g(T_3)}$$

The simplest functional relationship that can be chosen for g is that $g(T) = T$; thus

$$\frac{Q_1}{Q_3} = \frac{T_1}{T_3}$$

or in general for any externally reversible heat engine

$$\frac{Q_H}{Q_L} = \frac{T_H}{T_L} \qquad (4\text{--}25)$$

Equation 4–25, originally proposed by Lord Kelvin, constitutes a definition of an absolute thermodynamic temperature scale.

Equation 4–25 contains two temperatures, hence it does not completely specify the scale. Suppose we were to specify the size of the degree on this scale by writing

$$T_{\text{steam point}} - T_{\text{ice point}} = 100° \qquad (4\text{--}26)$$

Then if Q_H and Q_L could be measured for a Carnot engine operating with T_H equal to the normal boiling point and T_L equal to the normal freezing point, this ratio would be 1.366; thus

$$\frac{Q_H}{Q_L} = 1.366 = \frac{T_{\text{steam point}}}{T_{\text{ice point}}} \qquad (4\text{--}27)$$

Solving eqs. 4–26 and 4–27 simultaneously yields $T_{\text{steam point}} = 373.15$ and $T_{\text{ice point}} = 273.15$. We recognize these temperatures as values from the Kelvin scale which apparently satisfies the requirements of the absolute thermodynamic temperature scale. This method of obtaining the scale is of little practical significance since heat transfers had to be measured for a Carnot engine. You should recall that this cycle consists of externally reversible processes and in practice complete reversibility cannot be achieved.

With the determination of an absolute temperature scale the thermal efficiency of a Carnot engine can be determined as

$$\eta_{\text{th}} = \frac{W}{Q_H} = \frac{Q_H - Q_L}{Q_H} = 1 - \frac{Q_L}{Q_H} = 1 - \frac{T_L}{T_H} \qquad (4\text{--}28)$$

Thus all externally reversible heat engines have an efficiency given by eq. 4–28. This efficiency is called the Carnot efficiency. A knowledge of the temperature at which heat is added and the temperature at which heat is removed is sufficient to determine the thermal efficiency for externally reversible heat engines.

No heat engine can have a thermal efficiency higher than the Carnot efficiency when operating between the same two reservoirs. The proof of this statement is obtained by the same general technique used to prove that all externally reversible heat engines have the same efficiency when operating between the same two reservoirs. First an *assumption* is made that a heat engine can have a thermal efficiency higher than that of a Carnot engine. This higher efficiency engine is used to drive a reversed Carnot engine with the result that a violation of the Kelvin–Planck statement of the second law is obtained. Since this is an impossibility the assumption was in error. The details of this proof are left as an exercise for the student.

4.2.5   The Clausius Inequality

For the Carnot cycle of Fig. 4–10 we have

$$\oint_C \frac{\delta Q}{T} = \frac{Q_H}{T_H} - \frac{Q_{LC}}{T_L} \tag{4-29}$$

From eq. 4–25, for a Carnot cycle,

$$\frac{Q_H}{T_H} = \frac{Q_{LC}}{T_L}$$

Therefore

$$\oint_C \frac{\delta Q}{T} = 0 \tag{4-30}$$

For the irreversible cycle

$$\oint_A \frac{\delta Q}{T} = \frac{Q_H}{T_H} - \frac{Q_{LA}}{T_L} \tag{4-31}$$

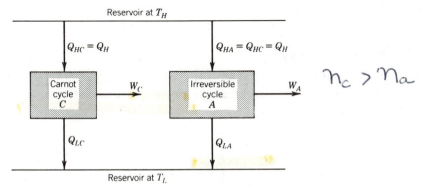

Figure 4-10   Reversible and irreversible heat engines operating between the same reservoirs and with the same heat supplied.

Since both cycles receive the same amount of heat Q_H from the same temperature T_H and reject heat to the same temperature T_L, the Carnot cycle will produce more work than the irreversible cycle (it has a higher thermal efficiency) so that

$$W_C > W_A \tag{4–32}$$

Application of the first law to the cycles yields

$$Q_H - Q_{LC} = W_C$$

$$Q_H - Q_{LA} = W_A$$

Hence it follows that

$$Q_{LA} > Q_{LC} \tag{4–33}$$

Thus

$$\frac{Q_H}{T_H} - \frac{Q_{LA}}{T_L} < \frac{Q_H}{T_H} - \frac{Q_{LC}}{T_L}$$

and

$$\oint_A \frac{\delta Q}{T} < \oint_C \frac{\delta Q}{T} = 0$$

In general then we have

$$\oint \frac{\delta Q}{T} \leq 0 \tag{4–34}$$

where the equal sign applies to externally reversible cycles and the less than sign to other cycles. Equation 4–34 is called the *inequality of Clausius*. It applies to all cycles, power cycles as well as heat pump cycles, although we did not show its applicability to the latter. The procedures for doing so are similar to those shown here for the heat engine.

4.2.6 The Property Entropy

The value of the cyclic integral of any property is zero, a fact previously noted in eq. 4–11. We can therefore conclude that $(\delta Q/T)_{Rev}$ is the differential of some thermodynamic property of the system. We will call this property *entropy*, and define it as

$$dS \equiv \left(\frac{\delta Q}{T}\right)_{Rev} \tag{4–35}$$

where δQ is the heat transfer to $(+)$ or from $(-)$ the system and T is the absolute temperature of the system. The units of entropy are kJ/K.

Changes in entropy can be evaluated by integrating eq. 4–35 to obtain

$$\int_1^2 dS = S_2 - S_1 = \int_1^2 \left(\frac{\delta Q}{T}\right)_{Rev} \tag{4-36}$$

If the process 1–2 is reversible, a direct evaluation of the integral on the right side of eq. 4–36 requires a knowledge of the relationship between temperature and heat transfer for that process. For example, if 300 kJ of heat is added reversibly to a system at a constant temperature of 300 K then

$$S_2 - S_1 = \frac{1}{T}\int_1^2 \delta Q = \frac{{}_1Q_2}{T} = \frac{300}{300} = 1.0 \text{ kJ/K}$$

Entropy is a property and the change in entropy is independent of the path taken during the process. In Fig. 4–11 an irreversible (actual) process between states 1 and 2 is shown by a broken line. The entropy change, $S_2 - S_1$, can be evaluated by computing it along a reversible path such as 1–a–2, 1–b–c–2, 1–c–2, or any other reversible path for which the integral can be evaluated. Note that the Q and T are for the reversible path used between the end states and are not the Q and T for the irreversible process.

Only *changes* in entropy can be calculated from this definition. This situation is analogous to energy, E, which was defined by the first law. For most of the discussion in this text where chemical reactions are not considered, absolute values are not required, and an arbitrary datum may be established. This was done in Tables A–1, the tabular properties for water. More complete texts on thermodynamics[1–5] contain discussions of the third law of thermodynamics and absolute values of entropy.

Since entropy is a new concept and cannot be directly measured with an entropy gauge, you may have difficulty obtaining a physical "feel" for the property. However, there are other properties that cannot be directly measured, but have proven useful. Internal energy is such a property. The fact that entropy cannot be directly measured does not make it any less real or any less important. Its value as a property lies in its usefulness in applying the second law of thermodynamics to real problems, and this is a welcome assistance. Certainly the classical statements of the second law are not particularly easy to apply to the

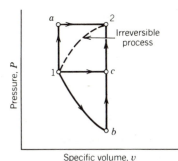

Figure 4-11   Calculation of entropy change.

variety of systems engineers wish to consider. As you use the property entropy you will develop a better understanding of its significance as well as the significance of the second law.

The entropy, S, is an extensive property of a system. The specific entropy, defined as $s = S/M$, is also used frequently. Tables of properties for pure substances (e.g., the tables for water in this text) list this property. In the two-phase region (liquid–vapor), specific entropy is computed via a relationship similar to that used to calculate the specific volume and specific internal energy

$$s = (1 - x)s_f + xs_g \qquad (4\text{-}37)$$

where x is the quality of the mixture, and subscripts f and g refer to saturated liquid and saturated vapor states, respectively.

Since entropy is a property, it is a very useful coordinate for diagrams representing processes and cycles. It should be clear from the definition of entropy that the entropy change for a reversible adiabatic process is equal to zero. A reversible adiabatic process is an isentropic process (one of constant entropy). The Carnot cycle consists of two reversible adiabatic processes and two reversible isothermal processes so that it appears as a rectangle on a temperature–entropy (T–S) diagram, as shown in Fig. 4–12.

The T–S diagram is also useful for showing the amount of heat transferred during a reversible process. From the definition of entropy, eq. 4–35, we have

$$\delta Q_{\text{Rev}} = T\,dS$$

or

$$\int_1^2 \delta Q_{\text{Rev}} = {}_1Q_{2\text{Rev}} = \int_1^2 T\,dS \qquad (4\text{-}38)$$

Thus the area on a T–S diagram under the path of a reversible process is proportional to the heat transferred on that process. From Fig. 4–12 we can see that Q_L is less than Q_H since Q_L = area 3–4–a–b–3 and Q_H = area 1–2–b–a–1. The area enclosed by the cycle is therefore proportional to $Q_H - Q_L$, the

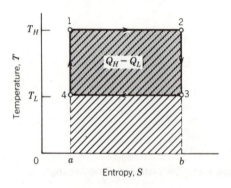

Figure 4-12   Temperature–entropy diagram for a Carnot cycle with any working substance.

net cycle heat transfer, and from the first law this net cycle heat transfer is equal to the net cycle work.

4.2.7 The Effect of Irreversibilities on Entropy

The irreversible cycle shown in Fig. 4–13 consists of irreversible process 1–2 and the reversible process 2–1. The Clausius inequality, eq. 4–34, for this cycle states that

$$\oint \frac{\delta Q}{T} < 0$$

$$\oint \frac{\delta Q}{T} = \int_1^2 \left(\frac{\delta Q}{T}\right)_{Irr} + \int_2^1 \left(\frac{\delta Q}{T}\right)_{Rev} < 0 \tag{4-39}$$

But process 2–1 is reversible; hence from the definition of entropy

$$\int_2^1 \left(\frac{\delta Q}{T}\right)_{Rev} = S_1 - S_2 \tag{4-40}$$

Substituting eq. 4–40 into eq. 4–39 and rearranging yields

$$S_2 - S_1 > \int_1^2 \left(\frac{\delta Q}{T}\right)_{Irr} \tag{4-41}$$

For a reversible process we know that

$$S_2 - S_1 = \int_1^2 \left(\frac{\delta Q}{T}\right)_{Rev} \tag{4-42}$$

so that we can write a general equation (combining eqs. 4–41 and 4–42) as

$$S_2 - S_1 \geq \int_1^2 \left(\frac{\delta Q}{T}\right) \tag{4-43}$$

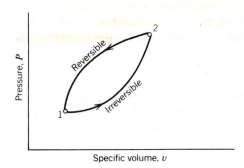

Figure 4-13 Irreversible cycle.

In differential form this can be written as

$$dS \geq \frac{\delta Q}{T} \qquad (4\text{-}44)$$

and in rate form as

$$\frac{dS}{dt} \geq \frac{\dot{Q}}{T} \qquad (4\text{-}45)$$

The equal sign applies to reversible processes and the greater than sign applies to irreversible processes. These equations are valid for heat addition (δQ positive), heat removal (δQ negative), and adiabatic ($\delta Q = 0$) processes.

The inequalities expressed in eqs. 4–43 through 4–45 can be removed by defining a term I, called *irreversibility*, which represents an entropy production due to irreversibilities. Thus we can convert eq. 4–44 to

$$dS = \frac{\delta Q}{T} + \delta I \qquad (4\text{-}46)$$

and eq. 4–45 to

$$\frac{dS}{dt} = \frac{\dot{Q}}{T} + \dot{I} \qquad (4\text{-}47)$$

The irreversibility term, δI, is always positive. In the limiting case of a reversible process it is zero. It is a function of the path and is not a property, but represents the entropy produced due to the irreversibility. $\dot{I}$ is the rate of entropy production due to the irreversibility.

Equation 4–46 points out several important aspects of entropy changes:

1. The entropy of a system can increase in only two ways, either by heat *addition* or by the presence of an irreversibility.

2. The entropy of a system can decrease in *only one way*, that is, by heat *removal.*

3. The entropy of a system cannot decrease during an adiabatic process.

4. The entropy change for an isolated system cannot be negative, that is,
$$dS_{\text{isolated system}} = \delta I \geq 0$$

5. All reversible adiabatic processes are isentropic; however, all isentropic processes are not necessarily reversible and adiabatic. The entropy can remain constant on a process if the heat removal balances the irreversibility.

4.2.8   The Principle of the Increase of Entropy

A small quantity of heat is removed from the surroundings at temperature T_∞ and added to a system at a lower temperature T, Fig. 4–14. The change in entropy for this system can be obtained using eq. 4–44.

$$dS_{\text{sys}} \geq \frac{\delta Q}{T}$$

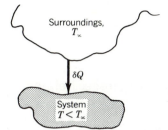

Figure 4-14   Irreversible heat transfer.

We will consider the term δQ to be a positive number so that when we apply eq. 4-44 to the surroundings we get

$$dS_{sur} = -\frac{\delta Q}{T_\infty}$$

The reversible form of eq. 4-44 can be used because the heat is removed from the surroundings at constant temperature, T_∞. The minus sign of course indicates that this heat is removed. The total change of entropy of the universe, where the universe is defined as the system plus surroundings, is

$$dS_{universe} = dS_{sys} + dS_{sur} \geq \frac{\delta Q}{T} - \frac{\delta Q}{T_\infty}$$

$$dS_{universe} \geq \delta Q\left(\frac{1}{T} - \frac{1}{T_\infty}\right) \tag{4-48}$$

Since T_∞ must be greater than or in the limit equal to T, the quantity $(1/T) - (1/T_\infty)$ is always positive (or in the limit zero). Since δQ was also considered to be positive, it follows from eq. 4-48 that

$$dS_{universe} \geq 0 \tag{4-49}$$

The equal sign applies for externally reversible processes, while the greater than sign applies for processes that are not externally reversible.

Equation 4-49 is often referred to as the *principle of the increase of entropy*. It indicates that the natural direction for processes to occur is in the direction where an increase of entropy of the universe takes place. Processes that would create a decrease in the entropy of the universe do not occur.

EXAMPLE 4-6

A system of 2.0 kg water is initially saturated liquid water at a pressure of 10.0 MPa (state 1). Heat from a reservoir at a temperature of 800 K is added to this system at constant pressure until the final state (state 2) of the system is saturated vapor at a pressure of 10.0 MPa. Compute ΔS for the system and ΔS for the universe for this process.

SOLUTION

$$\Delta S_{sys} = S_2 - S_1 = M(s_2 - s_1)$$

s_1 and s_2 are obtained from Table A–1.2 as s_f and s_g, respectively, at a pressure of 10 MPa.

$$\Delta S_{sys} = 2.0(5.6141 - 3.3596) = 4.509 \text{ kJ/K}$$

A second method for obtaining ΔS_{sys} is to use the first law to calculate $_1Q_2$ and use the definition of entropy to compute ΔS_{sys}. The first law yields

$$_1Q_2 - _1W_2 = U_2 - U_1$$

$$_1Q_2 = U_2 - U_1 + \int_1^2 P \, dV = U_2 - U_1 + P_2V_2 - P_1V_1 = H_2 - H_1$$

From Table A–1.2, h_1 and h_2 are obtained at $P = 10$ MPa as h_f and h_g, respectively.

$$_1Q_2 \quad = H_2 - H_1 = M(h_2 - h_1) = 2.0(2724.7 - 1407.56)$$

$$_1Q_2 \quad = 2634.28 \text{ kJ}$$

$$\int_1^2 dS = \int_1^2 \left(\frac{\delta Q}{T}\right)_{Rev} = S_2 - S_1 = \Delta S_{sys} = \frac{_1Q_2}{T}$$

From Table A–1.2 at $P = 10.0$ MPa, $T = 311.06 \,°C = 584.21$ K

$$\Delta S_{sys} = \frac{2634.28}{584.21} = 4.509 \text{ kJ/K}$$

This is the same answer as that previously obtained in a more simple manner.

To evaluate ΔS_{sur} we again use the definition of entropy

$$\int_1^2 dS_{sur} = \int_1^2 \left(\frac{\delta Q_{sur}}{T}\right)_{Rev} = \Delta S_{sur} = \left(\frac{_1Q_2}{T}\right)_{sur}$$

We have determined that 2634.28 kJ were added to the system from the surroundings (the reservoir); thus $(_1Q_2)_{sur} = -2634.28$ kJ. T_{sur} was given as 800 K.

$$\Delta S_{sur} = \frac{-2634.28}{800} = -3.293 \text{ kJ/K}$$

$$\Delta S_{universe} = \Delta S_{sys} + \Delta S_{sur} = 4.509 - 3.293 = 1.216 \text{ kJ/K}$$

COMMENT

Since $\Delta S_{universe}$ is greater than zero, this process is not *externally* reversible. There are no irreversibilities *in* the system (we assumed the process reversible (internally) in calculating ΔS_{sys} by the second method) and there are no irreversibilities in the reservoir (we again assumed reversibility in the reservoir in calculating ΔS_{sur}). However, there is an irreversibility at the boundary between the system and the surroundings where we have heat transfer across a finite temperature difference (from 800 to 584.21 K). Thus the process is not externally reversible although it is internally reversible.

4.3 THE *T–dS* EQUATIONS FOR A SIMPLE COMPRESSIBLE SUBSTANCE

The first law for a system that consists of a simple compressible substance with no motion or gravitational effects present is

$$\delta Q - \delta W = dU$$

If the process this system undergoes is reversible, then

$$\delta Q = T \, dS$$

and

$$\delta W = P \, dV$$

so that the first law yields

$$T \, dS - P \, dV = dU$$

or

$$T \, dS = dU + P \, dV \tag{4–50}$$

From the definition of enthalpy

$$H = U + PV$$

Hence

$$dH = dU + P \, dV + V \, dP$$

This equation, when solved for dU and substituted into eq. 4–50, yields

$$T \, dS = dH - V \, dP \tag{4–51}$$

Equation 4–50 is called the first $T \, dS$ equation and eq. 4–51 is called the second $T \, dS$ equation. They are property relationships which are extremely useful for calculating entropy changes for any process, reversible as well as irreversible, for a simple compressible substance in a stationary system. Although the assumption of a reversible process was made in the derivation, these equations involve only properties that are independent of path and can be used for any process between two end states. This is essentially the same concept as fitting a reversible process between the end states to evaluate the entropy change between those states independent of the actual process that took place.

The first and second $T\,dS$ equations, written in terms of the intensive properties, are

$$T\,ds = du + P\,dv$$

and

$$T\,ds = dh - v\,dP$$

EXAMPLE 4–7

A system of 10 kg of saturated liquid water at 10 kPa has its pressure increased to 1.0 MPa during a reversible adiabatic process. If the liquid is assumed incompressible, what is the change in internal energy and what is the change in enthalpy for this process?

SOLUTION

The process is stated to be reversible and adiabatic so that $ds = 0$. The assumption of an incompressible liquid means that the density and specific volume do not change, thus $dv = 0$.
Using the first $T\,ds$ equation we have

$$T\,ds = du + P\,dv$$

$$0 = du + 0$$

Therefore $u_2 - u_1 = 0 = U_2 - U_1$.
Using the second $T\,ds$ equation we have

$$T\,ds = dh - v\,dP$$

$$0 = dh - v\,dP$$

$$dh = v\,dP$$

$$\int_1^2 dh = = h_2 - h_1 = \int_1^2 v\,dP = v\int_1^2 dP = v(P_2 - P_1)$$

$$v_f = 0.001010 \text{ m}^3/\text{kg at 10 kPa (Table A–1.2)}$$

$$h_2 - h_1 = 0.001010(1000 - 10) = 0.9999 \text{ kJ/kg}$$

$$H_2 - H_1 = M(h_2 - h_1) = 10(0.9999) = 9.999 \text{ kJ}$$

For an ideal gas

$$du = c_v \, dT$$

and

$$P = \frac{RT}{v}$$

so that

$$\int_1^2 ds = \int_1^2 c_v \frac{dT}{T} + \int_1^2 R \frac{dv}{v} \tag{4–52}$$

If c_v is assumed constant, this equation can be easily integrated to yield

$$s_2 - s_1 = c_v \ln\left(\frac{T_2}{T_1}\right) + R \ln\left(\frac{v_2}{v_1}\right) \tag{4–53}$$

Similarly the second $T \, ds$ equation can be used to obtain for an ideal gas with constant specific heats

$$s_2 - s_1 = c_p \ln\left(\frac{T_2}{T_1}\right) - R \ln\left(\frac{P_2}{P_1}\right) \tag{4–54}$$

If the variation in specific heats for the process is too great to be ignored, then the functional relationship between c_v and T must be used before integration of eq. 4–52.

EXAMPLE 4–8

5.0 kg of air expands isothermally from a volume of 1.0 m^3 to a volume of 5.0 m^3. Assuming air an ideal gas with constant specific heats, compute the change in entropy of the air during the process.

SOLUTION

Equation 4–53 is directly applicable here since the assumption of an ideal gas with constant specific heats is made.

$$s_2 - s_1 = c_v \ln\left(\frac{T_2}{T_1}\right) + R \ln\left(\frac{v_2}{v_1}\right) = 0 + R \ln\left(\frac{V_2 M}{M V_1}\right)$$

From Table A–6 obtain a value of R for air of 0.287 kJ/kgK.

$$s_2 - s_1 = 0.287 \ln\left(\frac{5}{1}\right) = 0.4619 \frac{kJ}{kg \cdot K}$$

$$S_2 - S_1 = M(s_2 - s_1) = 5(0.4619) = 2.310 \frac{kJ}{K}$$

COMMENT

The entropy increased due to the heat addition.

4.4 TEMPERATURE–ENTROPY DIAGRAMS

Temperature–entropy diagrams are useful for showing certain processes and it is helpful to know the paths of these processes, particularly for ideal gases. For an ideal gas $du = c_v \, dT$, so that the first $T \, ds$ equation can be written as

$$T \, ds = du + P \, dv = c_v \, dT + P \, dv$$

For a constant-volume process $dv = 0$, thus

$$T \, ds = c_v \, dT$$

or

$$\left. \frac{dT}{ds} \right|_{v=c} = \frac{T}{c_v} \tag{4–55}$$

Equation 4–55 shows that the slope of a constant-volume line on a T–s diagram increases as the temperature increases. For a constant-pressure process $dP = 0$ and for an ideal gas $dh = c_p \, dT$, so that the second $T \, ds$ equation yields

$$T \, ds = dh - v \, dP = c_p \, dT - v \, dP$$

$$T \, ds = c_p \, dT \tag{4–56}$$

$$\left. \frac{dT}{ds} \right|_{p=c} = \frac{T}{c_p}$$

Thus the slope of a constant-pressure line on a T–s diagram also increases as the temperature increases, but at any given temperature the slope is less than that of a constant-volume line since c_p is always greater than c_v (recall that $c_p - c_v = R$). Figure 4–15 shows that on a T–s diagram the constant-volume line is more steep than the constant-pressure line for an ideal gas.

A temperature–entropy diagram for a pure substance in the liquid–vapor region is shown schematically in Fig. 4–16. A T–s diagram for water, drawn to scale, is included as Fig. A–6.

4.4.1 Process Efficiency

The T–s diagrams are useful for comparing actual processes to ideal processes. In many processes, such as the compression of a gas in a piston-cylinder device, the actual process involves very little heat transfer so that the process is essentially adiabatic. Thus the idealization of this type of process is a reversible adiabatic process. The actual process is then compared to the ideal process by a parameter

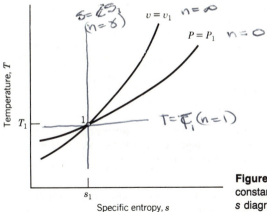

Figure 4-15 Constant specific volume and constant-pressure lines for an ideal gas on a T–s diagram.

called the adiabatic compression efficiency

$$\eta_{AC} \equiv \frac{W_i}{W_a} \qquad (4\text{--}57)$$

where W_a is the actual work of compression and W_i is the ideal (reversible adiabatic) work for compressing between the same pressures. The actual and ideal compression processes are illustrated in Fig. 4–17. The actual process is illustrated with a broken line since the path is not really known (but the initial and final states are known). We do, however, know that the entropy at state 2 is greater than at state 1 if the actual process is adiabatic. The reversible adiabatic process is of course isentropic, and state 2s is at the same pressure as state 2 and the same entropy as state 1.

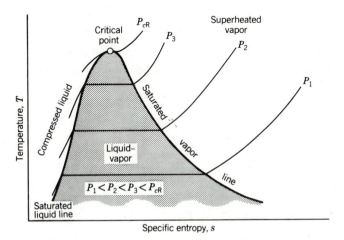

Figure 4-16 Temperature–entropy diagram showing the compressed liquid, liquid–vapor, and vapor regions.

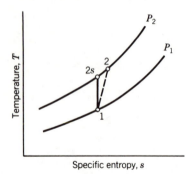

Figure 4-17 Temperature–entropy diagram for an ideal gas showing the actual and reversible adiabatic compression processes.

The efficiency defined by eq. 4–57 is a process efficiency and compares an actual process to an ideal process. It should not be confused with the thermal efficiency which was previously defined for a heat engine as the net work divided by the heat supplied. The efficiency of various processes can be defined in a similar manner by comparing the actual work or useful output (or input) of some process to the comparable output (or input) for an idealized process which closely approximates the real situation.

EXAMPLE 4–9

A system of 2 kg of air expands adiabatically from a pressure of 0.8 MPa and a temperature of 200 °C to a pressure of 0.1 MPa. If the adiabatic expansion efficiency is 85%, what is the air temperature after expansion and how much work is actually done?

SOLUTION

We will assume a stationary system and air to be an ideal gas with constant specific heats. Since the process is adiabatic and the system is stationary, the first law is

$$_1Q_2 - {_1}W_2 = U_2 - U_1$$

$$0 - {_1}W_2 = U_2 - U_1$$

$$_1W_2 = U_1 - U_2$$

and for an ideal gas

$$_1W_2 = Mc_v(T_1 - T_2)$$

The actual adiabatic process is shown as 1–2 in Fig. E4–9, while the ideal (reversible adiabatic) process is shown as 1–2s. State 2 must have a higher entropy than 2s since the process is adiabatic. The adiabatic expansion efficiency is defined as

$$\eta_{AE} = \frac{W_{actual}}{W_{ideal}} = \frac{{}_1W_2}{{}_1W_{2s}} = 0.85$$

Thus

$${}_1W_2 = 0.85\,{}_1W_{2s}$$

but

$${}_1W_{2s} = U_1 - U_{2s} = Mc_v(T_1 - T_{2s})$$

where T_{2s} can be calculated from the ideal gas P–T relationship for a reversible adiabatic process (eq. 3–41)

$$\left(\frac{P_2}{P_1}\right)^{(\gamma - 1)/\gamma} = \left(\frac{T_{2s}}{T_1}\right)$$

$$T_{2s} = T_1\left(\frac{P_2}{P_1}\right)^{(\gamma - 1)/\gamma} = 473.2\left(\frac{0.1}{0.8}\right)^{0.4/1.4} = 261.2 \text{ K}$$

From Table A–7 we get $c_v = 0.7165$ kJ/kg·K.
Then

$${}_1W_{2s} = 2.0\,(0.7165)(473.2 - 261.1) = 303.8 \text{ kJ}$$

and

$${}_1W_2 = 0.85(303.8) = 258.2 \text{ kJ}$$

We find T_2 from

$${}_1W_2 = U_1 - U_2 = Mc_v(T_1 - T_2)$$

$$258.2 = 2.0(0.7165)(473.2 - T_2)$$

$$T_2 = 293.0 \text{ K}$$

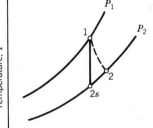

Temperature, T

Specific entropy, s

Figure E4-9 Temperature–entropy diagram for an ideal gas showing actual and reversible adiabatic expansion processes.

REFERENCES

1. Holman, J. P., *Thermodynamics*, 3rd ed., McGraw-Hill, New York, 1980.
2. Hatsopoulos, G. N., and Keenan, J. H., *Principles of General Thermodynamics*, Wiley, New York, 1965.
3. Morse, P. M., *Thermal Physics*, 2nd ed., Benjamin/Cummings Publishing, Reading, Mass., 1969.
4. Reynolds, W. C., *Thermodynamics*, McGraw-Hill, New York, 1965.
5. VanWylen, G. J., and Sonntag, R. E., *Fundamentals of Classical Thermodynamics*, 2nd ed., Revised Printing, SI Version, Wiley, New York, 1978.

PROBLEMS

4-1   A rigid tank containing 10 kg of air is being heated. If it is desired to increase the temperature of the air 100 °C in a period of 1000 s, at what rate must heat be added?

4-2   A system containing 3.0 kg of air is operated in a cycle consisting of the following three processes:

    1-2   A constant-volume heat addition: $P_1 = 0.1$ MPa, $T_1 = 20$ °C, and $P_2 = 0.2$ MPa

    2-3   A constant-temperature heat addition

    3-1   A constant-pressure heat rejection

    (a) Sketch this cycle on a p–v diagram.
    (b) Compute the work on each of the three processes 1–2, 2–3, and 3–1.
    (c) What is the net cycle work?
    (d) What is the net cycle heat transfer?
    (e) What is the cycle thermal efficiency?

4-3   A system consists of a flywheel with a mass of 70 kg and a mass moment of inertia of 5.096 kg · m² about its axis of rotation. If there is no heat transfer, how much work is done when the system changes from an elevation of 1.0 to 11.0 m while its speed of rotation changes from 10 to 5.64 revolutions per second? The temperature of the system remains constant.

4-4   Compute the rate of temperature change of 0.1 kg of air as it is being compressed adiabatically with a power input to the air of 1.0 kW.

4-5   Air, assumed to be an ideal gas with constant specific heats, is compressed in a closed piston–cylinder device in a reversible polytropic process with $n = 1.27$. The air temperature before compression is 30 °C and after compression is 130 °C. Compute the heat transferred on the compression process.

4-6   100 kJ of heat at 1000 K is added to a Carnot cycle. The cycle rejects heat at 300 K.
    (a) How much work does the cycle produce?
    (b) How much heat does the cycle reject?

4–7   A large utility power station produces 1000 MW of electrical power while operating with a cycle thermal efficiency of 40%. At what rate is heat rejected to the environment by this station?

4–8   A Brayton cycle is supplied with heat at a rate of 180 kW and rejects heat at a rate of 110 kW. The maximum cycle temperature is 1000 °C and the minimum cycle temperature is 100 °C. Compute the power produced and the cycle thermal efficiency.

4–9   A certain house requires a heating rate of 12 kW when the outside air is at −10 °C and the inside temperature is 21 °C.
(a) What is the minimum amount of power required to drive a heat pump to supply this heat at these conditions?
(b) List the factors that would cause the actual power requirement to be greater than this minimum

4–10   A device removes 10 kJ of heat from a low-temperature reservoir at 5 °C and delivers heat to a high-temperature reservoir at 25 °C. The work into the device is 5 kJ. Determine the coefficient of performance of this device if it is
(a) A heat pump.
(b) A refrigerator.

4–11   Paddle wheel work is done on a system containing 2.5 kg of air in a rigid tank to increase the air temperature from 20 to 200 °C. The tank is well insulated so that no heat transfer occurs.
(a) How much work is done?
(b) What is the irreversibility of the process?

4–12   100 kJ of heat is removed from a system containing 10 kg of steam while the system undergoes a reversible isothermal process at a temperature of 400 K. The heat is transferred from the system to the surroundings which are at a temperature of 300 K.
(a) What is the specific entropy change of the system?
(b) What is the total entropy change of the surroundings?
(c) What is the entropy change of the universe?

4–13   The entropy of a system containing 5.0 kg of helium is decreased by 6.5 kJ/K while the system undergoes a constant-pressure quasiequilibrium process. The initial temperature of the helium is 100 °C.
(a) What is the final temperature?
(b) What is the heat transfer for this process?

4–14   At a constant pressure of 0.1 MPa carbon dioxide sublimates at a temperature of −78.7 °C. The change in enthalpy for this sublimation process is 572.2 kJ/kg. Compute the entropy change for this sublimation process.

4–15   An incompressible liquid ($dv = 0$) experiences a reversible adiabatic process where the pressure is increased from 0.1 to 10 MPa. Compute the change in specific enthalpy for this process if the specific volume of the liquid is 0.001 m³/kg.

4–16   State whether the following processes are reversible, irreversible, or impossible, for liquid water (assumed incompressible) undergoing an adiabatic process.
(a) $u_2 - u_1$ is positive.
(b) $u_2 - u_1 = 0$.
(c) $u_2 - u_1$ is negative.

4–17   Saturated water vapor at $P = 0.40$ MPa is expanded reversibly and adiabatically in a piston–cylinder device to a pressure of $P = 0.1$ MPa.

(a) Sketch the process on a T–s diagram. Show the saturated vapor line.

(b) What is the quality of the water at state 2?

(c) How much work is done?

4–18   A system contains 0.1 kg of steam at a pressure of 1.0 MPa and a temperature of 250 °C. It expands adiabatically to a pressure of 0.15 MPa while producing 26 kJ of work.

(a) What is the actual quality of state 2?

(b) Calculate the adiabatic expansion efficiency defined as

$$\eta_{AE} = \frac{{}_1W_2 \text{ actual}}{{}_1W_2 \text{ reversible adiabatic}}$$

where the ideal work assumes the expansion process is both reversible and adiabatic from the initial state to the final pressure.

(c) Sketch the actual and ideal processes on a T–s diagram (show the saturated liquid and saturated vapor lines).

4–19   Sketch on a T–s diagram paths for an ideal gas undergoing reversible polytropic processes with $n = 0, 1, \gamma, \infty$.

4–20   A water-jacketed fireplace designed for heating water to be circulated through the baseboard units of a home heating system contains 0.035 m³ of water. An estimate of the amount of energy released when the unit ruptures is desired. Assume the unit is initially completely filled with saturated liquid at a pressure of 0.3 MPa and choose this mass as the system. The final state of this system is a combination of saturated liquid and saturated vapor at atmospheric pressure (0.10135 MPa). (When the pressure is reduced due to the rupture some of the liquid will evaporate into steam.)

(a) Compute the volume of the system at the final state.

(b) Although the actual rupture would most probably be extremely rapid like an explosion, estimate the energy released by *assuming* that the system does work on the atmosphere and this work can be computed by integrating $P\,dV$ with P equal to a constant (atmospheric).

5

Control Volume Analysis

5.1 INTRODUCTION

In Chapters 2, 3, and 4 we restricted ourselves to the discussion of thermodynamic systems, which means by definition that there is no flow or mass transport of energy across the system boundary. In this chapter we consider cases in which there is fluid flow and, hence, the mass transport of energy. This leads us to the thermodynamic analysis of control volumes.

As noted in Chapter 2 a thermodynamic analysis can be performed on a region in space through which mass flows. This region is called a control volume and its boundary, the control surface. As the mass flows into and out of the control volume, it transports energy. This mass transport of energy as well as the transfer of heat and work through the control surface must be considered when conducting a thermodynamic analysis. The study of an aircraft jet engine that has air flowing into its inlet and leaving through its exit nozzle is most conveniently accomplished using a control volume analysis. In this case the surface of the control volume consists of the inner walls of the engine and imaginary planes across the flow at the inlet and exit of the engine. At a particular instant of time, the control surface encloses all of the air within the engine and forms the control volume, Fig. 5–1. As we will see, the selection of the control surface is arbitrary, with the "best" control surface being dependent upon the information that is known about the particular problem being studied.

When we studied the transfer of energy to a system by heat transfer and work in Chapters 2 to 4, we were concerned with changes in the properties of the system, that is, changes in the pressure, temperature, specific volume, internal

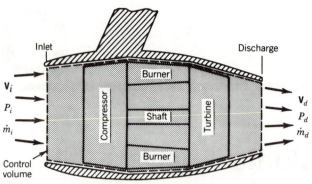

Figure 5-1   Control volume describing a jet engine. V = velocity (m/s), P = pressure (pa), $\dot{m}$ = massflow rate (kg/s)

energy, and entropy of the system. In considering a control volume we will still be concerned with property changes, but will also be concerned with the forces acting on the fluid moving through the control surface, or with the reaction of this fluid force on the control surface. The force on the fluid multiplied by the velocity of the fluid motion gives the power required to produce the motion.

It is necessary that we develop the tools that permit us to estimate the forces acting on the fluid. To do this we will first derive a general relationship called the *Reynolds transport theorem* (*RTT*) which relates the characteristics of a system to the characteristics of a control volume. We will continue to use a macroscopic approach to the study of energy and therefore will define the fluid as a continuum, meaning that it is continuous rather than being composed of a number of individual particles. The system characteristics we will investigate are the laws describing the conservation of mass, momentum, energy (the first law of thermodynamics), and the second law of thermodynamics.

We have discussed the first and second laws for a system in the previous chapters. You have studied the conservation of mass and momentum of a system in your physics courses. Before proceeding, however, these system conservation laws will be reviewed.

5.1.1   Conservation of Mass of a System

The law of the conservation of mass for a system states that the mass of the system, M, is always constant. Therefore, the rate of change of M with time is zero

$$\frac{dM}{dt} = 0 \qquad (5\text{--}1)$$

This is a scalar equation and is therefore independent of any coordinate direction.

5.1.2   Conservation of Momentum of a System

Newton's second law states that if a net force $\Sigma\mathbf{F}$ acts upon the system, the mass of the system will experience an acceleration $\mathbf{a}$ in the direction of $\Sigma\mathbf{F}$. The force $\Sigma\mathbf{F}$ and the acceleration $\mathbf{a}$ are related by

$$\Sigma\mathbf{F} = M\mathbf{a} = M\frac{d\mathbf{V}}{dt} \tag{5-2}$$

Since

$$\frac{dM}{dt} = 0$$

it follows that

$$\Sigma\mathbf{F} = \frac{d}{dt}(M\mathbf{V}) \tag{5-3}$$

The quantity $M\mathbf{V}$ is termed the linear momentum of the system in the direction of the velocity $\mathbf{V}$. Equation 5-3 states that the time rate of change of the linear momentum in the direction of $\mathbf{V}$ equals the resultant of *all* forces acting on the system in the direction of $\mathbf{V}$. The linear momentum equation is a vector equation and is therefore dependent upon a set of coordinate directions.

The term "linear" momentum indicates that the system motion proceeds along a continuous path in space. While it will not be considered here, there is also a law of the conservation of angular momentum which considers the rotation of the system around an axis. It states that the rate of change of angular momentum around an axis equals the sum of the torques around the axis.

$$\Sigma \text{ torque} = I\frac{d\Omega}{dt} \tag{5-4}$$

where I is the moment of inertia and Ω the angular velocity about the axis.

5.1.3   Conservation of Energy of a System

Section 4.1 presents the first law of thermodynamics and states that it is represents the conservation of energy of a system. This conservation law states that the algebraic (scalar) sum of all the energy transfers across the system's boundary must equal the change in energy of the system. Since heat and work are the only forms of energy which can be transferred across a system boundary, we have seen that, eq. 4-1,

$$\delta Q - \delta W = dE \tag{4-1}$$

We can also write this conservation law as a rate equation as long as the quasi-equilibrium assumption holds.

$$\dot{Q} - \dot{W} = \frac{dE}{dt} \qquad (4\text{--}7)$$

When heat is added to the system $\dot{Q} > 0$. When work is extracted from the system, the system produces work and $\dot{W} > 0$.

5.1.4   Second Law of Thermodynamics

The second law of thermodynamics for a system can be expressed in a variety of forms, but one of the most general statements is that given by eq. 4–45,

$$\frac{dS}{dt} \geq \frac{\dot{Q}}{T} \qquad (4\text{--}45)$$

This equation states that the rate of entropy change of a system will be greater than (or for a reversible process equal to) the rate of heat transfer divided by the system temperature. The entropy increase in excess of that due to the heat transfer is created by the irreversibility in the system.

5.2   REYNOLDS TRANSPORT THEORY (RTT)

The study of fluid dynamics requires the determination of the forces, velocities, and pressures which result from the movement of a fluid past an object or the movement of an object through a fluid. An example is the determination of the energy required to move a fluid through a piping system and the rate at which the fluid flows (the mass per unit time). Experience has shown that when determining either the energy required or the velocity of the fluid, it is easier to consider the motion of the fluid through the pipe (a control volume fixed in space) than it is to consider a fixed mass of fluid as it moves through the pipe (a system of fixed identity). It is therefore desirable to relate the system conservation laws given by eqs. 5–1, 5–3, 4–7, and 4–45 to a form that is applicable to a control volume fixed in space. The Reynolds transport theorem (RTT) provides such a relationship.

Rather than develop independent system/control volume relationships for each of the conservation laws, a general relationship for an arbitrary fluid property Φ will be developed. This arbitrary property will then be used to represent the mass (M), linear momentum ($M\mathbf{V}$), energy (E), and entropy (S) of the fluid. The arbitrary property per unit mass will be designated as $\varphi = \Phi/M$.

We will consider an arbitrary control volume of unit depth as shown by the solid line in Fig. 5–2. This control volume is fixed to an inertial coordinate system. At a time t this control volume contains a fluid system whose arbitrary fluid property is Φ_{sys}. This states that the system boundary and the boundary of

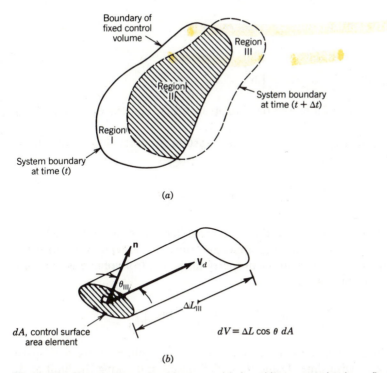

Figure 5-2 Control volume fixed in space. (a) An arbitrary control volume fixed in space. (b) Volume element of region III.

the control volume are identical at time t, or

$$[\Phi_{sys}]_t = [\Phi_{cv}]_t = \left[\iiint_{CV} \rho\varphi \, dV\right]_t \tag{5-5}$$

where

$$\rho = \text{density of the fluid in the control volume}$$

$$dV = \text{differential volume}$$

The mass of the fluid in the control volume is

$$M = \iiint_{CV} \rho \, dV$$

Since the fluid has motion relative to the control volume, at a time $t + \Delta t$, the boundary enclosing the arbitrary system property Φ_{sys} and the boundary of the control volume will no longer be coincident. The boundary enclosing Φ_{sys} at $t + \Delta t$ is shown by the broken line in Fig. 5–2 and encloses a volume that can be described as the sum of regions II and III. At time $t + \Delta t$ the arbitrary system property can therefore be expressed as

$$[\Phi_{sys}]_{t+\Delta t} = [\Phi_{II} + \Phi_{III}]_{t+\Delta t} \tag{5-6}$$

$$= [\Phi_{cv} - \Phi_{I} + \Phi_{III}]_{t+\Delta t}$$

since the volume enclosed by the fixed control surface consists of regions I and II.

The *total rate of change* of the arbitrary system property Φ_{sys} follows from the definition of a derivative

$$\frac{D\Phi_{sys}}{Dt} = \lim_{\Delta t \to 0} \frac{1}{\Delta t} \{[\Phi_{sys}]_{t+\Delta t} - [\Phi_{sys}]_t\}$$

The total derivative, also termed the substantial derivative, must be used since the system property can change with time and, since it is moving, with position. The rate of change of the arbitrary property in the control volume is due only to its change with time. Hence,

$$\frac{\partial \Phi_{cv}}{\partial t} = \lim_{\Delta t \to 0} \frac{1}{\Delta t} \{[\Phi_{cv}]_{t+\Delta t} - [\Phi_{cv}]_t\}$$

where the partial derivative is used since the control volume is fixed in space and will only vary with time. Using these definitions and taking the limit as $\Delta t \to 0$ of the difference between eqs. 5–5 and 5–6, gives

$$\frac{D\Phi_{sys}}{Dt} = \underbrace{\frac{\partial \Phi_{cv}}{\partial t}}_{①} - \underbrace{\lim_{\Delta t \to 0} \left\{ \frac{[\Phi_{I}]_{t+\Delta t}}{\Delta t} \right\}}_{②} + \underbrace{\lim_{\Delta t \to 0} \left\{ \frac{[\Phi_{III}]_{t+\Delta t}}{\Delta t} \right\}}_{③} \tag{5-7}$$

Physically, the terms on the right hand side of this equation represent

Term ①   Time rate of change Φ in the control volume (cv)

Term ②   Flux of Φ carried by the mass flow into the control volume

Term ③   Flux of Φ carried by the mass flow out of the control volume

Equation 5–7 relates the total time rate of change of an arbitrary property of a system, Φ_{sys}, to the same arbitrary property of a control volume, Φ_{cv}. This is the Reynolds transport theorem, but further consideration must be given to terms ①, ②, and ③ before it can be conveniently employed.

The amount of the arbitrary property in the control volume at any instant of time can be written as

$$\Phi_{cv} = \iiint_{cv} \rho \varphi \, dV \tag{5-8}$$

Thus term ① becomes

$$\text{Term ①} = \frac{\partial \Phi_{\text{cv}}}{\partial t} = \frac{\partial}{\partial t} \left\{ \iiint_{\text{cv}} \rho \varphi \, dV \right\} \tag{5-9}$$

If the control volume boundary is rigid then its volume will not change with time and

$$\frac{\partial \Phi_{\text{cv}}}{\partial t} = \iiint_{\text{cv}} \frac{\partial}{\partial t} [\rho \varphi] dV \tag{5-9a}$$

Term ③ of eq. 5–7 represents the flux of Φ out of the fixed control volume and into region III. A small volume element of region III, Fig. 5–2b, has a length ΔL_{III} in the direction of the fluid velocity V_d. The velocity is in a direction that makes an angle Θ_{III} to the control surface's differential area element dA. The volume of differential element can be written as

$$dV_{\text{III}} = \Delta L_{\text{III}} \cos \Theta_{\text{III}} \, dA$$

Term ③ can then be expressed as

$$\text{Term ③} = \lim_{\Delta t \to 0} \left\{ \iiint_{V_{\text{III}}} \rho \frac{\varphi}{\Delta t} \, dV_{\text{III}} \right\}$$

Substituting the above expressions for dV_{III} reduces the volume integral to a surface integral.

$$\text{Term ③} = \lim_{\Delta t \to 0} \left\{ \iint_{A_{\text{III}}} \rho \varphi \frac{\Delta L_{\text{III}}}{\Delta t} \cos \Theta_{\text{III}} \, dA \right\}$$
$$= \iint_{A_{\text{III}}} \rho \varphi |V_d| \cos \Theta_{\text{III}} \, dA \tag{5-10}$$

where A_{III} is the portion of the control surface which separates region II and region III, and V_d is the velocity of the flow through A_{III}.

Similarly, term ② can be written as

$$\text{Term ②} = \iint_{A_I} \rho \varphi |V_i| \cos \Theta_I \, dA \tag{5-11}$$

where V_i is the velocity of the flow through A_I.

A unit vector $\mathbf{n}$ is defined which is always normal to the control surface, Fig. 5–2b, and is positive in a direction away from the surface of the control volume. The fluid velocity relative to the control surface is V and is, in general, variable in both magnitude and direction over the control surface. The difference between terms ③ and ② is, by the definition of a dot (or scalar) product;

$$\text{Term ③} - \text{Term ②} = \iint_{S_{CV}} \rho\varphi[|V_d|\cos\Theta_{III} - |V_i|\cos\Theta_I]\,dA \tag{5–12}$$

$$= \iint_{S_{CV}} \rho\varphi[\mathbf{V}\cdot\mathbf{n}]\,dA$$

The integration is now a surface integral over the entire surface of the fixed control volume, the sum of regions I and III.

Substituting eqs. 5–9 and 5–12 into eq. 5–7 gives the general integral form of the Reynolds transport theorem.

$$\frac{D\Phi_{sys}}{Dt} = \frac{\partial}{\partial t}\iiint_{CV} \rho\varphi\,dV + \iint_{S_{CV}} \rho\varphi[\mathbf{V}\cdot\mathbf{n}]\,dA \tag{5–13}$$

Total rate of change of Φ_{sys}	Time rate of change of Φ within the control volume	Net efflux of Φ through the surface of the control volume

If the control volume is rigid, then from eq. 5-9a

$$\frac{D\Phi_{sys}}{Dt} = \iiint_{CV} \frac{\partial}{\partial t}[\rho\varphi]\,dV + \iint_{S_{CV}} \rho\varphi[\mathbf{V}\cdot\mathbf{n}]\,dA \tag{5–13a}$$

Most of the discussion in this book will assume the flow to be a steady state, steady flow (SSSF), which makes the first term on the right of eqs. 5–13 and 5–13a identically zero.

5.2.1 Averaged or Uniform RTT

The general form of the Reynolds transport theorem (RTT), eq. 5–13, requires the evaluation of an integral over the surface of the control volume in order to determine the net flux of Φ through the control surface. The specification of a simple control volume shape can simplify the evaluation of this surface integral. In some cases, however, a simplified shape can not be specified and it is easier to assume that the fluid properties are uniform, or averaged, over the inlet (i) and discharge (d) portions of the control surface. This means that they are constant over a projection of the control surface area on a plane surface normal to the averaged velocity.

To illustrate this assumption, consider the flow through a duct shown in Fig. 5–3a. The duct has a constant width b in the y direction and variations in velocity and density are functions of x and z. The control volume for this case is chosen to be coincident with the inner walls of the duct and to be plane surfaces at the

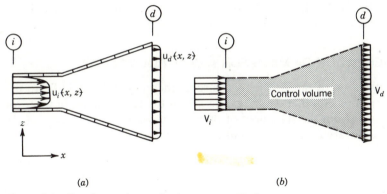

(a)   *(b)*

Figure 5-3 Pipe flow and control volume representation.

inlet (i) and discharge (d) as shown in Fig. 5–3b. The fluid velocities through the inlet and outlet planes are chosen to be *normal* to these planes and *uniform, or constant, over each plane*. The magnitude of these uniform velocities, which now vary only in the x direction in this example, are defined as

$$V_i \equiv \frac{1}{A_i} \iint_{A_i} u_i(y,z)\, dA_i$$

$$V_d \equiv \frac{1}{A_d} \iint_{A_d} u_d(y,z)\, dA_d$$

(5–14)

Since uniform fluid properties are defined by taking an average over an area, they become one dimensional. If the property is a vector, such as the velocity, it is possible to define three components of the averaged velocity, but each is averaged over a different area. Therefore each component varies only in one direction, but in a different direction from the other two. Consider a scalar fluid property such as the density. If $\rho_i = \rho_d$ the flow is termed incompressible. If $\rho_i \neq \rho_d$ the flow is a one dimensional, compressible flow since ρ is a function of only one spatial coordinate. The characteristics of a one-dimensional, compressible flow will be discussed in Chapter 6.

With these definitions, the RTT, eq. 5–13, with uniform or averaged fluid properties becomes

$$\frac{D\Phi_{\text{sys}}}{Dt} = \frac{\partial}{\partial t} \iiint_{\text{cv}} \rho\varphi\, dV + \rho_d\, \varphi V_d A_d - \rho_i\, \varphi V_i A_i$$

The product $\rho V A$, where V is normal to the surface of area A, is the fluid *mass flow rate* $\dot{m}$ (kg/s). Thus, the uniform or one-dimensional RTT is written as

$$\frac{D\Phi_{\text{sys}}}{Dt} = \frac{\partial}{\partial t} \iiint_{\text{cv}} \rho\varphi\, dV + [\varphi\dot{m}]_d - [\varphi\dot{m}]_i$$

(5–15)

This form of the RTT will be used almost exlusively in the remainder of this book, rather than the general form given by eq. 5–13.

5.3 CONSERVATION OF MASS FOR A CONTROL VOLUME

The first application of the Reynolds transport theorem (RTT) will be to determine the relationship that describes the law of the conservation of mass for a control volume from the statement of the same law for a system, eq. 5–1. In this case the arbitrary properties Φ and φ are replaced by

$$\Phi \equiv M$$

and

$$\varphi \equiv \frac{\Phi}{M} = 1$$

Therefore the one-dimensional RTT, eq. 5–15, becomes

$$\frac{DM_{sys}}{Dt} = \frac{\partial}{\partial t} \iiint_{cv} \rho \cdot 1 \, dV + [1 \cdot \dot{m}]_d - [1 \cdot \dot{m}]_i \qquad (5\text{–}16)$$

Total rate of change of $M_{sys}\,(= 0)$	Time rate of change of mass in the control volume	Net flux of mass through the surface of the control volume

Equation 5–1 states that the total rate of change of mass of the system is zero. If we restrict ourselves to a steady state, steady flow (SSSF), then the partial derivative with time is zero and eq. 5–16 reduces to

$$\dot{m}_d \qquad = \qquad \dot{m}_i \qquad (5\text{–}17)$$

Rate of mass flow leaving the control volume (kg/s)	Rate of mass flow entering the control volume (kg/s)

This relationship can also be expressed in terms of the average velocity $\mathbf{V}$ and density ρ over the inlet and discharge of the control volume by substituting $\Phi = M$, $\varphi = 1$, and eq. 5–1 into eq. 5–13.

$$[\rho(\mathbf{V} \cdot \mathbf{n})A]_d = [\rho(\mathbf{V} \cdot \mathbf{n})A]_i$$

or

$$[\rho V_n A_n]_d = [\rho V_n A_n]_i \qquad (5\text{–}18)$$

where V_n is the magnitude of velocity *normal* to the area A_n.

If the flow through the control volume is incompressible, $\rho_d = \rho_i$, then

$$[V_n A_n]_n \qquad = \qquad [V_n A_n]_i \qquad\qquad (5\text{--}19)$$

Volume flow	Volume flow
rate leaving	rate entering
the control	the control
volume (m³/s)	volume (m³/s)

While eq. 5–19 has many applications, it is best to remember eq. 5–18 since it is more general and is valid for both compressible and incompressible flows.

If the flow through a control volume is incompressible, ρ = const, eqs. 5–18 and 5–19 can provide an immediate qualitative description of the velocity at the inlet and discharge by inspection of the control volume. Consider the control volume shown in Fig. 5–3. Assuming an incompressible flow, the velocity at the discharge, V_d, must be less than the velocity at the inlet, V_i, since $A_d > A_i$. When we consider the SSSF energy equation in a later section, we will see that for an incompressible flow this means that the uniform pressure at the discharge P_d is greater than the uniform pressure P_i at the inlet. Such a control volume represents a *diffuser*. If the control volume is such that $A_d < A_i$, then $V_d > V_i$ and $P_d < P_i$, if $\rho_d = \rho_i$. This type of control volume represents a *nozzle*.

These qualitative findings are also valid for compressible flow if the velocity through the control volume is subsonic, less than the acoustic velocity (speed of sound) in the fluid. When the fluid velocity is greater than the acoustic velocity (supersonic), this simple logic is not valid although eq. 5–18 still holds. This case will be discussed in Section 6.5.5.

EXAMPLE 5–1

Air flows at standard conditions (P = 101 kPa and T = 20 °C) through a porous walled pipe as shown in Fig. E5–1. At these conditions the density ρ is constant. What is the average (uniform) velocity at the exit of the pipe? The porosity of

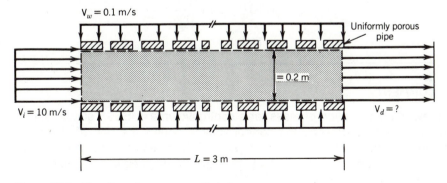

Figure E5-1   Flow through a porous-walled pipe.

the pipe walls is assumed to be uniform and the inside diameter of the pipe, d, constant.

SOLUTION

If the pipe shown is taken as the entire control volume

$$\dot{m}_{entering} = \dot{m}_{leaving}$$

or

$$\rho_i V_i A_i + \rho_w V_w A_w = \rho_d V_d A_d$$

where

$$A_i = \text{pipe inlet cross section area}$$
$$A_d = \text{pipe exit cross section area}$$
$$A_w = \text{pipe porous wall surface area} = \pi\, dL$$
$$V, V_w = \text{uniform velocities through}$$
$$\text{the pipe and porous walls, respectively}$$

Since $\rho = \text{const}$,

$$V_i \frac{\pi d^2}{4} + V_w \pi\, dL = V_d \frac{\pi d^2}{4}$$

or

$$V_d' = V_i + V_w 4 \frac{L}{d}$$

$$= 10.0 + 0.1 \frac{4(3)}{0.2}$$

$$= 16 \text{ m/s}$$

COMMENT

The law of conservation of mass for a control volume in SSSF states that the rate of mass flow entering the control volume equals the rate of mass flow leaving the control volume. Thus, the sum of the rates of mass flow from the sources of entering flow, the inlet and porous wall, must equal the rate of mass flow leaving the pipe discharge.

EXAMPLE 5–2

Flow enters the pipe shown in Fig. E5–2 at station 1 and exits at stations 2 and 3. The mass flow rate leaving the pipe at station 3 is one-quarter that entering at station 1. The diameter of the pipe at station 2 is $d_2 = 0.5d_1$, and the uniform velocity at station 3, $V_3 = 0.5V_1$. Determine the uniform velocity at station 2 in terms of V_1 and determine the pipe diameter at station 3 in terms of d_1. The fluid is incompressible and, therefore, the density of the fluid is constant, throughout the fluid.

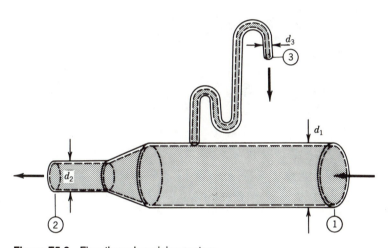

Figure E5-2   Flow through a piping system.

SOLUTION

A control volume is chosen which includes the inside contour of the entire piping system and whose surface is coincident with the flow cross-sectional areas at stations 1, 2, and 3. The law of conservation of mass states that

$$\dot{m}_1 = \dot{m}_2 + \dot{m}_3$$

the mass flow rate entering the control volume equals the mass flow rate leaving. Thus,

$$\dot{m}_2 = \dot{m}_1 - \dot{m}_3 = \dot{m}_1 - 0.25\dot{m}_1$$

$$= 0.75\,\dot{m}_1$$

But,

$$\dot{m}_1 = \rho V_1 \frac{\pi d_1^2}{4}$$

$$\dot{m}_2 = \rho V_2 \frac{\pi d_2^2}{4}$$

$$\dot{m}_3 = \rho V_3 \frac{\pi d_3^2}{4}$$

Therefore,

$$\rho V_2 \frac{\pi d_2^2}{4} = 0.75 \left[\rho V_1 \frac{\pi d_1^2}{4} \right]$$

or

$$\frac{V_2}{V_1} = 0.75 \frac{d_1^2}{d_2^2} = 0.75 \frac{d_1^2}{(0.5 d_1)^2}$$

$$= 3$$

Also,

$$\rho V_3 \frac{\pi d_3^2}{4} = 0.25 \left[\rho V_1 \frac{\pi d_1^2}{4} \right]$$

or

$$\frac{d_3}{d_1} = \sqrt{0.25 \frac{V_1}{V_3}} = \sqrt{0.25 \frac{V_1}{(0.5 \, V_1)}}$$

$$= 0.707$$

COMMENT

Employing the law of conservation of mass for a control volume, the ratio of inlet to discharge velocity and control surface area (diameter) can be determined. In this way the uniform velocity at different stations in a duct of varying size, or the duct dimensions to give a certain value of velocity, can be determined.

5.4 CONSERVATION OF LINEAR MOMENTUM FOR A CONTROL VOLUME

5.4.1 One-Dimensional Linear Momentum Equation

Equation 5–3 is an expression of Newton's second law for a system of mass, M. It states that if a force is applied to the mass the force is equal to the change in momentum $(M\mathbf{V})$ of the system. The one-dimensional form of this conservation law for a control volume can be determined for a fluid passing through a control volume by substituting

$$\Phi \equiv M\mathbf{V}$$

and

$$\varphi \equiv \frac{\Phi}{M} = \mathbf{V}$$

into the one-dimensional form of the Reynolds transport theorem (RTT), eq. 5–15. Thus,

$$\frac{D(M\mathbf{V})_{sys}}{Dt} = \frac{\partial}{\partial t}\iiint_{cv} \rho\mathbf{V}\,dV + [\mathbf{V}\dot{m}]_a - [\mathbf{V}\dot{m}]_i \qquad (5\text{–}20)$$

Equation 5–3 states that the total rate of change of the momentum of a system equals the sum of *all* the forces on the system. Therefore

$$\Sigma\mathbf{F}_{sys} = \frac{\partial}{\partial t}\iiint_{cv} \rho\mathbf{V}\,dV + [\mathbf{V}\dot{m}]_a - [\mathbf{V}\dot{m}]_i \qquad (5\text{–}21)$$

where $\mathbf{V}$ is the velocity *relative* to the control volume.

Since in the derivation of the RTT the control volume and system are assumed to be coincident at time t, Section 5.2, the resultant force on the system, $\Sigma\mathbf{F}_{sys}$, is equal to the resultant force on the control volume, $\Sigma\mathbf{F}_{cv}$. Thus, eq. 5–21 states that the sum of *all* of the forces on the control volume, or the fluid in the control volume, is equal to the change in linear momentum of the fluid as it flows through the control volume. Therefore, for a steady state, steady flow (SSSF) eq. 5–21 becomes

$$\Sigma\mathbf{F}_{cv} \qquad = \qquad [\mathbf{V}\dot{m}]_a \qquad - \qquad [\mathbf{V}\dot{m}]_i \qquad (5\text{–}22)$$

| Resultant force on fluid in the control volume (N) | Fluid momentum leaving the control volume (kg· m/s²) | Fluid momentum entering the control volume (kg· m/s²) |

The terms on the right side of this equation express the change in uniform linear momentum of the fluid as it passes through the control volume. Note that eq. 5–22 is a vector equation and, therefore, has a component in each of the coordinate directions. The right side of this equation is called the fluid inertia force. It represents the tendency of the fluid to remain in motion unless acted upon by an external force, represented by the left side of eq. 5–22. A discussion of the force term $\Sigma \mathbf{F}_{CV}$ and the individual forces contributing to this term is given below.

5.4.2   Forces Acting on the Control Volume

The use of eq. 5–22 permits the determination of the resultant force acting on a control volume by determining the change of fluid momentum as it passes through the control volume. There are problems in which it is also desirable to determine the various contributions to this resultant force.

The forces acting on a control volume can be one of two types; forces that act on the surface of the control volume, surface forces, and body forces that are related to the mass of fluid within the control volume. *Body forces*, $\mathbf{F}_B$, are those that result from the existence of an external gravitational, electrical, or magnetic force field. The only external force field to be considered here is that due to the gravitational attraction of the earth. The body force on a fluid element of volume V and density ρ is then

$$\mathbf{F}_{grav} = \rho \mathbf{g} V \tag{5–23}$$

On the surface of the earth the average magnitude of $\mathbf{g}$ at sea level is equal to 9.807 m/s^2. The force $\mathbf{F}_{grav}$ is then equal to the weight of the fluid element.

The *surface forces* on a control volume occur because of the pressure and viscous forces that act on the surface of the control volume.

Figure 5–4 is a two-dimensional representation of the surface of an arbitrary control volume showing a distribution of pressure over the surface. The pressure is a scalar quantity and therefore acts in all directions at a point in space. This means that the external pressure force always acts normal to the control surface and toward the inside of the control volume. If dA is an increment of the surface

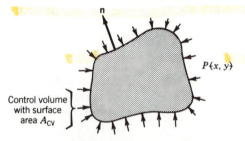

Control volume
with surface
area A_{CV}

$P\{x, y\}$

Figure 5-4   Pressure acting on an arbitrary control volume.

area of the control volume, then the surface force due to the pressure is

$$\mathbf{F}_{pres} = \iint_{CS} (-\mathbf{n})P \, dA \tag{5-24}$$

$\mathbf{n}$ is a unit vector normal to the surface and is defined to be positive in direction away from the control surface.

The viscous force on the surface occurs because the fluid has viscosity and is moving past the control surface. The *no-slip condition* discussed in Chapter 1 states that there can be no relative motion between the moving fluid and the control volume at the surface of the control volume. Shear stresses occur on the control volume surface as a result of the no-slip condition. These shear stresses then result in a surface force on the control volume which takes the following form when summed over the entire control surface area.

$$\mathbf{F}_{vis} = \iint_{CS} \boldsymbol{\tau} \, dA \tag{5-25}$$

The resultant or total force on the control volume is therefore

$$\Sigma \mathbf{F}_{CV} = \mathbf{F}_{grav} + \mathbf{F}_{pres} + \mathbf{F}_{vis} \tag{5-26}$$

$$= \rho g V + \iint_{CS} (-\mathbf{n})P \, dA + \iint_{CS} \boldsymbol{\tau} \, dA$$

Figure 5–5 is a representation of the components of the resultant force on the arbitrary control volume depicted by Fig. 5–4. It shows how these components add vectorially to produce the resultant force.

It is necessary to discuss each of these contributors to the total force on the control volume if the maximum amount of information is to be obtained from the linear momentum equation, eq. 5–22. The contribution due to gravity is straightforward. The contribution due to the pressure is discussed in the following section. We will delay the discussion of $\mathbf{F}_{vis}$ until Chapters 7 and 8.

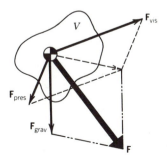

Figure 5-5   Total force on an arbitrary control volume.

5.4.2.A Contribution of Pressure to the Forces on the Control Volume

Equation 5–24 shows that the contribution of the pressure acting on a control volume is the integral of the pressure over the surface of the control volume. Thus, if an object is surrounded by a constant pressure, such as that due to the atmosphere, there will be no contribution from the pressure to the resultant force on the object,

$$\mathbf{F}_{pres} = \iint_{CS} (-\mathbf{n}) P_{atm} \, dA = 0 \tag{5–27}$$

This is illustrated by the two-dimensional control volume representing a nozzle shown in Fig. 5–6. For this case there is no flow through the nozzle.

If the nozzle of Fig. 5–6 has a flow through it, an additional pressure P_G will act on the control volume representing the nozzle. This pressure will vary in magnitude over the control volume, $P_G (x, y) \neq$ const, since the flow velocity will vary through the nozzle. If the velocity is subsonic, eq. 5–18 indicates $V_{ad} < V_{bc}$ since $A_{ad} > A_{bc}$. Figure 5–7 depicts the additional pressure on the nozzle control volume when there is flow. The total pressure force $\mathbf{F}_{pres}$, due to the flow and the atmospheric pressure, is

$$\mathbf{F}_{pres} = \iint_{CS} (-\mathbf{n}) [P_G + P_{atm}] \, dA$$

using eq. 5–27,

$$\mathbf{F}_{pres} = \iint_{CS} (-\mathbf{n}) P_G \, dA \tag{5–28}$$

The sum of the pressure P_G, the *gauge pressure,* and the atmospheric pressure P_{atm} is called the *absolute pressure,* $P = P_G + P_{atm}$. The gauge pressure is then the pressure that would be measured by a differential pressure sensor which uses P_{atm} as its reference. However, the pressure that exists at a point is always the absolute pressure P.

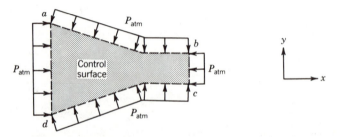

Figure 5-6 Pressure on a nozzle control volume with no flow (top view)

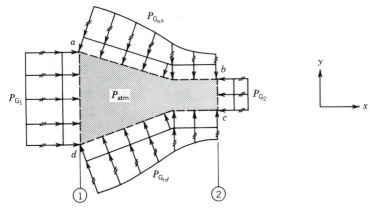

Figure 5-7   Pressure on a nozzle control volume with flow.

5.4.2.B   Resultant Force on the Control Volume

What is the resultant force that acts on the nozzle shown in Fig. 5–7? To determine this force the control volume is chosen so that it coincides with the walls of the nozzle, sides *ab cd*, Fig. 5–8, and is normal to the flow direction along sides *ad* and *bc*. The free-body diagram of this control volume is shown in Fig. 5–8 assuming that the velocity, pressure, and density are uniform on sides *ad* and *bc*, which are denoted as stations 1 and 2, respectively. The force $\mathbf{F_N} = \mathbf{F_{NP}} + \mathbf{F_{NS}}$ on the control volume is the reaction to the force on the nozzle.

The contribution of the pressure to the total force on the control volume consists of a pressure force due to the gauge pressure $P_{G_1} = P_1 - P_{atm}$ acting on *ad* and due to $P_{G_2} = P_2 - P_{atm}$ acting on *bc*, plus the pressure force on the control volume surfaces *ab* and *cd* which is denoted as $\mathbf{F_{NP}}$, the reaction to the pressure forces on the nozzle.

Also acting on *ab* and *cd* is the force due to the shear stress between the control volume and the nozzle wall, $\mathbf{F_{NS}}$. There is an additional force on the

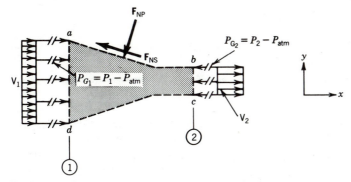

Figure 5-8   Free-body diagram of a nozzle control volume.

control volume in the negative z direction (vertical) which is a body force due to the fluid within the control volume and equals the weight of the fluid. The components of the pressure and shear forces on *ab* and *cd* in the y direction and z direction are equal in magnitude but opposite in direction. This is due to the symmetry of the control volume about the x direction. Therefore, the component of the resultant force on the control volume in the z direction is equal to the body force, F_{NZ},

$$F_{NZ} = \rho(\mathbf{k} \cdot \mathbf{g})(\text{nozzle volume})$$

and acts in the negative z direction. The resultant force in the y direction $F_{NY} = 0$ because of symmetry.

In the x direction the force of the nozzle on the control volume is

$$F_{N_x} = [\mathbf{F}_{NP} + \mathbf{F}_{NS}]_x \tag{5–29}$$

Thus, the x component of the momentum equation for the control volume becomes (assuming F_{N_x} to be acting in the negative x direction),

$$-F_{N_x} + (P_1 - P_{atm})A_1 - (P_2 - P_{atm})A_2 = \dot{m}_2 V_{x_2} - \dot{m}_1 V_{x_1} \tag{5–30}$$

or

$$F_{N_x} = \dot{m}(V_{x_1} - V_{x_2}) + P_1 A_1 - P_2 A_2 + (A_2 - A_1)P_{atm} \tag{5–31}$$
$$= \dot{m}(V_{x_1} - V_{x_2}) + P_{G_1} A_1 - P_{G_2} A_2$$

Since the momentum equation is a vector equation it is important to determine the direction of F_{N_x}. If the direction of F_{N_x} is not known, it is assumed to act in either the positive or negative x direction, and the assumption is carried throughout the problem. The correct direction will then be determined when numerical values are substituted into the momentum equation. If the assumed direction is correct, the numerical value will be positive.

The resultant force on the nozzle is the reaction of the resultant force on the control volume (or fluid).

$$\mathbf{F}_{\text{nozzle}} = -\mathbf{F}_{\text{control volume (fluid)}} \tag{5-32}$$

EXAMPLE 5–3

Water flows steadily through a circular 90° reducing elbow which discharges to the atmosphere, Fig. E5–3. The elbow is a part of a horizontal piping system (x, y plane) and is connected to the remainder of the pipe by a flange. Determine the force on the elbow flange in the x and y directions, if the mass flow rate through the elbow is 40.0 kg/s ($\rho_{water} = 998$ kg/m³ at 20° C). The positive directions in the x, y plane are shown.

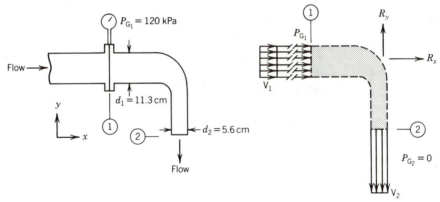

Figure E5-3   Flow through a 90° reducing elbow.

SOLUTION

The control volume is chosen to enclose the flow in the elbow and to be coincident with the inlet (1) and exit (2) faces of the elbow. The forces acting on the control volume in the horizontal (x, y) plane include the gauge pressure of the water at (1) and (2) and the reaction to the force on the elbow and flange, $\mathbf{R} = \mathbf{i}R_x + \mathbf{j}R_y$. The linear momentum equation can be written in the x and y directions as

$$P_{G_1}\frac{\pi d_1^2}{4} + R_x = \dot{m}_2 V_{x_2} - \dot{m}_1 V_{x_1} = \dot{m}(V_{x_2} - V_{x_1})$$

$$P_{G_2}\frac{\pi d_2^2}{4} + R_y = \dot{m}_2(-V_{y_2}) - \dot{m}_1 V_{y_1} = \dot{m}(-V_{y_2} - V_{y_1})$$

since $\dot{m}_1 = \dot{m}_2 = \dot{m}$ by the conservation of mass. Since $P_{G_2} = 0$, $V_{x_2} = 0$ and $V_{y_1} = 0$, the momentum equation in the x and y directions becomes (note that velocity is a vector and has a sign depending upon its direction)

$$P_{G_1}\frac{\pi d_1^2}{4} + R_x = -\dot{m}V_{x_1}$$

$$R_y = \dot{m}(-V_{y_2})$$

The velocities V_{x_1} and V_{y_2} are determined by the conservation of mass as

$$V_{x_1} = \frac{4\dot{m}}{\rho \pi d_1^2} = \frac{4(40)}{998\pi(0.113)^2} = 4.0 \text{ m/s}$$

$$V_{y_2} = \frac{4\dot{m}}{\rho \pi d_2^2} = \frac{4(40)}{998\pi(0.056)^2} = 16.3 \text{ m/s}$$

Hence,

$$R_x = -\dot{m}V_{x_1} - P_{G1}\frac{\pi d_1^2}{4}$$

$$= -(40)4 - (120{,}000)\frac{\pi(0.113)^2}{4}$$

$$= -1363.5 \text{ N } \{\text{minus sign indicates } R_x \text{ acts in}$$

the negative x direction}

$$R_y = \dot{m}(-V_{y_2})$$

$$= 40(-16.3)$$

$$= -652.0 \text{ N } \{\text{minus sign indicates } R_y \text{ acts in}$$

the negative y direction}

The force on the elbow is the reaction to the force of the elbow on the control volume.

$$R_{elbow} = \{(-R_x)^2 + (-R_y)^2\}^{1/2}$$

$$= \{(1364.5)^2 + (652)^2\}^{1/2}$$

$$= 1511.4 \text{ N}$$

$$\theta = \tan^{-1}\left(\frac{R_y}{R_x}\right)_{elbow}$$

$$= \tan^{-1}\left(\frac{652}{1363.5}\right) = 25.6°$$

COMMENT

The components R_x and R_y are assumed to act in the positive x and y directions, respectively. The numerical values of these components are both negative which indicates that the assumed direction was wrong. The force on the elbow is the reaction to the force on the fluid and is therefore in the opposite direction. This example uses the law of conservation of linear momentum to determine the force components on the fluid, but must also use the law of conservation of mass.

EXAMPLE 5–4

A curved vane is mounted on wheels and moves in the x direction at a steady velocity U = 8 m/s as a result of being struck by a jet of water (ρ = 998 kg/m³) from a stationary nozzle, Fig. E5–4. The velocity of the water leaving the

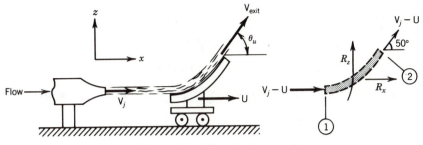

Figure E5-4 Jet of water discharged against a moving curved vane.

nozzle V_j = 25 m/s. As the water first strikes the vane it is moving only in the x direction. When the water leaves the vane it has been turned to a direction which is 50° above the x direction. The exit area of the nozzle is 0.0025 m². Neglecting body forces, what is the force of the water on the moving vane?

SOLUTION

The control volume is chosen as the fluid on the vane and is fixed to the vane, meaning that the control volume moves in the positive x direction with a velocity U = 8 m/s. The velocity of the water entering the control volume is the velocity *relative* to the moving control volume, V_j − U, in the x direction. The magnitude of the velocity leaving the control volume is also V_j − U, as determined by the law of conservation of mass, but in a direction of 50° from the x direction. The force on the vane is the reaction to the force the vane exerts on the fluid and has components R_x and R_z. There is no pressure force contribution since the entire surface of the control volume experiences atmospheric pressure. Body forces, the weight of the water, will be neglected. The linear momentum equation has two components

$$R_x = \dot{m}(V_{x_2} - V_{x_1})$$

$$R_z = \dot{m}(V_{z_2} - V_{z_1})$$

$$\dot{m} = \rho(V_j - U)A_j; \qquad V_{x_1} = V_j - U; \qquad V_{x_2} = (V_j - U)\cos\theta$$

$$V_{z_1} = 0; \qquad V_{z_2} = (V_j - U)\sin\theta$$

Therefore,

$$R_x = \rho(V_j - U)^2 A_j[\cos\theta - 1]$$

$$= 998(25 - 8)^2(0.0025)[\cos 50° - 1]$$

$$= -257.6 \text{ N } \{\text{acts on fluid in negative } x \text{ direction}\}$$

$$R_z = \rho(V_j - U)^2 A_j \sin \theta$$

$$= 998(25 - 8)^2(0.0025) \sin 50°$$

$$= 552.4 \text{ N \{acts in positive } z \text{ direction\}}$$

The force on the vane is the reaction to the force on the control volume.

$$-R_{x_{CV}} = R_{x_{vane}} = 257.6 \text{ N}$$

$$-R_{z_{CV}} = R_{z_{vane}} = -552.4 \text{ N}$$

$$R_{vane} = [R_x^2 + R_z^2]^{1/2}_{vane}$$

$$= 609.5 \text{ N} \left\{ \begin{array}{l} \text{neglecting the weight of the} \\ \text{water on the vane} \end{array} \right\}$$

$$\theta = \tan^{-1}\left(\frac{R_z}{R_x}\right) = \tan^{-1}\left(\frac{-552.4}{257.6}\right)$$

$$= -65°$$

COMMENT

Since the force on the vane is to be calculated, the control volume must be fixed to the vane. The flow enters the control volume at the velocity *relative* to the control volume. This relative velocity is $V_j - U$ in the x direction since both the jet velocity and the control volume velocity are in the same direction. If the vane were moving toward the jet as the result of an applied external force, the relative velocity would be $(V_j + U)$. Imagine yourself setting on the vane in both cases. What would be the velocity of the water approaching you?

5.5   CONSERVATION OF ENERGY (FIRST LAW OF THERMODYNAMICS) FOR A CONTROL VOLUME

5.5.1   One-Dimensional Energy Equation

The first law of thermodynamics for a system was presented and discussed in considerable detail in Chapter 4. As with the laws for the conservation of mass and linear momentum, we must formulate the first law for a control volume if we wish to analyze problems in which there is fluid flow. This is done using the one-dimensional Reynolds transport theorem (RTT), eq. 5–15, with the arbitrary system property

$$\Phi \equiv E$$

and

$$\varphi \equiv \frac{\Phi}{M} = e$$

Thus,

$$\frac{DE_{sys}}{Dt} = \frac{\partial}{\partial t} \iiint_{CV} \rho e \, dV + [e\dot{m}]_a - [e\dot{m}]_i \tag{5–33}$$

The time rate of change of system energy, E_{sys}, can be related to the rate at which energy is transferred across the system boundary as heat, $\dot{Q}$, or as power, $\dot{W}$. This relationship has been given in eq. 4–7 as

$$\dot{Q}_{sys} - \dot{W}_{sys} = \frac{dE_{sys}}{dt}$$

$\dot{Q}$ and $\dot{W}$ are not properties of the fluid but rather are rates at which energy that is not associated with mass transfer (fluid flow) is being transferred across a boundary. $\dot{Q}$ occurs because of a difference in temperature across the boundary and $\dot{W}$ occurs because of a difference in potential other than temperature. By the convention established in Chapter 2, $\dot{Q}$ is positive when heat is *added to* the system and $\dot{W}$ is positive when work is *done by* the system.

In addition to energy transferred as $\dot{Q}$ and $\dot{W}$ the control volume will experience energy transfer across its boundary because of the mass transfer across its boundary. The term $[(e\dot{m})_a - (e\dot{m})_i]$ represents a portion of the energy transfer due to mass transfer. An additional form of energy transfer is due to the stresses, pressure and viscous shear, acting on the surface of the control volume because of mass transfer and is not included in the term $[(e\dot{m})_a - (e\dot{m})_i]$ in eq. 5–33.

The rate of energy transfer which results due to mass transfer through the surface of the control volume is designated as $\mathfrak{E}_{sur}$. The energy transfer across the control volume boundary and the system boundary are equal since the control surface boundary and the system boundary are coincident at time (t).

$$\underbrace{\dot{Q}_{sys} - \dot{W}_{sys}}_{\substack{\text{Energy transfer} \\ \text{across system} \\ \text{boundary at} \\ \text{time } t}} = \underbrace{\dot{Q}_{cv} - \dot{W}_{cv} - \mathfrak{E}_{sur}}_{\substack{\text{Energy transfer} \\ \text{across control} \\ \text{volume boundary} \\ \text{at time } t}} \tag{5–34}$$

$\mathfrak{E}_{sur}$ is defined as being positive if energy is transferred *out of* the control volume.

Further discussion of the term $\mathfrak{E}_{sur}$ is required to determine its origin and why it is not included in the convective energy transfer terms on the right side of eq. 5–33. This additional rate of energy transfer occurs when the moving fluid crosses a control surface on which a force is acting in the flow direction. Since a force and a motion in the direction of the force are both involved, an energy transfer rate results. (Some authors would call this energy term work, but the definition of work used here precludes that terminology.) This energy rate term may be more readily visualized by thinking of the control surface moving through the fluid. The product of the force on the control surface in the direction of motion and the velocity of the control surface equals the power required to produce the motion.

The normal force due to the pressures acting on the control surface can be separated from the force due to viscosity on the control surface, so that

$$\mathfrak{E}_{sur} = \mathfrak{E}_{pres} + \mathfrak{E}_{vis} \tag{5-35}$$

The term $\mathfrak{E}_{vis}$ can be made equal to zero if the control volume is chosen so that the flow is normal to the surface of the control volume. Since the viscous forces are parallel to the control surface and normal to the velocity through the control surface, their contribution to the energy transfer rate is zero. In determining the one-dimensional energy equation it will be assumed that this is the case.

The other energy transfer rate term, $\mathfrak{E}_{pres}$, is the result of the pressure acting on the control surface and the relative motion between the fluid and this surface. The rate of this energy transfer equals the product of the pressure force $-P\,dA$, acting toward the center of the control volume, and the velocity normal to dA, $(\mathbf{V} \cdot \mathbf{n})$, integrated over the surface of the control volume.

$$\mathfrak{E}_{pres} = -\iint_{cs} P(\mathbf{V} \cdot \mathbf{n})\, dA$$

Assuming that the pressure is uniform over the inlet and outlet of the control volume, the one-dimensional pressure term is

$$\mathfrak{E}_{pres} = \left[\frac{P}{\rho}\, \dot{m}\right]_d - \left[\frac{P}{\rho}\, \dot{m}\right]_i \tag{5-36}$$

Since $\mathfrak{E}_{vis}$ is zero by the choice of the surface of the control volume,

$$\mathfrak{E}_{sur} = \left[\frac{P}{\rho}\, \dot{m}\right]_d - \left[\frac{P}{\rho}\, \dot{m}\right]_i$$

$\mathfrak{E}_{sur}$ represents the rate of energy required to move the fluid into, $(P\dot{m}/\rho)_i$, and out of, $(P\dot{m}/\rho)_d$, the control volume.

The first term on the right side of eq. 5–33 represents the time rate of change of the total energy in the control volume. Since the energy in the control volume is only a function of time, the partial derivative can be replaced by an ordinary derivative. Therefore,

$$\frac{\partial}{\partial t}\iiint_{cv} \rho e\, dV = \frac{d}{dt}\iiint_{cv} \rho e\, dV = \frac{d\,E_{cv}}{d\,t} \tag{5-37}$$

Combining eqs. 5–33, 5–34, 5–36, and 5–37, the uniform or one-dimensional energy equation for a control volume becomes

$$\dot{Q}_{cv} - \dot{W}_{cv} - \left\{ \left[\frac{P}{\rho} \dot{m} \right]_d - \left[\frac{P}{\rho} \dot{m} \right]_i \right\} = \frac{dE_{cv}}{dt} + [e\dot{m}]_a - [e\dot{m}]_i \quad (5\text{–}38)$$

| Rate of energy transfer as heat and work | Rate of energy transfer due to pressure forces on the control surface | Rate of change of energy in the control volume | Convective rate of energy transfer through the control surface |

The total energy transported across the control surface by virtue of the mass transfer is the combination of the convective terms, $e\dot{m}$, and the energy transfer due to forces on the surface through which there is flow, $\dot{m}P/\rho$. This combination can be written as

$$e\dot{m} + \frac{P}{\rho}\dot{m} = \left[e + \frac{P}{\rho} \right]\dot{m} = \left[h + \frac{V^2}{2} + gz \right]\dot{m}$$

since

$$e \equiv u + \frac{V^2}{2} + gz \quad \text{and} \quad h \equiv u + \frac{P}{\rho}$$

The energy equation for a control volume then becomes,

$$\dot{Q}_{cv} - \dot{W}_{cv} + \left[h + \frac{V^2}{2} + gz \right]_i \dot{m} - \left[h + \frac{V^2}{2} + gz \right]_d \dot{m} = \frac{dE_{cv}}{dt} \quad (5\text{–}39)$$

| Rate of energy transfer through control surface as heat (J/S) | Rate of energy transfer through control surface as work (J/S) | Rate of energy transfer through control surface as a result of mass transfer (N · m/kg · kg/s, J/S) | Rate of change of energy in control volume (N · m/s, J/S) |

Equation 5–39 is the general form of the first law of thermodynamics and is applicable to a wide variety of control volumes for uniform or averaged flows. If more than one inlet or discharge stream is present a transport term $[(h + V^2/2 + gz)\dot{m}]$ is included in this equation for each stream. Note that in the absence of any mass flow terms this equation reduces to the energy equation for a system (eq. 4–7).

If a uniform, steady state, steady flow (SSSF) through a rigid control volume is considered, $dE_{cv}/dt = 0$ and eq. 5–39 simplifies to

$$\frac{\dot{Q}_{cv}}{\dot{m}} - \frac{\dot{W}_{cv}}{\dot{m}} = \left[h + \frac{V^2}{2} + gz \right]_d - \left[h + \frac{V^2}{2} + gz \right]_i \quad (5\text{–}40)$$

| Heat added to control volume per unit mass flow (J/kg) | Work done by fluid in control volume per unit mass flow (J/kg) | Fluid energy leaving control volume per unit mass flow (N · m/kg, J/kg) | Fluid energy entering control volume per unit mass flow (N · m/kg, J/kg) |

All of the terms in eq. 5–40 describe the energy per unit mass flow *of the fluid*. Recall that a positive value of $\dot{W}_{cv}$ is defined as the work done by the fluid. In order to use the work done by the fluid it is necessary to convert it into mechanical work through a *turbine* or a piston-cylinder arrangement which outputs power by a rotating shaft. Such devices have irreversibilities and therefore produce less shaft power than the available fluid power.

Similarly, when work is done *on* a fluid the work rate $-\dot{W}_{cv}$ represents the rate at which work is transferred *to* the fluid. To produce this work transfer to the fluid a device (a *compressor* or *pump*) is required which converts mechanical shaft work into fluid work, again by a piston-cylinder arrangement or by a set of rotating blades. This conversion also involves irreversibilities so that more power must be applied to the compressor or pump shaft than is done on the fluid.

The existence of irreversibilities in this shaft-to-fluid power conversion results in turbine and pump efficiencies which relate the available or applied power, respectively, to the fluid power. The work terms in eqs. 5–39 and 5–40 represent the fluid work and *do not* include these efficiencies. A further discussion of energy conversion is given in Section 5.8.

5.5.2 Application of Uniform, One-Dimensional Energy Equation

The steady state, steady flow (SSSF) energy equation given by eq. 5–40 provides a valuable tool for the analysis of engineering flow problems in which heat and work transfer occur. Such problems include the analysis of power-generating machines (turbines), pumping machines (pumps and compressors) that transport a fluid between two locations, air conditioning and ventilating networks, for example.

When the form of the SSSF energy equation given in eq. 5–40 is used for the analysis of problems involving the flow of a ideal gas, the change in enthalpy per unit mass of the gas through the control volume can be expressed as a difference in temperature

$$h_d - h_i = c_P(T_d - T_i) \quad (5\text{–}41)$$

In this case, eq. 5–40 reduces to (note $\dot{W} \equiv \dot{W}_{cv}$ and $\dot{Q} = \dot{Q}_{cv}$)

$$\frac{\dot{Q}}{\dot{m}} - \frac{\dot{W}}{\dot{m}} = c_P(T_d - T_i) + \frac{1}{2}(V_d^2 - V_i^2) + g(z_d - z_i) \tag{5–42}$$

If the fluid being considered is a liquid, eqs. 5–40 or 5–42 are not the most convenient forms of the SSSF energy equation. This is because of the difficulty in measuring the small temperature differences that will occur in a liquid. For the case of liquid flows, eq. 5–40 is rearranged into the following form

$$-\frac{\dot{W}}{g\dot{m}} = \left[\frac{P}{g\rho} + \frac{V^2}{2g} + z\right]_d - \left[\frac{P}{g\rho} + \frac{V^2}{2g} + z\right]_i + \left[u_d - u_i - \frac{\dot{Q}}{\dot{m}}\right]\frac{1}{g} \tag{5–43}$$

Each of the terms in this representation has the dimensions of a length (m) and is called a *head*. Thus,

$$\underbrace{\left[\frac{P}{g\rho} + \frac{V^2}{2g} + z\right]_i}_{\substack{\text{Total fluid} \\ \text{head entering} \\ \text{the control} \\ \text{volume, } H_{T_i}}} = \underbrace{\left[\frac{P}{g\rho} + \frac{V^2}{2g} + z\right]_d}_{\substack{\text{Total fluid} \\ \text{head leaving} \\ \text{the control} \\ \text{volume, } H_{T_d}}} + \underbrace{\left[u_d - u_i - \frac{\dot{Q}}{\dot{m}}\right]\frac{1}{g}}_{\substack{\text{Fluid head} \\ \text{loss to friction} \\ \text{and heat} \\ \text{transfer, } h_L}} + \underbrace{\frac{\dot{W}}{g\dot{m}}}_{\substack{\text{Fluid head} \\ \text{equivalent of} \\ \text{work done} \\ \text{by fluid, } h_w}} \tag{5–44}$$

$$H_{T_i} = H_{T_d} + h_L + h_w \tag{5–45}$$

The *total fluid head* $(H_T = P/\rho g + V^2/2g + z)$ is defined as the sum of the pressure head $(P/\rho g)$, the velocity head $(V^2/2g)$, and the potential head (z) per unit mass flow rate of fluid.

The term head has its origin in the field of hydraulics and is a representation of the pressure in a liquid. For example, an atmospheric pressure of 101 kPa can be stated as an equivalent height of a column of a liquid. If the liquid is water with a density, $\rho = 998$ kg/m³.

$$P_{atm} = \rho_{water} g \, h_{water}$$

$$h_{water} = \frac{P_{atm}}{\rho_{water} g} = \frac{101,000}{998(9.807)} = 10.32 \text{ m of water}$$

Therefore, an atmospheric pressure of 101 kPa is equal to the force per unit area at the bottom of a column of water which is 10.32 m in height, or 1 atm pressure corresponds to a head of 10.32 m of water.

The estimation of h_L is discussed in considerable detail in Chapter 8. In the remainder of this chapter it will be assumed to be known.

EXAMPLE 5–5

A siphon system with an inside diameter $d = 0.075$ m is used to remove water from container A to container B, Fig. E5–5a. When operational the steady-flow

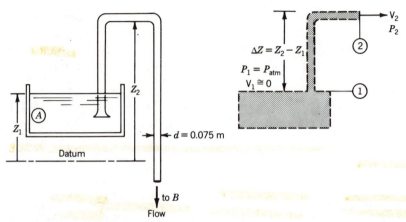

Figure E5-5a A siphon system.

rate through the siphon is 0.03 m³/s. The temperature of the water is 20° C and its density 998 kg/m³. Compute the elevation of the siphon above the water surface of reservoir A, Δz, at which the minimum pressure in the siphon equals the vapor pressure of the water. The head loss due to friction and heat transfer is assumed to be negligible, $h_L = 0$. Assume that the depth of the water in the reservoir is maintained constant.

SOLUTION

A control volume is chosen which includes the surface of the water in the reservoir and is perpendicular to the flow through the siphon at a point Δz above the reservoir. The SSSF, one-dimensional energy equation written between points 1 and 2 at the surface of the reservoir and the point of maximum elevation, respectively, is

$$\frac{P_1}{\rho g} + \frac{V_1^2}{2g} + z_1 = \frac{P_2}{\rho g} + \frac{V_2^2}{2g} + z_2 + h_L + \frac{\dot{W}}{g \dot{m}}$$

For this particular problem

$z_2 - z_1 = \Delta z$ (unknown)

$P_1 = P_{atm} = 101$ kPa (surface is exposed to atmosphere)

$V_1 = 0$ (reservoir is large compared to siphon)

$h_L = 0$ (stated assumption)

$\dot{W} = 0$ (there is no pump or turbine)

$P_2 = 2.3$ kPa (vapor pressure of water at 20° C, Table A.1.1)

$V_2 = \dot{m}/\rho A_2 = 0.03/\pi(0.075^2/4) = 6.791$ m/s

Thus, the energy equation becomes

$$\Delta z = \frac{P_{atm}}{\rho g} - \frac{P_2}{\rho g} - \frac{V_2^2}{2g}$$

$$= \frac{1}{9.807}\left[\frac{101,000}{998} - \frac{2300}{998} - \frac{(6.79)^2}{2}\right]$$

$$= 7.73 \text{ m}$$

COMMENT

The control volume inlet is chosen as the free surface of the reservoir since the velocity and pressure are known at that point. When the pressure at point 2 equals the liquid vapor pressure, bubbles of liquid vapor will start to form. This phenomena is called cavitation. Figure 5–5b is a picture taken in a water tunnel of tip vortex cavitation on a marine propeller. As the bubbles are transported into a region of higher pressure they will shrink in size and eventually

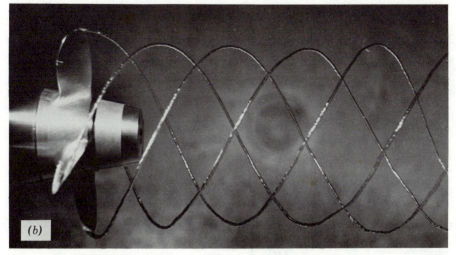

Figure E5-5b Cavitation of a marine propeller. (Courtesy of Applied Research Laboratory, The Pennsylvania State University.)

collapse. This collapsing produces noise and can lead to a pitting of the surface on which they collapse. This phenomena is termed cavitation erosion and is a very serious problem on ship propellers.

5.6 SELECTION OF A CONTROL VOLUME

There are a large number of control surface contours that can be selected for the solution of a particular problem. The best control surface contour is determined from experience and by a careful analysis of the known and unknown quantities for the particular problem.

To illustrate this, consider the problem of determining the amount of shaft power that is required to pump a liquid, say water, at a specified flow rate through a pipe of known diameter from one storage reservoir to another at a higher elevation, Fig. 5–9. The difference in elevation between the surfaces of the reservoirs is known, as is the length and diameter of the pipe through the piping network. Since the quantity to be calculated is the rate of work done *on* the fluid by the pump, $-\dot{W}$, the control volume selected must enclose the fluid in the pump. Two possible choices are shown in Fig. 5–9. Note that the control surfaces shown are meant to enclose only the fluid passing through the piping system. The representation shown simplifies the drawing of the control surface and is used in many texts, but the control surface encloses the fluid only and not the pipe.

For each control surface, the following information is available regarding the terms in the SSSF, one-dimensional energy equation.

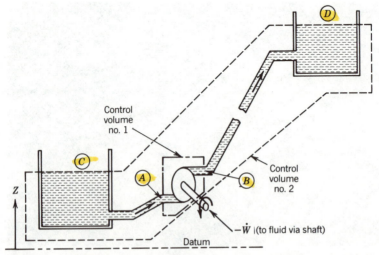

Figure 5-9 Liquid pumping system between two reservoirs. Note: Simplified control surface shown is meant to enclose only the fluid between Ⓐ and Ⓑ, Ⓒ and Ⓓ.

Control Volume No. 1 (between points A and B).

P_i and P_d	Unknown; can be determined only by measurement
V_i and V_d	Known; calculated from mass flow rate and pipe area, both of which are known
z_i and z_d	Unknown; could be measured from survey of terrain
h_L	Known; can be estimated knowing velocity, length of pipe in control volume and its diameter (see Chapter 8)
$\dot{W}$	Quantity to be determined

Result: five unknowns and three known quantities without additional measurements being performed.

Control Volume No. 2 (Between Points C and D).

P_i and P_d	Known; both pressures are equal to atmospheric
V_i and V_d	Known, both velocities are negligible since reservoirs are much larger than pipe.
z_i and z_d	Difference is known since location of reservoirs is known
h_L	Known; can be estimated knowing velocity, length of pipe in the control volume, and its diameter (see Chapter 8)
$\dot{W}$	Quantity to be determined

Result: one unknown and seven known quantities.

It is obvious that for this particular problem the best choice is control volume no. 2. Substituting the known quantities into eq. 5–44 gives the power or rate of work done on the fluid

$$-\dot{W} = \dot{m}g(z_d - z_i + h_L) \qquad (5\text{–}46)$$

($\dot{W}$ is positive if work is done *by* the fluid, negative if work is done on the fluid).

Since the energy equation for a control volume describes the energy of the fluid, $\dot{W}$ represents the work that is transferred to the fluid, but not the shaft work that must be applied to the pump shaft. In other words, it does not account for the irreversibilities in the pump. The power that must be delivered to the shaft of the pump is $-\dot{W}$ divided by the efficiency of the pump, η_p

$$-\dot{W}_s(\text{required shaft power}) = \frac{-\dot{W}}{\eta_p} \qquad (5\text{–}47)$$

According to eq. 5–46 this power goes to increase the potential energy of the water and to overcome the energy lost to friction and heat transfer in the entire piping system, $\dot{m}gh_L$. Note that the pump efficiency must be known.

EXAMPLE 5–6

The pipe network shown in Fig. 5–9 consists of a 15.24-cm diameter duct with a mass flow rate $\dot{m}$ = 140 kg/s. The level of water in the receiving reservoir is maintained at 122 m above the level of the supply reservoir. The head loss, in meters, in the network due to duct wall friction, elbows, and fittings varies with the square of the average velocity through the duct, h_L = 1.07V^2 (see Chapter 8), where V is in m/s. If the pump has an efficiency η_p = 0.90, determine the power that must be supplied to the *shaft* of the pump (ρ_{water} = 998 kg/m³).

From eq. 5–46 the rate at which work is done on the water by the pump is

$$-\dot{W} = \dot{m}g[z_d - z_i + h_L]$$

when control volume no. 2 is selected. The average velocity through the duct is

$$V = \frac{\dot{m}}{\rho A} = \frac{4(140)}{998\pi(0.1524)^2}$$

$$= 7.69 \text{ m/s}$$

and

$$h_L = 1.07 (7.69)^2$$

$$= 63.28 \text{ m}$$

Thus,

$$-\dot{W} = 140(9.807)[122 + 63.28]$$

$$= 0.2544 \text{ MW}$$

This is the power delivered to the water, or required in the pump shaft if the pump were 100% efficient. Since η_p = 0.90 the required shaft power is

$$-\dot{W}_s(\text{required shaft power}) = \frac{-\dot{W}}{\eta_p} = \frac{0.2544}{0.90}$$

$$= 0.2827 \text{ MW}$$

EXAMPLE 5–7

The pressures at the inlet and exit of a water turbine are 300 and 90 kPa, respectively. The volume flow rate through the turbine, Fig. E5–7, is 0.9 m³/s. If the turbine efficiency η_t = 0.82, what is the shaft output power of the turbine? Assume SSSF and ρ = 1000 kg/m³.

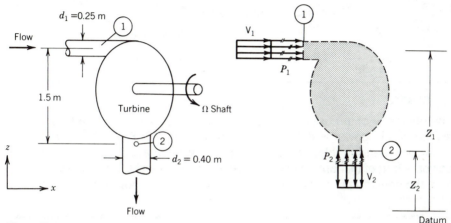

Figure E5-7   A water turbine.

SOLUTION

From eq. 5–44 and Fig. E5–7,

$$\left[\frac{P}{\rho g} + \frac{V^2}{2g} + z\right]_1 = \left[\frac{P}{\rho g} + \frac{V^2}{2g} + z\right]_2 + h_L + \frac{\dot{W}}{g\dot{m}}$$

$$P_1 = 300 \text{ kPa}; \quad P_2 = 90 \text{ kPa};$$

$$\dot{m} = \rho(0.9); \quad z_1 - z_2 = 1.5 \text{ m}$$

$$V_1 = \frac{\dot{m}}{\rho A_1} = \frac{0.90}{(\pi/4)(0.25)^2}$$

$$= 18.33 \text{ m/s}$$

$$V_2 = \frac{\dot{m}}{\rho A_2} = \frac{0.90}{(\pi/4)(0.40)^2}$$

$$= 7.162 \text{ m/s}$$

If h_L = 0, the available fluid power or maximum available turbine output (η_t = 1.0) can be calculated.

$$\dot{W} = \dot{m}\left[\frac{(P_1 - P_2)}{\rho} + \frac{(V_1^2 - V_2^2)}{2} + g(z_1 - z_2)\right]$$

$$= 0.90\left[\frac{(300,000 - 90,000)}{1000} + \frac{(18.33)^2 - (7.162)^2}{2} + 9.807(1.5)\right]$$

$$= 329.9 \text{ W}$$

$$\dot{W}_s(\text{available shaft power}) = \eta_t\dot{W}$$

$$= 0.82(329.9)$$

$$= 270.5 \text{ W}$$

COMMENT

The efficiency of the turbine expresses the head loss through the turbine. The available fluid power is determined by writing the SSSF energy equation, eq. 5–44, between the inlet and discharge of the turbine with $h_L = 0$. The shaft power output of the turbine must be less than the available water power. Therefore, the shaft output power equals the available water power times the turbine efficiency

---

EXAMPLE 5–8

A large air tank at a service station is being filled by an air compressor while a customer is using air from the tank to fill a tire. Calculate the rate of pressure rise in the tank at an instant of time when the following conditions prevail.

1. The compressor is delivering air to the tank at a flow rate of 0.01 kg/s and a temperature of 90° C.

2. The customer is using air from the tank at a rate of 0.001 kg/s.

3. The tank volume is 3.0 m³ and the temperature of the air in the tank is 40° C.

SOLUTION

Take the tank as the control volume and make the following assumptions:

(a) Air is an ideal gas ($PV = MRT$, $h = f\{T\}$) with constant specific heats.

(b) The tank is adiabatic and rigid ($\dot{Q} = 0$, $\dot{V} = 0$).

(c) The air in the tank is well mixed, it is at a uniform temperature, hence $T_d = T$.

(d) The kinetic and potential energies of the air streams entering and leaving the tank are negligible.

(e) The energy stored in the tank is only internal energy ($E = U$).

The general form of the first law for the control volume, eq. 5–39, with the above assumptions reduces to:

$$\dot{m}_i h_i - \dot{m}_d h_d = \frac{dU}{dt}$$

but $U = Mu$ so that

$$\frac{dU}{dt} = M\frac{du}{dt} + u\frac{dM}{dt} = Mc_v\frac{dT}{dt} + c_v T\frac{dM}{dt}$$

The equation of state can be used to obtain a dP/dt term and eliminate the dT/dt term as follows:

$$PV = MRT$$

Take the logarithm of this equation to obtain

$$\ln P + \ln V = \ln M + \ln R + \ln T$$

Take the derivative of this equation with respect to time

$$\frac{1}{P}\frac{dP}{dt} + 0 = \frac{1}{M}\frac{dM}{dt} + 0 + \frac{1}{T}\frac{dT}{dt}$$

Solve for

$$\frac{dT}{dt} = \frac{T}{P}\frac{dP}{dt} - \frac{T}{M}\frac{dM}{dt}$$

Substituting this expression into the dU equation yields

$$\frac{dU}{dt} = \frac{Mc_v T}{P}\frac{dP}{dt} - \frac{Mc_v T}{M}\frac{dM}{dt} + c_v T\frac{dM}{dt}$$

$$= \frac{Mc_v T}{P}\frac{dP}{dt} = \frac{c_v V}{R}\frac{dP}{dt}$$

Using this relationship in the first law and solving for dP/dt yields

$$\frac{dP}{dt} = \frac{R}{c_v V}[\dot{m}_i c_p T_i - \dot{m}_d c_p T_d] = \frac{c_p R}{c_v V}[\dot{m}_i T_i - \dot{m}_d T_d]$$

$$= \frac{1.4(0.287)}{3.0}[(0.01)363.2 - (0.001)313.2]$$

$$= 0.4445 \text{ kPa/s}$$

COMMENT

The pressure in the tank increases because the rate of energy transported into the tank is greater than the rate at which energy is transported out of the tank.

EXAMPLE 5–9

A tank containing nitrogen is being emptied at a rate of 0.01 kg/s. At what rate must heat be added to the nitrogen so that the temperature of the nitrogen remains constant at a value of 50° C?

SOLUTION

Take the nitrogen tank as the control volume (the inside of the tank wall forms the control surface) and make the following assumptions:

(a) The tank is rigid ($\dot{V} = 0$) and no kinetic or potential energy is stored in the tank ($E = U$).

(b) The fluid leaving the tank carries negligible kinetic and potential energy with it, $(h + V^2/2 + gz)_d = h_d$.

(c) Nitrogen is an ideal gas ($PV = MRT$, $h = f(T)$) with constant specific heats.

(d) The temperature is uniform throughout the control volume, thus $T_d = T$.

The first law equation for this control volume reduces to

$$\dot{Q} - \dot{m}_d h_d = \frac{dU}{dt}$$

$$\dot{Q} = \dot{m}_d h_d + M\frac{du}{dt} + u\frac{dM}{dt}$$

From continuity

$$\frac{dM}{dt} = -\dot{m}_d$$

Therefore

$$\dot{Q} = \dot{m}_d c_p T_d + Mc_v\frac{dT}{dt} - c_v T\dot{m}_d$$

But the temperature is constant so that $dT/dt = 0$

$$\dot{Q} = \dot{m}_d T(c_p - c_v) = \dot{m}_d RT$$

$$\dot{Q} = 0.01(0.2968)(323.2) = 0.9593 \text{ kW}$$

COMMENT

Note that this rate of heat addition is independent of the size of the tank, the amount of mass in the tank, and the pressure in the tank.

EXAMPLE 5–10

Compute the work during the intake stroke and the temperature of the air at the end of the intake stroke for a control volume consisting of the space inside the cylinder of an engine. Let state 1 represent the state of the control volume at the beginning and state 2 the state of the control volume at the end of the intake stroke. Given information:

$$P_1 = P_2 = P_i = P = 0.1 \text{ MPa}$$

$$T_1 = 150° \text{ C} = 423 \text{ K}$$

$$T_i = 20° \text{ C} = 293 \text{ K}$$

$$V_1 = 0.0008 \text{ m}^3$$

$$V_2 = 0.008 \text{ m}^3$$

SOLUTION

Assume the intake process is relatively slow so that the pressure inside the control volume remains constant during the intake stroke. (Obviously the pressure inside the cylinder must be less than that outside in order for flow to occur, but the assumption of constant pressure is made here to simplify the problem.)

$$\dot{W} = P\frac{dV}{dt}$$

$$\int_1^2 \dot{W}\, dt = {}_1W_2 = \int_1^2 P\frac{dV}{dt}\, dt = P\int_1^2 dV = P(V_2 - V_1)$$

$${}_1W_2 = 100(0.008 - 0.0008) = 0.72 \text{ kJ}$$

If we assume the control volume is adiabatic, an ideal gas with constant specific heats, and negligible kinetic energy entering the control volume (this is consistent with neglecting the pressure difference across the intake valve), the first law for the control volume reduces to

$$-\dot{W} + \dot{m}_i h_i = \frac{dE}{dt} = \frac{dU}{dt}$$

$$\int_1^2 -\dot{W}\, dt + \int_1^2 \dot{m}_i h_i\, dt = \int_1^2 \frac{dU}{dt}\, dt$$

$$-{}_1W_2 + h_i \int_1^2 \dot{m}_i\, dt = U_2 - U_1$$

From continuity

$$\frac{dM}{dt} = \dot{m}_i$$

Therefore

$$-{}_1W_2 + h_i \int_1^2 \frac{dM}{dt}\, dt = U_2 - U_1$$

$$-P(V_2 - V_1) + h_i(M_2 - M_1) = M_2 c_v T_2 - M_1 c_v T_1$$

$$-P(V_2 - V_1) + c_p T_i\left(\frac{P_2 V_2}{RT_2} - \frac{P_1 V_1}{RT_1}\right) = \frac{P_2 V_2}{RT_2} c_v T_2 - \frac{P_1 V_1}{RT_1} c_v T_1$$

$$-(V_2 - V_1) + \frac{c_p T_i}{R}\left(\frac{V_2}{T_2} - \frac{V_1}{T_1}\right) = \frac{c_v}{R}(V_2 - V_1)$$

$$(V_2 - V_1)\left(1 + \frac{c_v}{R}\right) = \frac{c_p T_i}{R}\left(\frac{V_2}{T_2} - \frac{V_1}{T_1}\right)$$

Using

$$\frac{c_v}{R} = \frac{1}{\gamma - 1}, \quad \frac{c_p}{R} = \frac{\gamma}{\gamma - 1},$$

and solving for T_2 yields

$$(V_2 - V_1)\left(\frac{\gamma}{\gamma - 1}\right) = \left(\frac{\gamma}{\gamma - 1}\right)\left[V_2\frac{T_i}{T_2} - V_1\frac{T_i}{T_1}\right]$$

$$T_2 = \frac{T_i}{[(V_2 - V_1)/V_2 + (T_i/T_1)(V_1/V_2)]}$$

$$= \frac{T_i}{[1 - (V_1/V_2) + (T_i/T_1)(V_1/V_2)]}$$

$$T_2 = \frac{T_i}{\{1 + (V_1/V_2)[(T_i/T_1) - 1]\}}$$

$$T_2 = \frac{293.2}{[1 + 0.1(293.2/423.2 - 1)]}$$

$$= 302.5 \text{ K}$$

COMMENT

The work is positive indicating work is done by the control volume as the volume increases. The final temperature is a function of T_i, T_1, and the volume ratio V_1/V_2.

EXAMPLE 5–11

Air enters a compressor at 20° C and leaves the compressor at 360° C. The mass flow rate of air of 10 kg/s, the velocity of the air entering the compressor is 10 m/s, and the velocity of the air leaving the compressor is 100 m/s. Assuming SSSF, an adiabatic compressor, and an ideal gas with constant specific heats, compute the power required to drive this compressor.

SOLUTION

Using the SSSF form of the first law, eq. 5–40,

$$\frac{\dot{Q}}{\dot{m}} - \frac{\dot{W}}{\dot{m}} = \left(h + \frac{V^2}{2} + gz\right)_d - \left(h + \frac{V^2}{2} + gz\right)_i$$

and neglecting changes in potential energy we have

$$\dot{W} = \dot{m}\left[\left(h + \frac{V^2}{2}\right)_i - \left(h + \frac{V^2}{2}\right)_d\right] = \dot{m}\left(h_i - h_d + \frac{V_i^2 - V_d^2}{2}\right)$$

$$\dot{W} = \dot{m}\left[c_p(T_i - T_d) + \frac{V_i^2 - V_d^2}{2}\right]$$

$$\dot{W} = 10\left[1.004(20 - 360) + \frac{100 - 10000}{2000}\right] = -3463 \text{ kW}$$

COMMENT

The power is negative indicating it is power transferred into the control volume. The change in kinetic energy in this case is not negligible, although it is less than 2% of the change in enthalpy. The assumption of a negligible change in potential energy is quite reasonable since a difference in elevation of 1 m between inlet and discharge would produce a change in power of only 0.1 kW. The specific heat of air has a mean value over this temperature range about 2.5% higher than the value used and a more precise calculation should account for this fact.

5.7 THE SECOND LAW OF THERMODYNAMICS FOR A CONTROL VOLUME

The same procedure used to formulate the first law of thermodynamics for a control volume can be used to obtain the second law of thermodynamics for a control volume. Either the system inequality relationship (eq. 4–45) or the equality relationship (eq. 4–47) can be used.

For a system, eq. 4–45 states

$$\frac{dS_{sys}}{dt} \geq \frac{\dot{Q}}{T}$$

Using eq. 5–15 with $\Phi = S$ and $\phi = \Phi/M = s$

$$\frac{dS_{sys}}{dt} = \frac{\partial}{\partial t}\iiint_{cv} \rho s \, dV + (\dot{m}s)_d - (\dot{m}s)_i \tag{5–48}$$

where

$$\frac{\partial}{\partial t}\iiint_{cv} \rho s \, dV = \frac{dS_{cv}}{dt} \tag{5–49}$$

since S_{cv} is a function of time alone.

Substituting eq. 5–49 into eq. 5–48 and then eq. 5–48 into eq. 4–45 yields

$$\frac{dS_{cv}}{dt} \geq \frac{\dot{Q}}{T} + (\dot{m}s)_i - (\dot{m}s)_d \tag{5–50}$$

If eq. 4–47 is used as the starting point rather than eq. 4–45, the second law of thermodynamics for a control volume becomes

$$\frac{dS_{cv}}{dt} = \frac{\dot{Q}}{T} + \dot{I} + (\dot{m}s)_i - (\dot{m}s)_d \tag{5–51}$$

Either eq. 5–51 or the inequality relationship, eq. 5–50, may be regarded as the statement of the second law for a control volume. For a reversible process the equal sign in eq. 5–50 applies and, of course, the term $\dot{I}$ equals zero in eq. 5–51 so that these two equations are identical for reversible processes. These equations are written for a control volume with only one inlet and one exit. If the control volume has more than one inlet or exit, a summation must be made of all inlets and exits. Similarly, if heat is added or removed at more than one temperature on the control surface the $\dot{Q}/T$ term must be obtained by integrating over the area,

$$\frac{\dot{Q}}{T} = \iint_{cs} \frac{\dot{Q}/A}{T} \, dA \tag{5–52}$$

Equations 5–50 and 5–51 demonstrate that

1. The entropy of a control volume can be decreased in only two ways—by heat removal or by the removal of mass.

2. The entropy of a control volume can be increased in three ways—by heat addition, by mass addition, and by irreversibilities.

Equation 5–51 can sometimes be used to evaluate the magnitude of an irreversibility by solving for $\dot{I}$. Since

$$S_{cv} = Ms \tag{5–53}$$

$$\frac{dS_{cv}}{dt} = s\frac{dM}{dt} + M\frac{ds}{dt}$$

The term dS_{cv}/dt can be calculated if dM/dt is known from continuity and ds/dt is known from a property relationship. For example, when an ideal gas is contained in the control volume the second $T\,ds$ equation, eq. 4–49, yields

$$T\frac{ds}{dt} = c_p\frac{dT}{dt} - v\frac{dP}{dt} \tag{5–54}$$

which can be solved for ds/dt if the rates of pressure and temperature change can be measured.

For steady state, steady flow, eq. 5–51 yields

$$0 = \dot{m}(s_i - s_d) + \frac{\dot{Q}}{T} + \dot{I} \tag{5-55}$$

If the SSSF process is reversible, $\dot{I} = 0$, and

$$\frac{\dot{Q}}{\dot{m}} = q = T(s_d - s_i) \tag{5-56}$$

If the SSSF process is adiabatic, $\dot{Q} = 0$, and

$$\dot{I} = \dot{m}(s_d - s_i) \tag{5-57}$$

If the SSSF process is reversible ($\dot{I} = 0$) and adiabatic ($\dot{Q} = 0$) (isentropic),

$$s_i = s_d \tag{5-58}$$

The first law equation for the SSSF process, eq. 5–40, can be written as

$$\dot{Q} - \dot{W} = \dot{m}\left[h_d - h_i + \frac{V_d^2 - V_i^2}{2} + g(z_d - z_i) \right]$$

when divided by $\dot{m}$ yields

$$q - w = h_d - h_i + \frac{V_d^2 - V_i^2}{2} + g(z_d - z_i)$$

For a process with negligible changes in kinetic and potential energy this reduces to

$$q - w = h_d - h_i \tag{5-59}$$

For a reversible process

$$\int T \, ds = \int \delta q = q \tag{5-60}$$

so that the second $T \, ds$ equation can be written as

$$\int_i^d T \, ds = \int_i^d dh - \int_i^d v \, dP$$

$$q = h_d - h_i - \int_i^d v \, dP \tag{5-61}$$

Substituting eq. 5–61 into eq. 5–59 yields

$$w = -\int v \, dP \tag{5-62}$$

Thus, in a reversible SSSF process the work per unit mass is $\int -v \, dP$, not $\int P \, dv$. This integral is proportional to the area on a P–v diagram as shown in Fig. 5–10. A reversible constant-pressure process involves no work in SSSF.

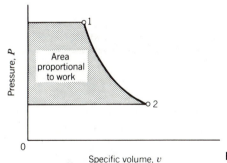

Figure 5-10   Work in a reversible SSSF process.

EXAMPLE 5–12

Calculate the rate of entropy production due to irreversibilities, $\dot{I}$, for a control volume consisting of the water side of a heat exchanger. Saturated liquid water at a pressure of 1.0 MPa enters and saturated vapor at a pressure of 0.95 MPa leaves the heat exchanger. Steady-state, steady-flow conditions prevail and the mass flow rate is 1.0 kg/s. The differences in inlet and discharge kinetic and potential energy may be assumed negligible.

SOLUTION

Use eq. 5–51.

$$\frac{dS_{cv}}{dt} = \frac{\dot{Q}}{T} + \dot{I} + (\dot{m}s)_i - (\dot{m}s)_d$$

For SSSF conditions $dS_{cv}/dt = 0$ and $\dot{m}_i = \dot{m}_d = \dot{m}$. The first law, eq. 5–57, with negligible changes in kinetic energy and potential energy, can be used to obtain the expression for $\dot{Q}$ since $\dot{W} = 0$

$$\dot{Q} = \dot{m}(h_d - h_i) = 1.0(2776.1 - 762.81) = 2013.3 \text{ kW}$$

The values of h_d and h_i are found in Table A–1.2. For T use the arithmetic average since T_d and T_i are very nearly equal;

$$\bar{T} = \frac{179.91 + 177.69}{2} = 178.8° \text{ C} = 452 \text{ K}$$

$$\dot{I} = \dot{m}(s_d - s_i) - \frac{\dot{Q}}{T} = 1.0(6.6041 - 2.1387) - \frac{2013.3}{452}$$

$$\dot{I} = 0.01120 \text{ kW/K}$$

COMMENT

This irreversibility is due to the pressure drop in the heat exchanger.

5.8 ENERGY CONVERSION

Most of the energy used is in the form of either mechanical or electrical work. All transportation systems use mechanical work since a force and a displacement are involved. The electrical work produced by large generators at the electrical utilities is mechanical shaft work before it is converted to electrical work. Manufacturing plants use mechanical and electrical work in forming and shaping the products they produce. Most air conditioning and refrigeration systems use mechanical and electrical work to move energy in the form of heat from a low-temperature to a high-temperature region.

Where does this work we use come from? It is energy converted to work from some source. There are three principal sources of this energy: fossil fuels, nuclear fuels, and solar. The fossil fuels include coal, oil, and natural gas; the nuclear fuels consist mainly of fissionable materials; and solar includes hydropower, wind, ocean thermal energy, biomass, as well as the direct use of the sun's rays. There are of course other sources (geothermal, salinity gradients, etc.) but these do not contribute a significant portion of the total energy we use today. How is this energy converted from its source form into the usable form, work? It is done by two distinctly different methods: either through the use of certain processes or through the use of cycles.

5.8.1 Energy Conversion by Processes

Devices that convert other forms of energy into work using specific processes are familiar in everyday living. Batteries and fuel cells convert the energy from a chemical reaction to electrical work using a nearly isothermal process. Internal combustion engines such as spark ignition, diesel, and gas turbine engines convert the chemical energy from the combustion of a fossil fuel in air into work using a series of processes. While spark ignition and diesel engines operate in a *mechanical* cycle (piston motion is repetitive), they do not operate in a *thermodynamic* cycle because the fluid leaving the device is not identical in composition to the fluid entering. All these devices that operate on a process, or a series of processes that do not constitute a thermodynamic cycle, are not heat engines. They generally do not convert heat to work, thus they are not limited in efficiency by the Carnot efficiency. There may in fact be no heat transfer involved during any of the processes. The second law still applies however, and a theoretical maximum possible work output attainable from these processes can be computed. We shall examine only several of these processes briefly in this text but other texts are completely devoted to studies of devices that convert energy using a process or series of processes.[1,2,3]

5.8.1.A   Nozzles

A nozzle is a device that converts the enthalpy of a flowing fluid into fluid kinetic energy. The device can be extremely simple, consisting only of a channel of variable area in the flow direction (see Fig. 5–6). This device was previously analyzed in Section 5.4 from a momentum point of view; an energy analysis will now be performed. If the nozzle is chosen as the control volume there is no mechanical work involved, the heat transfer and changes in potential energy are usually negligible, and the flow is usually assumed to be SSSF. Thus a first law analysis of the nozzle as the control volume yields

$$\dot{Q}_{cv} - \dot{W}_{cv} + \left[\dot{m}\left(h + \frac{V^2}{2} + gz\right)\right]_i - \left[\dot{m}(h + \frac{V^2}{2} + gz)\right]_d = 0$$

$$h_i + \frac{V_i^2}{2} = h_d + \frac{V_d^2}{2}$$

$$V_d = \sqrt{2(h_i - h_d) + V_i^2} \tag{5–63}$$

Frequently the inlet kinetic energy can be neglected because the inlet velocity is low.

A device that performs the inverse function, converting kinetic energy into enthalpy, is called a diffuser. A first law analysis on the diffuser yields the same relationship, eq. 5–63, obtained for the nozzle.

5.8.1.B   Turbines and Piston Engines

Turbines and piston engines convert the kinetic energy or enthalpy of a flowing fluid, or the heat added to the fluid, into work. Occasionally an engine device will have a geometry different from the conventional piston–cylinder geometry, for example, a Wankel engine, but the energy conversion processes in such geometries are similar to those in a piston–cylinder device. The control surface around the device is usually chosen so that there is no change in potential energy and the flow is SSSF. In a piston–cylinder engine, the flow is pulsating near the intake and exhaust valves; however, at a considerable distance upstream and downstream from the valves, the pulsations have damped appreciably and the flow may be assumed SSSF. If the heat transfer is negligible and $z_i = z_d$, a first law analysis yields

$$\dot{Q}_{cv} - \dot{W}_{cv} + \left[\dot{m}\left(h + \frac{V^2}{2} + gz\right)\right]_i - \left[\dot{m}\left(h + \frac{V^2}{2} + gz\right)\right]_d = 0$$

$$\dot{W}_{cv} = \dot{m}\left(h_i - h_d + \frac{V_i^2 - V_d^2}{2}\right) \tag{5–64}$$

The power can be divided by the mass flow rate to obtain the work per unit mass of fluid flowing.

$$\frac{\dot{W}_{cv}}{\dot{m}} = w_{cv} = h_i - h_d + \frac{V_i^2 - V_d^2}{2} \qquad (5\text{-}65)$$

If the heat transfer is not negligible, the first law yields

$$\dot{W}_{cv} = \dot{m}\left(h_i - h_d + \frac{V_i^2 - V_d^2}{2} \right) + \dot{Q}_{cv} \qquad (5\text{-}66)$$

or

$$\frac{\dot{W}_{cv}}{\dot{m}} = w_{cv} = h_i - h_d + \frac{V_i^2 - V_d^2}{2} + q_{cv} \qquad (5\text{-}67)$$

EXAMPLE 5–13

Determine the work obtained per unit mass of air flowing in the SSSF of air through a turbine where the inlet conditions are $P_i = 1.0$ MPa and $T_i = 500$ K and the discharge pressure is 0.1 MPa. First conduct the computation for a reversible adiabatic process, then for a reversible isothermal process, assuming a negligible change in kinetic energy in both cases.
Assume the air is an ideal gas.

SOLUTION

(a) Reversible adiabatic process, eq. 5–65.

$$w_{cv} = h_i - h_d + \frac{V_i^2 - V_d^2}{2}$$

$$w_{cv} = h_i - h_d = c_p(T_i - T_d)$$

$$\frac{T_d}{T_i} = \left(\frac{P_d}{P_i}\right)^{(\gamma - 1)/\gamma} = 0.1^{0.4/1.4} = 0.5179$$

$$T_d = 0.5179(500) = 258.9\,\text{K}$$

$$w_{cv} = 1.004(500 - 258.9) = 242.1\,\text{kJ/kg}$$

(b) Reversible isothermal process, eq. 5–67.

$$w_{cv} = h_i - h_d + \frac{V_i^2 - V_d^2}{2} + q_{cv}$$

$$w_{cv} = c_p(T_i - T_d) + q_{cv}$$

$$w_{cv} = q_{cv}$$

From the definition of entropy, $q_{rev} = \int T\,ds$, and from the second $T\,ds$ equation, $T\,ds = dh - v\,dP$, obtain

$$w_{cv} = q_{cv} = \int_i^d T\,ds = \int_i^d -v\,dP$$

$$= RT\int_i^d \frac{-dP}{P} = -RT\ln\left(\frac{P_d}{P_i}\right) = RT\ln\left(\frac{P_i}{P_d}\right)$$

$$w_{cv} = 0.287(500)\ln 10 = 330.4\,kJ/kg$$

COMMENT

There are two points that merit comment in this example. First, the reversible isothermal work is observed to be greater than the reversible adiabatic work for the expansion of an ideal gas between fixed pressures. This is true in general. If the process had been a compression process between fixed pressures, the reversible isothermal process would involve less work input than the reversible adiabatic process. Thus, in both cases the reversible isothermal process would be preferable over the reversible adiabatic process. However, in practice the isothermal process is difficult, if not impossible, to achieve because of the required high rate of heat transfer.

The second point is that here we have a *process* (the reversible isothermal process) where all the heat added to the control volume at 500 K has been converted into work. This is not a violation of the second law since this is a *process*, not a *cycle*. This process required air at 1.0 MPa and 500 K. If there happened to be an infinite supply of such air, this process would have tremendous practical significance, but such an energy source does not exist. In practice, the air would have to be compressed to that high pressure and that compression process would require a work input, reducing significantly the net work output. Note that in a *process* there is no prohibition against converting all the heat supplied into work. In fact, the reversible adiabatic process had zero heat supplied, yet work was produced.

Devices that perform the inverse function of turbines and piston engines are called compressors if the fluid is compressible or pumps if the fluid is incompressible. The analysis of these devices produces equations for the work which are identical to eq. 5–65 and 5–67.

5.8.1.C   The Gas Turbine Engine

A gas turbine internal combustion engine is shown schematically in Fig. 5-11. The air from the atmosphere is compressed in the compressor, flows into the burner where fuel is mixed with the air, and combustion occurs. The products of the combustion reaction expand through the turbine producing power to drive the compressor as well as to provide some power to the load. The load could

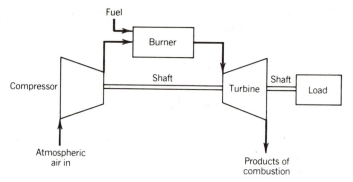

Figure 5-11 Schematic diagram of a gas turbine engine.

be an electrical generator or the propeller of an aircraft or ship. This device does not operate in a thermodynamic cycle, since the composition of the mass returned to the atmosphere is different from that of the mass removed from the atmosphere.

A temperature-specific entropy diagram showing the processes taking place in the gas turbine is shown in Fig. 5-12. Both the compressor and turbine are assumed to be reversible and adiabatic so that isentropic processes occur through these components. The combustion process is assumed to be a constant-pressure process where the temperature increases significantly due to the chemical reaction. A simplified first law analysis can be performed on the compressor and turbine, assuming: SSSF, negligible kinetic and potential energy, and ideal gases with constant specific heats, to yield

Compressor:

$$-\frac{\dot{W}_{cv}}{\dot{m}} = -w_{cv} = w_c = h_2 - h_1 = c_p(T_2 - T_1) \qquad (5\text{--}68)$$

Turbine:

$$\frac{\dot{W}_{cv}}{\dot{m}} = w_{cv} = w_t = h_3 - h_4 = c_p(T_3 - T_4) \qquad (5\text{--}69)$$

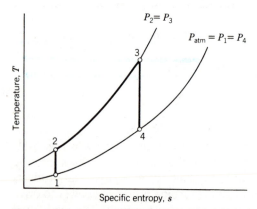

Figure 5-12 Temperature–specific entropy diagram for a gas turbine engine.

and the net work is given by

$$W_{net} = w = w_t - w_c = c_p[(T_3 - T_4) - (T_2 - T_1)] \qquad (5\text{–}70)$$

From the T–s diagram it can be seen that

$$(T_3 - T_4) > (T_2 - T_1)$$

Thus the net work is positive. A more detailed thermodynamic analysis can be made which also shows this to be true. Thus the gas turbine engine is a device that, operating through a series of processes, converts some of the energy released in the chemical reaction into work.

5.8.2 Energy Conversion by Cycles—Heat to Work

As noted in Chapter 4, where the Carnot cycle was studied, power-producing cycles take heat from a high-temperature source, let it run "downhill" to a low-temperature sink, and on the way convert into work some of the heat taken from the high-temperature source. The Carnot efficiency represents the maximum percentage of the heat added which can be converted into work. All externally reversible cycles achieve this Carnot efficiency, while lower efficiencies are obtained when the cycle is not externally reversible. In the real world there are no externally reversible processes so that we would expect actual cycles to have an efficiency lower than the Carnot efficiency.

Most practical cycles involve a fluid that experiences a series of processes that constitute the cycle. The fluid, called the cycle working fluid, may be a gas, a liquid, or it may involve both in a phase change process. The fluid is usually circulated through devices or components where heat is added, then through devices or components where work is extracted, and then through devices or components where heat is removed from the fluid. Most cycles also involve a process where some work must be added to the fluid so that devices or components such as pumps or compressors are also involved. Obviously the amount of work added to the fluid must be less than that extracted from the fluid for it to be of interest as a power-producing cycle. If net work is added to the cycle (i.e., more is added than extracted) it would be a heat pump or refrigeration cycle. Vapor compression refrigeration cycles are of considerable importance since most refrigerators, freezers, and home air conditioners operate on such a cycle. The components of the heat pump or refrigeration cycles are similar to those of the power-producing cycles and we shall discuss them after we have discussed the power-producing cycles.

In addition to using fluids as the working substance in thermodynamic cycles, solid working substances may be used. Both rubber and metallic bands or wires have been used as working substances in power-producing cycles. These cycles will produce power but at a level usually too low to merit practical use. Thermoelectric and thermionic devices also convert heat to work continuously. These devices are static in nature; they involve no moving parts such as pistons, or turbine blades. They are of course limited by the Carnot efficiency and in practice do not approach that theoretical limit. They are not widely used and are limited to selected applications. Further discussion of the performance of these devices can be found in Refs. 2 and 4.

5.8.3 The Rankine Cycle

There are many systems in use today which convert heat to work through a thermodynamic cycle. The most common cycle is the Rankine cycle. Utilities use this cycle to produce power from both fossil and nuclear energy sources. The working fluid used in these cycles is water, although Rankine cycles can and have operated on other phase change fluids such as mercury, potassium, ammonia, and a variety of organic fluids such as the freons. We will describe the basic Rankine cycle along with the hardware or components used to carry out this cycle. We will also discuss modifications that are made to the basic cycle and the reasons for making these modifications.

The components of the basic Rankine cycle are shown in Fig. 5-13. Water is used as the working fluid in this cycle. Water leaves the condenser and enters the pump at state 1. Power is supplied to the pump which increases the pressure of the water before it enters the boiler at state 2. Heat is added to the water in the boiler and the water leaves the boiler and enters the turbine as a vapor (state 3). Power is produced by the turbine as the steam expands through the turbine and enters the condenser (state 4). Heat is removed to condense the steam into the liquid phase after which the water enters the pump to complete the cycle.

The processes shown in Fig. 5-14 are idealizations of the actual processes that occur in the system components shown in Fig. 5-13. The process through the pump (1–2) is idealized as a reversible adiabatic (isentropic) process; the actual process is in fact extremely close to being adiabatic but some irreversibilities are present and the actual process is different from that shown. The process through the boiler (2–3) is idealized as a constant-pressure heating process, but actually there would be a pressure drop as the fluid flows through the boiler (due both to fluid viscosity and probably a momentum increase). The expansion through

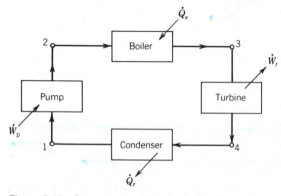

Figure 5-13   Components of the simple Rankine cycle.

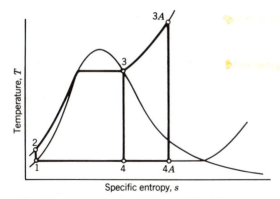

Figure 5-14   Temperature–specific entropy diagram for the ideal Rankine cycle.

the turbine (3–4) is idealized as a reversible adiabatic expansion. Again the process is very close to being adiabatic, but some irreversibilities are present and a slight increase in entropy will actually occur. The *T–s* diagram in Fig. 5-14 also shows an expansion process through a turbine when superheated vapor (3*A*) rather than saturated vapor (3) enters the turbine. The boiler must obviously provide the additional heat (from 3 to 3*A*) when superheated steam is supplied to the turbine. The turbine exhaust state is 4*A* when the turbine inlet state is at 3*A*. In most actual Rankine cycle systems the steam is supplied to the turbine in the superheated state. The process through the condenser (4–1) or (4A–1) is idealized as a constant-pressure heat removal process while actually a small pressure drop would occur.

5.8.3.A   Ideal Rankine Cycle Analysis

Before discussing the more complex cycle, a simple first law analysis of the ideal Rankine cycle shown in Figs. 5-13 and 5-14 will be performed. The cycle is called ideal because all the processes are assumed reversible; there are no pressure drops in the heat exchangers and no irreversibilities (losses) in the pump and turbine. We will *assume steady state, steady flow for all the components* and analyze each component separately as a control volume. The SSSF form of the first law, eq. 5-40, can be written as

$$\dot{Q}_{cv} - \dot{W}_{cv} + \left[\dot{m}\left(h + \frac{V^2}{2} + gz\right)\right]_i - \left[\dot{m}\left(h + \frac{V^2}{2} + gz\right)\right]_d = 0$$

The Pump.   When eq. 5–40 is applied to the control volume representing the pump, along with the assumptions that:

1.   The pump is adiabatic ($\dot{Q} = 0$) and reversible ($s_2 = s_1$).
2.   The changes in kinetic and potential energy are negligible.

3. **The fluid is incompressible** $(v_2 = v_1)$.
   It reduces to

$$\dot{W}_{cv} = \dot{m}(h_i - h_d) = \dot{m}(h_1 - h_2) \qquad (5\text{–}71)$$

$$\frac{\dot{W}_{cv}}{\dot{m}} = w_{cv} = h_1 - h_2 \qquad (5\text{–}72)$$

from the second $T\, ds$ equation

$$T\, ds = dh - v\, dP = 0$$

$$\int_1^2 dh = \int_1^2 v\, dP \qquad (5\text{–}73)$$

$$h_2 - h_1 = v(P_2 - P_1)$$

Therefore,

$$w_{cv} = h_1 - h_2 = -v(P_2 - P_1) \qquad (5\text{–}74)$$

or

$$w_p = -w_{cv} = v(P_2 - P_1) = (h_2 - h_1) \qquad (5\text{–}75)$$

The subscript p denotes that the work is for a pump and all pumps require a work input. Thus, whenever this term is used without the subscript p in a thermodynamic equation the negative sign must be returned to the work quantity.

The Turbine.   When the first law is applied to the turbine along with assumptions 1 and 2 listed above for the pump, the following equation is obtained.

$$\dot{W}_{cv} = \dot{m}(h_i - h_d) = \dot{m}(h_3 - h_4)$$

$$\frac{\dot{W}_{cv}}{\dot{m}} = w_{cv} = w_t = h_3 - h_4 \qquad (5\text{–}76)$$

In many actual steam turbines the change in kinetic energy across the turbine is not negligible, but in this ideal analysis the kinetic energy change is being ignored since we have no specific means of determining inlet and discharge velocities.

The Boiler.   When the first law is applied to the boiler along with the assumptions that:

1. Changes in kinetic and potential energy are negligible.
2. There is no work crossing the control surface.

we obtain

$$\dot{Q}_{cv} = \dot{m}(h_a - h_i) = \dot{m}(h_3 - h_2) \tag{5–77}$$

$$\frac{\dot{Q}_{cv}}{\dot{m}} = q_{cv} = h_3 - h_2 = q_s$$

where the subscript s denotes supplied.

The Condenser.   The assumptions listed for the boiler are also applied to the condenser so that the first law for the condenser yields

$$\dot{Q}_{cv} = \dot{m}(h_a - h_i) = \dot{m}(h_1 - h_4)$$

$$\frac{\dot{Q}_{cv}}{\dot{m}} = q_{cv} = h_1 - h_4 \tag{5–78}$$

$$q_r = -q_{cv} = h_4 - h_1 \tag{5–79}$$

The subscript r denotes rejected, thus q_r is written as a positive quantity.

Net Cycle Work and Thermal Efficiency.   The net cycle work is determined as the turbine work minus the pump work,

$$w = w_t - w_p = (h_3 - h_4) - (h_2 - h_1) \tag{5–80}$$

The net cycle heat transfer, which is also equal to the net cycle work, is

$$q_s - q_r = (h_3 - h_2) - (h_4 - h_1) = w_t - w_p \tag{5–81}$$

The cycle thermal efficiency is then given by

$$\eta_{th} = \frac{w}{q_s} = \frac{w_t - w_p}{q_s} \tag{5–82}$$

This efficiency is less than the Carnot efficiency if the maximum and minimum Rankine cycle temperatures are used in the Carnot efficiency expression.

EXAMPLE 5-14

Steam in a Rankine cycle enters a turbine at $P_3 = 10$ MPa and $T_3 = 500°$ C and leaves the turbine at $P_4 = 10$ kPa. For the ideal Rankine cycle:

(a) Sketch the cycle on T–s coordinates.
(b) Compute the cycle thermal efficiency.
(c) How much net power is produced if the steam flow rate is 10 kg/s?

SOLUTION

(a) The T–s diagram in Fig. 5-14 is an adequate representation of the data given here if states 3A and 4A are used.

(b) Obtain the properties at states 3 and 1 where $P_1 = P_4$. From Table A-1.3 at $T_3 = 500°$ C and $P_3 = 10$ MPa: $h_3 = 3373.7$ kJ/kg, $s_3 = 6.5966$ kJ/kg·K.

From Table A-1.2 at $P_1 = 10$ kPa: $h_1 = 191.83$ kJ/kg, $s_1 = 0.6493$ kJ/kgK, $v_1 = 0.001010$ m³/kg, $T_1 = 45.81°$ C. A first law analysis on the pump yields (eq. 5–75)

$$w_p = v(P_2 - P_1) = 0.00101(10,000 - 10) = 10.09 \text{ kJ/kg}$$

and

$$h_2 = h_1 + w_p = 191.83 + 10.09 = 201.91 \text{ kJ/kg}$$

Since the process in the turbine is assumed reversible and adiabatic

$$s_4 = s_3 = 6.5966 \text{ kJ/kg}$$

and

$$P_4 = 10 \text{ kPa}$$

From Table A-1.2 at 10 kPa: $s_g = 8.1502$ kJ/kg·K, $s_f = 0.6493$ kJ/kg·K, $h_g = 2584.7$ kJ/kg, $h_f = 191.83$ kJ/kg.

$$s_4 = (1 - x)s_f + xs_g$$

$$x = \frac{s_4 - s_f}{s_g - s_f} = \frac{6.5966 - 0.6493}{8.1502 - 0.6493} = 0.7929$$

$$h_4 = (1 - x)h_f + xh_g = (.2071)191.83$$

$$+ \ 0.7929(2584.7) = 2089.1 \text{ kJ/kg}$$

$$w_t = h_3 - h_4 = 3373.7 - 2089.1 = 1284.6 \text{ kJ/kg}$$

$$w = w_t - w_p = 1274.5 \text{ kJ/kg}$$

$$q_s = h_3 - h_2 = 3373.7 - 201.9 = 3171.8 \text{ kJ/kg}$$

$$\eta_{th} = \frac{w}{q_s} = \frac{1274.5}{3171.8} = 0.4018$$

(c) $$\dot{W} = \dot{m}w = 10(1274.5) = 12,745 \text{ kW}.$$

COMMENT

Note that the pump work is very much smaller than the turbine work for a Rankine cycle. The irreversibilities present in a real system will cause the cycle efficiency for a real system to be on the order of 30% for the given conditions.

5.8.3.B REGENERATION

The simple Rankine cycle does not have a very high thermal efficiency. The efficiency of the ideal cycle under the conditions given in Example 5–13 was found to be only 40%, yet the Carnot efficiency between these temperature limits is

$$\eta_c = \frac{T_H - T_L}{T_H} = \frac{773.2 - 319.0}{773.2} = 0.5874$$

The difference is due to the fact that the Rankine cycle is not an externally reversible cycle. Heat is added to the Rankine cycle at temperatures below the maximum cycle temperature. A logical question then is, can any modifications be made to the simple Rankine cycle to improve its efficiency and bring it closer to the theoretical maximum? The answer is yes, a concept called regeneration will improve the simple Rankine cycle efficiency.

In practice the concept of regeneration for the Rankine cycle consists of extracting steam from the turbine and passing this steam through a heat exchanger (called a feedwater heater) to heat the water before it enters the boiler. The extracted steam is condensed in this heat exchanger and the liquid is returned to the cycle (in the condenser). This extraction may occur at only one point or it may occur at many points along the expansion process. In large power-generating facilities there may be up to nine extraction points. The exact number is a trade-off between improved thermal efficiency (reduced fuel cost) and increased capital cost of the hardware.

Since the extracted steam can no longer produce work in the turbine, the power output of the turbine (hence the net output) is reduced. However, the amount of heat that must be supplied is reduced by a larger amount, so that a net increase in efficiency results. With regeneration the heat being supplied from the external source is being supplied at a higher average temperature and the Carnot efficiency expression tells us that the efficiency should therefore be higher. The computation of the improvement in efficiency and reduction in net power will not be carried out in this text, but such computations can be found in thermodynamics texts(6) and power systems texts.(4, 5)

5.8.3.C Reheat

The Rankine cycle used in Example 5–13 had steam leaving the turbine with a quality of about 80%. A liquid (moisture) content of 20% is so high that the small liquid droplets would impinge on the blades causing serious turbine blade

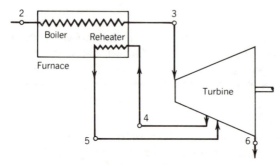

Figure 5-15 Schematic sketch of cycle with reheat.

erosion as well as a reduction in turbine efficiency. An examination of the *T–s* diagram, Fig. 5–14, for this cycle shows that this condition could be avoided if the turbine inlet pressure were reduced. This would move state 3A to the right on the *T–s* diagram and move state 4A to the right by the same amount, thus reducing the liquid content in the mixture at state 4A. This reduction of turbine inlet pressure, while solving the moisture problem, would reduce the cycle efficiency. A more desirable solution to the problem is to add reheat, that is, expand the steam to a pressure well above the condenser pressure, then route the steam back to the furnace where heat is added to increase its temperature, then bring it back to the turbine and continue the expansion until the condenser pressure is reached. Reheat is illustrated in the schematic of Fig. 5–15 and the *T–s* diagram of Fig. 5–16.

Reheat significantly increases the turbine work, but it also increases the required heat supplied from the external source. The cycle efficiency is not altered significantly for a given turbine inlet pressure; however, the turbine moisture problem can be eliminated. On the other hand, for a given tolerable moisture content in the turbine exhaust, reheat allows the use of a much higher turbine inlet pressure with a resulting higher cycle efficiency.

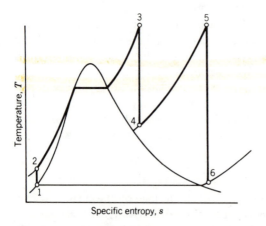

Figure 5-16 Temperature–specific entropy diagram for cycle with reheat.

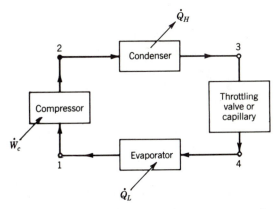

Figure 5-17 Schematic diagram of a vapor compression refrigeration cycle.

5.8.4   Actual Power Cycles

The major components of the actual cycles are those described here under ideal cycles; however there are irreversibilities present which are ignored in the ideal case. Pressure losses occur in piping and heat losses also occur from the high-temperature steam. There are also pressure losses in the heat exchangers and the pumps and turbines involve losses due to friction and flow losses. In addition to these losses in the major components, actual cycles contain additional components such as feedwater heaters to improve the thermal efficiency and to improve the overall operation of the system.

5.8.5   Power Absorbing Cycles

Most domestic refrigeration and heat pump systems use a vapor compression cycle to pump heat from a low-temperature region to a high-temperature region. The cycle is very much like the Rankine cycle run in reverse. It looks quite similar both on a schematic (Fig. 5–17) and on a T–s diagram (Fig. 5–18), with the major difference being that the turbine (where the expansion process occurs) has been replaced by a throttling valve or a capillary tube. The cycle working

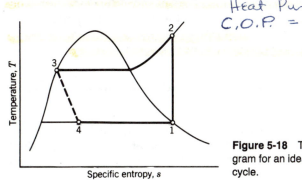

$$\text{Heat Pump,} \quad \text{Refrig.}$$
$$C.O.P. = \frac{\dot{Q}_H}{\dot{W}_c} \quad = \quad \frac{\dot{Q}_L}{\dot{W}_c}$$
$$(3\text{-}4)$$

Figure 5-18 Temperature–specific entropy diagram for an ideal vapor compression refrigeration cycle.

dotted-irreversible

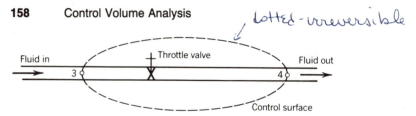

Figure 5-19 Schematic diagram of the throttling process.

fluid (refrigerant) can be any substance that has the desired thermodynamic properties (so that it changes phase at the desired temperatures and reasonable pressures), and has other satisfactory characteristics related to toxicity, corrosiveness, and cost. A number of compounds are used as refrigerants with Freon 12 being one of the more common for domestic applications. Properties of Freon 12 are included in Table A–2.

The analysis of the ideal vapor compression refrigeration cycle is similar to the Rankine cycle analysis. The same assumptions are made and each component is analyzed separately. The only new component is the throttling valve. A throttling valve is shown schematically in Fig. 5–19. The fluid flows through a restriction represented by a valve where a pressure drop occurs. A discussion of this pressure drop and why it occurs will be presented in Chapter 8. We assume for this throttling process that the flow is SSSF, the control volume shown is adiabatic and work free, and there are negligible changes in the kinetic and potential energies of the fluid. While there may be a slight change in velocity between states 3 and 4, the velocities are relatively low and the change in kinetic energy can be very, very small. Note that state 4 is located a considerable distance downstream from the valve in a region where the flow has become more uniform. The first law applied to the control volume yields

$$\dot{Q}_{cv} - \dot{W}_{cv} + \left[\dot{m}\left(h + \frac{V^2}{2} + gz\right)\right]_3 - \left[\dot{m}\left(h + \frac{V^2}{2} + gz\right)\right]_4 = 0$$

$$0 - 0 + \dot{m}(h_3 - h_4) = 0$$

$$h_3 = h_4 \tag{5–83}$$

Thus the enthalpy after the throttling process is identical to that before the throttling process. This is true whether a valve or a capillary tube performs the throttling (pressure drop) function. The throttling process is inherently irreversible and an entropy increase occurs across the valve so that $s_4 > s_3$.

EXAMPLE 5–15

Compute the coefficient of performance and the heating rate for an ideal vapor–compression cycle operating as a heat pump and using Freon 12 as the refrigerant. The refrigerant temperature in the evaporator is $-20°$ C and in the condenser is $50°$ C. The refrigerant flow rate is 0.05 kg/s.

SOLUTION

From Table A–2.1, at $T = T_3 = 50°$ C: $h_3 = h_f = 84.868$ kJ/kg, $P_3 = P_2 = 1.2193$ MPa. At $T = T_4 = T_1 = -20°$ C: $h_1 = h_g = 178.610$ kJ/kg, $s_1 = s_g = 0.7165$ kJ/kgK, $P_4 = P_1 = 0.1509$ MPa. $s_2 = s_1 = 0.7165$ and $P_2 = 1.2193$ MPa. Therefore from Table A–2.2 using double interpolation,

$$T_2 = 65.21° C \quad \text{and} \quad h_2 = 218.64 \text{ kJ/kg}$$

The first law applied to the control volume representing the compressor yields

$$w_c = h_2 - h_1 = 218.64 - 178.61 = 40.03 \text{ kJ/kg}$$

The first law applied to the control volume representing the condenser yields

$$q_H = h_2 - h_3 = 218.64 - 84.87 = 133.8 \text{ kJ/kg}$$

$$\beta = \frac{q_H}{w_c} = \frac{133.8}{40.03} = 3.342$$

Heating rate $= \dot{m}q_H = 0.05 (133.8) = 6.690$ kW.

COMMENT

The coefficient of performance of 3.342 means that a power of $6.690/3.342 = 2.002$ kW is required to provide the heating rate of 6.690 kW.

REFERENCES

1. Campbell, A. S., *Thermodynamic Analysis of Combustion Engines*, Wiley, New York, 1979.
2. Angrist, S. W., *Direct Energy Conversion*, Allyn & Bacon, Boston, 1976.
3. Lichty, L. C., *Combustion Engine Processes*, McGraw-Hill, New York, 1967.
4. Culp, Jr., A. W., *Principles of Energy Conversion*, McGraw-Hill, New York, 1979.
5. *Steam, Its Generation & Use*, Babcock & Wilcox, New York, 1978.
6. Van Wylen, G. J., and Sonntag, R. E., *Fundamentals of Classical Thermodynamics*, Wiley, New York, 1978.

PROBLEMS

5–1   The velocity of a two-dimensional flow field can be written in the xy plane as

$$\mathbf{V} = \mathbf{i}ax + \mathbf{j}b$$

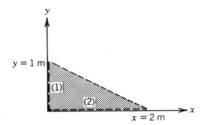

Figure P5-1 Control volume.

A control volume of width b in the z direction (perpendicular to the xy plane) has the shape shown in fig. P5–1. The area elements of this cv can be written as

$$d\mathbf{A} = -c[\mathbf{i}\ dy - \mathbf{j}\ dx]$$

where the negative sign means that $d\mathbf{A}$ is in a direction *outward* from the cv.
   Determine:

(a) $\mathbf{V} \cdot d\mathbf{A}$ (c) $\int_{A_{(1)}} \mathbf{V} \cdot d\mathbf{A}$

(b) $\mathbf{V} \cdot (\mathbf{V} \cdot d\mathbf{A})$ (d) $\int_{A_{(2)}} \mathbf{V} \cdot d\mathbf{A}$

5–2 Engine oil at a temperature of 300 K flows through a circular duct. The diameter of the duct changes along its length. If the flow rate through the duct is 0.147 m³/ s, calculate (a) the average velocity of the oil where the duct diameter is 0.25 m and (b) the mass flow rate. Repeat these calculations where the diameter is 0.50 m.

5–3 Steam enters a constant diameter duct at $P = 1$ MPa and a quality $x = 0.80$ with an average velocity of 25 m/s. Within the duct the steam is heated to a final state of $P = 0.8$ MPa and $T = 400°$ C. What is the average velocity of the steam at the exit of the duct?

5–4 The velocity profile in a circular duct is sometimes parabolic and obeys the equation

$$u = u_0[1 - r^2/R^2]$$

where

$$u_0 = \text{centerline velocity } (r = 0)$$

$$R = \text{duct radius}$$

$$r = \text{radius measured from centerline}$$

Determine:
(a) The volume flow rate (m³/s) in a duct with $R = 0.10$ m and $u_0 = 10$ m/s.
(b) The average velocity V of the flow in the duct.

5–5 A cylindrical tank 0.4 m in diameter discharges water through a hole in its bottom. Determine the rate at which the water depth in the tank is changing (falling) at the instant when the water depth is 1.0 m and the mass flow rate through the hole is 6.0 kg/s. The water temperature is 10° C.

5–6 In an air conditioning unit, Fig. P5–6, two air streams are brought together, mixed and cooled as shown. The states of the air at the inlet and exit are given. Determine

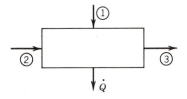

Figure P5-6   Air conditioning unit.

the volume flow rate (m³/s) at the exit

$$\dot{V}_1 = 2.5 \text{ m}^3\text{/s} \qquad \dot{V}_2 = 6.0 \text{ m}^3\text{/s} \qquad \dot{V}_3 = ?$$
$$P_1 = 100 \text{ kPa} \qquad P_2 = 100 \text{ kPa} \qquad P_3 = 95 \text{ kPa}$$
$$T_1 = 50° \text{ C} \qquad T_2 = 70° \text{ C} \qquad T_3 = 20° \text{ C}$$

5–7 A jet engine produces a total thrust of 231.3 kN when run on a test stand. Fuel enters vertically at the top of the engine at a rate equal to 3.0% of the mass flow rate of inlet air which enters (1) in the horizontal direction. The air is accelerated through the engine and exits (2) in the horizontal direction. Calculate the mass flow rate through the exit of the engine if

$$T_2 = 15°\text{C} \qquad P_2 = P_{atm} = 101,394 \text{ Pa}$$
$$V_2 = 366 \text{ m/s} \qquad A_1 = 6.6 \text{ m}^2$$
$$V_1 = 152.5 \text{ m/s} \qquad P_{G_1} = -4784.0 \text{ Pa}$$

5–8 A 7.5-cm diameter jet of kerosene (specific gravity = 0.82) moves at a velocity of 33.0 m/s in the horizontal direction. The jet is symmetrically deflected by a triangular wedge which has a total included angle of 120° and is moving at a velocity of 12 m/s in the opposite direction. Determine the magnitude of the horizontal force required to maintain the velocity of the wedge.

5–9 The hose connecting a pumper truck and the fire nozzle has an internal diameter of 7.5 cm. The pressure at the inlet to the nozzle is 375.0 kPa. The nozzle has an exit diameter of 3.0 cm and discharges the water at 10° C and a velocity of 25.0 m/s to the atmosphere, P_{atm} = 101.3 kPa. What is the net force on the nozzle and what is its direction?

5–10 The siphon system described in Example 5–5 will experience a head loss, although the head loss in the example was assumed to be zero. If the exit of the siphon is placed at a distance of 10.0 m below the surface of the reservoir, determine the head loss h_L through the siphon.

5–11 Water flows at the rate of 2.2 m³/s through a horizontal duct that has an internal diameter of 3.1 m. The duct has a sudden contraction to a diameter of 1.2 m. Experience with similar contractions indicates that the head loss across the contraction is given as $h_L = 0.36 \, V_a^2/2g$, where V_a is the velocity after the contraction (see Fig. 8–7). Determine the pressure drop, $P_1 - P_2$, across the contraction.

5–12 A horizontal jet of water strikes a stationary curved blade and is deflected upward through an angle of 60°. The jet velocity is 25 m/s, its area 0.010 m², and the temperature of the water 10° C. If the jet velocity is constant, what is the resultant force exerted on the vane?

5–13 A water jet pump is configured as shown in Fig. P5–13. The cross-sectional area of the jet is A_j = 0.01 m², and the jet velocity 35 m/s. The velocity of the secondary

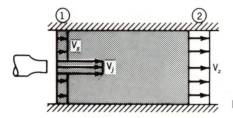

Figure P5-13   Water jet pump.

stream of water $V_s = 4$ m/s and the total cross-sectional area of the duct is 0.08 m². Assume that the jet and secondary flows are mixed thoroughly and leave the pump at station (2) as a uniform flow. The pressure in the jet and secondary flows are equal at the inlet, station (1). Determine the exit velocity, V_2, and the pressure rise, $P_2 - P_1$. The water temperature is 25° C. (Neglect the effect of viscous forces.)

5–14   A block of aluminum weighs 10 N and is constrained in a circular duct as shown in Fig. P5–14. A jet of water is directed upward at the block from a vertical nozzle

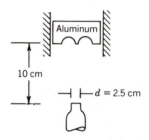

Figure P5-14   Jet of water striking block of aluminum.

whose diameter is 2.5 cm. What is the required jet velocity to hold the aluminum block 10 cm above the jet? The water temperature is 283 K.

5–15   A 1.0-kw electric motor is required to drive a ventilating fan. The fan delivers a stream of air (temperature = 300 K) which is 0.75 m in diameter and has an average velocity of 12 m/s. Determine the efficiency of the ventilating system if the pressure in the inlet room and at the exit of the ventilating duct are equal to atmospheric pressure, Fig. P5–15.

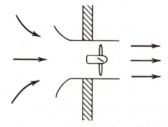

Figure P5-15   Ventilating fan.

5–16   A tank initially containing 1.0 kg of air at 0.1 MPa and 25° C is connected to a high-pressure air line where the pressure and temperature may be assumed to remain constant at 2.0 MPa and 30° C, respectively. A valve between the tank and the line is opened for a very short period of time and then closed when the mass in the tank reaches 10 kg. What is the pressure and temperature of the air in the tank at the instant the valve is closed? Assume the filling process is fairly rapid so that the heat transfer is negligible, assume the air to be an ideal gas, and $h = c_p T$ and $u = c_v T$.

5–17   Compute the power produced by an adiabatic steam turbine operating in steady-state, steady-flow conditions. The steam enters the turbine at a flow rate of 500 kg/s, a pressure of 5.0 MPa, a temperature of 600° C, and a velocity of 30 m/s. The steam leaves the turbine at a pressure of 7.5 kPa, a quality of 0.95, and a velocity of 100 m/s.

5–18   Compute the power required to drive a compressor which is compressing air at 2.0 kg/s in steady-state, steady-flow operation. Air enters the compressor at a pressure of 0.1 MPa, a temperature of 20° C, and a velocity of 5 m/s. The air leaves the compressor at 1.5 MPa, 400° C, and a velocity of 50 m/s. Assume the heat loss to be negligible.

5–19   Air flows reversibly and steadily through an adiabatic diffuser, entering at a pressure of 1.0 MPa, a temperature of 27° C and a velocity of 180 m/s. If the discharge velocity is 15 m/s:
(a)   What is the discharge temperature?
(b)   What is the discharge pressure?
(c)   What is the area ratio A_d/A_i?

5–20   Calculate $\dot{I}$, the rate of entropy production due to the irreversibilities, for:
(a)   Problem 5–17.
(b)   Problem 5–18.
(c)   Problem 5–19.

5–21   For a reversible steady-state, steady-flow process with negligible changes in kinetic and potential energy, the work per unit mass flow rate is given by eq. 5–62. Compare this work to the system work for compressing a unit mass of air between the same pressure limits (starting with air at a given initial state) for:
(a)   A reversible isothermal process.
(b)   A reversible adiabatic process.

5–22   At a certain instant in time, liquid water is entering the pressurizer of a nuclear reactor power system at a rate of 100 kg/s and the time rate of change of mass in the pressurizer is −20 kg/s. Compute the flow rate out of the pressurizer.

5–23   Air is heated in a constant diameter tube in a steady-state, steady-flow process. At the tube entrance the air is at $P_i = 300$ kPa, $T_i = 200°$ C, and has an average velocity of $V_i = 20$ m/s. If the air flow rate is 0.5 kg/s, what is the diameter of the tube?

5–24   Water at 30° C, assumed to be an incompressible liquid, is pumped in a reversible adiabatic steady-state, steady-flow process from a pressure of 0.1 MPa to a pressure of 1.0 MPa. The velocity entering the pump is 1.0 m/s and the velocity leaving the pump is 20 m/s. Compute the power required to drive the pump if the flow rate is 10 kg/s and changes in potential energy are negligible.

5–25   Freon-12 enters a compressor as a saturated vapor at 0° C and is compressed to a pressure of 1.60 MPa. The compressor is assumed to be reversible and adiabatic and is operating in steady state, steady flow with negligible changes in kinetic and potential energy. Compute the energy added as work to each kilogram of Freon as it flows through the compressor.

5–26 Steam in a steady-state, steady-flow process enters a nozzle with a velocity of 50 m/s, a pressure of 1.2 MPa, and a temperature of 400° C. The steam leaves the nozzle at a pressure of 0.4 MPa. Compute the nozzle exit velocity if:
  (a) The process is reversible and adiabatic.
  (b) The nozzle efficiency is 95%, where

$$\eta_n = \frac{\Delta h_{actual}}{\Delta h_{isentropic}}$$

5–27 Using the *T–s* diagram for water, Fig. A–6 in the appendix, estimate the quality after saturated liquid water is throttled from a pressure of 150 bars (1 bar = 10^5 Pa) to a pressure of 1 bar.

6

Special Flows and Differential Form of Conservation Laws

6.1 INTRODUCTION

The purpose of this chapter is to describe the application of the laws of the conservation of mass, momentum, and energy to special fluid flows. These special flows permit several assumptions to be made which simplify the solutions of certain problems. The three cases to be discussed are: inviscid, incompressible flows (effects of viscosity are neglected for a constant-density fluid), fluids at rest (fluid velocity is zero) and inviscid, and compressible flows (effects of viscosity are neglected but fluid density is not constant).

This chapter also introduces the differential forms of these conservation laws. These differential equations are of major importance in determining the distribution of pressure, velocity, and density throughout the flow and are therefore applicable to the study of two- and three-dimensional flows. The closed form solution of the complete fluid flow differential equations is usually difficult. While closed form solutions of the complete set of fluid flow differential equations do exist for some cases, most practical solutions are obtained using numerical techniques. Therefore, the differential equation forms represent the basis for the increasingly important field of computational fluid dynamics.

This chapter is divided into required and optional sections. The optional sections are designated by asteriks (*) and are recommended if a complete introduction to thermal sciences is to be presented. The required sections cover the basic phenomena but are restricted to the study of one-dimensional flows in which the flow properties are uniform over the inlet and exit of the control volume being considered.

*6.2 FLOWS WITH NEGLIGIBLE VISCOUS SHEAR STRESSES; EULER'S EQUATIONS OF MOTION

6.2.1 Differential Control Volume

Although all fluids have viscosity and the relative motion of fluid particles results in the generation of viscous shear stresses, there are regions in many flows where the viscous shear stresses are very, very small and can be neglected for purposes of engineering analysis. Such flows are called *inviscid flows*.

The viscous shear stresses are proportional to the velocity gradient in the direction normal to the flow. As discussed in Chapter 1, the shear stress in a Newtonian fluid in the x direction can be expressed as

$$\tau_x = \mu \frac{du}{dy}$$

where μ is the dynamic viscosity of the fluid and u is the velocity in the x direction. The velocity gradient normal to the flow, du/dy, is not negligible in a region adjacent to a solid surface. Outside of this boundary layer region du/dy is very small, and τ_x is approximately zero. Therefore, the study of inviscid flows excludes the flow near a solid surface or boundary. A further discussion of boundary layer flows will be presented in Chapters 7 and 8.

To take advantage of the special flow regions where viscous stresses are negligible, we develop special tools, for example, the law of conservation of linear momentum with no viscous forces present. We will do this for a differential-sized control volume, dV, whose dimensions are dx, dy, dz. As with the larger arbitrary control volume discussed in Section 5–4, the law of the conservation of linear momentum states that for a steady state, steady flow (SSSF).

$$\begin{bmatrix} \text{The sum of all the forces} \\ \text{(now only due to gravitational and} \\ \text{pressure forces) on } dV \end{bmatrix} = \begin{bmatrix} \text{Change of fluid momentum} \\ \text{as it passes through } dV \end{bmatrix}$$

$$\Sigma d\mathbf{F}_{grav} + \Sigma d\mathbf{F}_{pres} = [\rho(\mathbf{V} \cdot \mathbf{n}) \, dA \, \mathbf{V}]_d - [\rho(\mathbf{V} \cdot \mathbf{n}) \, dA \, \mathbf{V}]_i$$

Similarly, the law of the conservation of mass for the differential control volume states

$$[\text{Mass flow rate entering } dV] = [\text{mass flow rate leaving } dV]$$

$$\dot{m}_i = \dot{m}_d$$

Figure 6–1 shows the differential control volume $dV = dx \, dy \, dz$ with a pressure applied to each face, Fig. 6–1a, and the velocities entering and leaving this control volume, Fig. 6–1b. The pressure and velocity at the center of the control volume, the origin of the coordinate system, are designated as P and $\mathbf{V} = \mathbf{i}u + \mathbf{j}v + \mathbf{k}w$, respectively. At the inlet and discharge faces of the control volume

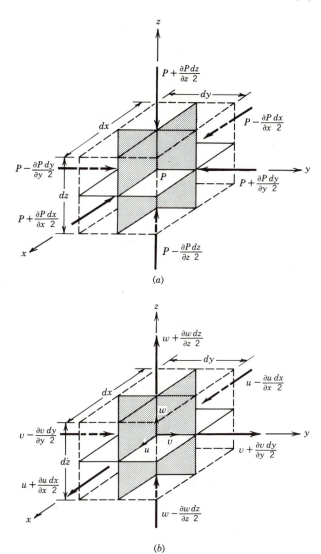

Figure 6-1   Differential control volume. (*a*) Pressures acting on faces of *CV*. (*b*) velocity components entering and leaving *CV*.

perpendicular to the x, y, and z directions, the pressure and velocity will have the following values:

	x Direction	y Direction	z Direction
Inlet:	$P - \dfrac{\partial P}{\partial x}\dfrac{dx}{2}$	$P - \dfrac{\partial P}{\partial y}\dfrac{dy}{2}$	$P - \dfrac{\partial P}{\partial z}\dfrac{dz}{2}$
	$u - \dfrac{\partial u}{\partial x}\dfrac{dx}{2}$	$v - \dfrac{\partial v}{\partial y}\dfrac{dy}{2}$	$w - \dfrac{\partial w}{\partial z}\dfrac{dz}{2}$
Discharge:	$P + \dfrac{\partial P}{\partial x}\dfrac{dx}{2}$	$P + \dfrac{\partial P}{\partial y}\dfrac{dy}{2}$	$P + \dfrac{\partial P}{\partial z}\dfrac{dz}{2}$
	$u + \dfrac{\partial u}{\partial x}\dfrac{dx}{2}$	$v + \dfrac{\partial v}{\partial y}\dfrac{dy}{2}$	$w + \dfrac{\partial w}{\partial z}\dfrac{dz}{2}$

6.2.2 Equation of Continuity

The law of the conservation of mass for a SSSF with constant density ρ through the differential control volume $dx\,dy\,dz$ can be written as

$$\rho\left(u - \frac{\partial u}{\partial x}\frac{dx}{2}\right)dy\,dz + \rho\left(v - \frac{\partial v}{\partial y}\frac{dy}{2}\right)dx\,dz + \rho\left(w - \frac{\partial w}{\partial z}\frac{dz}{2}\right)dx\,dy$$

$$\underbrace{\phantom{\rho\left(u - \frac{\partial u}{\partial x}\frac{dx}{2}\right)dy\,dz + \rho\left(v - \frac{\partial v}{\partial y}\frac{dy}{2}\right)dx\,dz + \rho\left(w - \frac{\partial w}{\partial z}\frac{dz}{2}\right)}}_{\text{Mass flow rate entering } dV} \tag{6-1}$$

$$= \rho\left(u + \frac{\partial u}{\partial x}\frac{dx}{2}\right)dy\,dz + \rho\left(v + \frac{\partial v}{\partial y}\frac{dy}{2}\right)dx\,dz + \rho\left(w + \frac{\partial w}{\partial z}\frac{dz}{2}\right)dx\,dy$$

$$\underbrace{\phantom{= \rho\left(u + \frac{\partial u}{\partial x}\frac{dx}{2}\right)dy\,dz + \rho\left(v + \frac{\partial v}{\partial y}\frac{dy}{2}\right)dx\,dz}}_{\text{Mass flow rate leaving } dV}$$

Collecting terms reduces this equation to the continuity equation for an incompressible flow

$$\frac{\partial u}{\partial x} + \frac{\partial v}{\partial y} + \frac{\partial w}{\partial z} = 0 \tag{6-2}$$

Defining the vector operator

$$\nabla \equiv \mathbf{i}\frac{\partial}{\partial x} + \mathbf{j}\frac{\partial}{\partial y} + \mathbf{k}\frac{\partial}{\partial z} \tag{6-3}$$

Equation 6-2 becomes $\nabla \cdot \mathbf{V} = 0$. If the flow is compressible, $\rho \neq$ const, the equation of continuity for a SSSF becomes

$$\nabla \cdot (\rho\mathbf{V}) = 0 \tag{6-4}$$

The derivation of this equation is left as a student exercise.

6.2.3 Euler's Equation of Fluid Motion

The effect of the gravitational field acting on a fluid is to produce a force that acts at the center of the mass of the differential control volume and in the minus z direction since the positive z direction is defined as upward. Therefore, the resultant force acting on the control volume has components in the x, y, and z

directions which are

$$F_x = \left(P - \frac{\partial P}{\partial x} \frac{dx}{2} \right) dy\ dz - \left(P + \frac{\partial P}{\partial x} \frac{dx}{2} \right) dy\ dz \tag{6-5a}$$

$$= - \frac{\partial P}{\partial x} dx\ dy\ dz$$

$$F_y = \left(P - \frac{\partial P}{\partial y} \frac{dy}{2} \right) dx\ dz - \left(P + \frac{\partial P}{\partial y} \frac{dy}{2} \right) dx\ dz \tag{6-5b}$$

$$= - \frac{\partial P}{\partial y} dx\ dy\ dz$$

$$F_z = \left(P - \frac{\partial P}{\partial z} \frac{dz}{2} \right) dx\ dy - \left(P + \frac{\partial P}{\partial z} \frac{dz}{2} \right) dx\ dy - \rho g\ dx\ dy\ dz \tag{6-5c}$$

$$= - \frac{\partial P}{\partial z} dx\ dy\ dz - \rho g\ dx\ dy\ dz$$

The resultant force on $dx\ dy\ dz$ is therefore

$$\mathbf{F} = \mathbf{i}F_x + \mathbf{j}F_y + \mathbf{k}F_z$$

$$= - \left[\mathbf{i} \frac{\partial P}{\partial x} + \mathbf{j} \frac{\partial P}{\partial y} + \mathbf{k} \left(\frac{\partial P}{\partial z} + \rho g \right) \right] dx\ dy\ dz \tag{6-6}$$

Introducing the vector operator ∇

$$\mathbf{F} = - \left[\nabla P + \mathbf{k}\rho g \right] dx\ dy\ dz \tag{6-7}$$

This expression for the resultant force on dV shows that the pressure contributes to this force only if there is a change, or gradient, of pressure through the control volume. Also, the direction of the force due to pressure is in the opposite direction of the *pressure gradient*, that is, if the pressure increases in the positive x direction, the resulting force is in the minus x direction.

To complete the derivation of the linear momentum equation for the control volume shown in Fig. 6–1, we must determine the change in fluid momentum between the inlet and discharge of the control volume. To do this we must find the difference between the product of the mass flow rate and the velocity entering and leaving in the x, y, and z directions. The sum of the mass flow rates entering each face of the control volume, $\Sigma \dot{m}_i$, must be equal to the sum leaving the discharge faces of the control volume, $\Sigma \dot{m}_d$. In turn these mass flow rates must be equal to that crossing the three perpendicular planes at the center of the control volume, $\Sigma \dot{m}_0$, Fig. 6–1. The mass flow rate associated with each component direction at the center of the control volume is

$$\dot{m}_{x=0} = \rho u\ dy\ dz$$

$$\dot{m}_{y=0} = \rho v\ dx\ dz \tag{6-8}$$

$$\dot{m}_{z=0} = \rho w\ dy\ dz$$

in the x, y, and z directions, respectively.

The change in momentum in the x, y, and z directions can then be written as

$$(\Delta \text{ momentum})_x = \dot{m}_x \Delta u = \rho u \, dy \, dz \left[\left(u + \frac{\partial u}{\partial x} \frac{dx}{2} \right) - \left(u - \frac{\partial u}{\partial x} \frac{dx}{2} \right) \right] \tag{6-9a}$$

$$= \rho u \frac{\partial u}{\partial x} dx \, dy \, dz$$

$$(\Delta \text{ momentum})_y = \dot{m}_y \Delta v = \rho v \, dx \, dz \left[\left(v + \frac{\partial v}{\partial y} \frac{dy}{2} \right) - \left(v - \frac{\partial v}{\partial y} \frac{dy}{2} \right) \right] \tag{6-9b}$$

$$= \rho v \frac{\partial v}{\partial y} dx \, dy \, dz$$

$$(\Delta \text{ momentum})_z = \dot{m}_z \Delta w = \rho w \, dx \, dy \left[\left(w + \frac{\partial w}{\partial z} \frac{dz}{2} \right) - \left(w - \frac{\partial w}{\partial z} \frac{dz}{2} \right) \right] \tag{6-9c}$$

$$= \rho w \frac{\partial w}{\partial z} dx \, dy \, dz$$

Adding these vector components gives the total change in momentum through the control volume

$$[\dot{m}\mathbf{V}]_d - [\dot{m}\mathbf{V}]_i = \rho \left[\mathbf{i} u \frac{\partial u}{\partial x} + \mathbf{j} \, v \frac{\partial v}{\partial y} + \mathbf{k} w \frac{\partial w}{\partial z} \right] dx \, dy \, dz$$

$$= \frac{\rho}{2} \left[\mathbf{i} \frac{\partial u^2}{\partial x} + \mathbf{j} \frac{\partial v^2}{\partial y} + \mathbf{k} \frac{\partial w^2}{\partial z} \right] dx \, dy \, dz \tag{6-10}$$

$$= \rho \, (\mathbf{V} \cdot \nabla) \, \mathbf{V} \, dx \, dy \, dz$$

The law of the conservation of linear momentum states that eq. 6–7 must equal eq. 6–10. Therefore,

$$\mathbf{F} = [\dot{m}\mathbf{V}]_d - [\dot{m}\mathbf{V}]_i$$

or

$$- \nabla P - \rho g \nabla z = \rho (\mathbf{V} \cdot \nabla) \mathbf{V} \tag{6-11}$$

where $\mathbf{k} = \nabla z$. In expanded form,

$$- \mathbf{i} \frac{\partial P}{\partial x} - \mathbf{j} \frac{\partial P}{\partial y} - \mathbf{k} \left(\frac{\partial P}{\partial z} + \rho g \right) = \frac{\rho}{2} \left[\mathbf{i} \frac{\partial u^2}{\partial x} + \mathbf{j} \frac{\partial v^2}{\partial y} + \mathbf{k} \frac{\partial w^2}{\partial z} \right]$$

Note that this equation, called *Euler's equation of motion*, assumes an incompressible, steady state, steady flow (SSSF) with no viscous or friction forces, an inviscid flow.

6.3 BERNOULLI'S EQUATION

A special class of fluid flow is one in which the effects of viscous shear stresses (fluid friction) are negligible. Such flows occur at a sufficient distance from a surface or wall to *exclude* the boundary layer on the surface and, hence, the flow region in which the viscous shear stresses are concentrated. A further discussion of the boundary layer region is presented in Chapters 7 and 8.

Section 6–2 derives the differential equation of motion, eq. 6–11, for an incompressible steady state, steady flow (SSSF) with no viscous shear stresses. This differential equation, called Euler's equation of motion, can be integrated to determine the distribution of the velocity and pressure in the flow outside of the boundary layer. It can also be integrated in the flow direction, along a streamline, or normal to a streamline to give a very important fluid dynamic relationship, *Bernoulli's equation.*

A *streamline* is defined as a line in the flow along which the fluid velocity vectors are tangent at a given instant of time. When the flow is a steady state, steady flow (SSSF) the streamlines also represent the path of the fluid particles. Examples of experimentally determined streamlines are shown in Fig. 6–2.

Figure 6-2 Examples of streamline patterns. (From *Illustrated Experiments in Fluid Mechanics* (The NCFMF Book of Film Notes), National Committee for Fluid Mechanics Films, Educational Development Center, Inc., Copyright © 1972.) (*a*) Smoke flow past airfoil at zero angle of attack. (*b*) High-speed photograph of boundary layer on airfoil undergoing transition. Large eddies are formed prior to breakdown into turbulence. (*c*) Same airfoil at large angle of attack with flow separation.

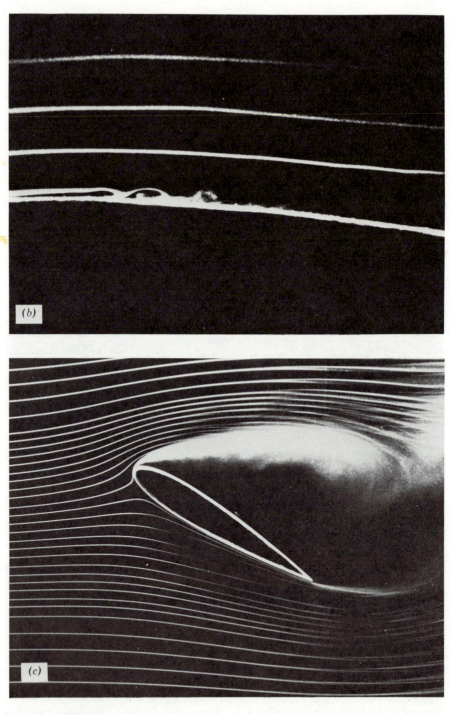

Figure 6-2 Continued

Since Section 6–2 is an optional section, the integration of eq. 6–11 to obtain Bernoulli's equation will also be in an optional section. For those students not studying these optional sections, Bernoulli's equation will be *deduced* from the incompressible SSSF energy equation, eq. 5–44. This is done by making a number of simplifying assumptions before applying the incompressible, SSSF energy equation along a streamline. It must be emphasized that this approach is not the usual way to obtain Bernoulli's equation; it merely gives the same result once these restrictions have been made. This approach is restrictive since it is not possible to determine Bernoulli's equation between streamlines, as can be done by the integration of Euler's equations of motion.

6.3.1 Special Case of the SSSF Energy Equation

The incompressible, SSSF energy equation stated in eq. 5–44 describes the law of the conservation of energy for a rigid control volume with uniform, or averaged, fluid properties entering and leaving the control volume. We assume that:

1. The flow is reversible and adiabatic, that is, it is inviscid and the heat transfer to the control volume is negligible, $h_L = 0$.
2. There is no work done by or on the fluid in the control volume, $\dot{W} = 0$.

With these assumptions eq. 5–44 reduces to

$$\left[\frac{P}{\rho g} + \frac{V^2}{2g} + z\right]_i = \left[\frac{P}{\rho g} + \frac{V^2}{2g} + z\right]_d \qquad (6\text{–}12)$$

If the control volume is reduced to a differential streamtube which includes a streamline of the flow whose inlet (i) and discharge (d) occur at two points, say 1 and 2, along the streamline, Fig. 6–3, eq. 6–12 becomes

$$\frac{P_1}{\rho g} + \frac{V_1^2}{2g} + z_1 = \frac{P_2}{\rho g} + \frac{V_2^2}{2g} + z_2 \qquad (6\text{–}13)$$

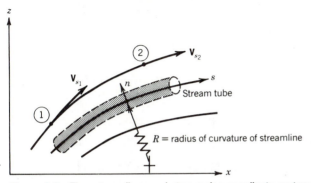

Figure 6-3 Flow streamlines and streamwise coordinate system.

This is equivalent to Bernoulli's equation and states that the total head, $H_T = P/\rho g + V^2/2g + z$, remains constant between two points in a flow if, between the two points:

1. The flow is incompressible.
2. The flow is a steady state, steady flow.
3. The flow is inviscid with no heat transfer (reversible and adiabatic).
4. No work is done by or on the fluid.
5. The two points are along the same streamline.

   While these restrictions preclude a great number of flows, there are many flows where this equation is valid. One must determine if the restrictions can be satisfied before using eq. 6–13.

*6.3.2 Integration of Euler's Equation of Motion

In Section 6–2, the differential equation describing the motion of a fluid particle in an incompressible SSSF with no viscous shear stresses was derived, eq. 6–11.
   This equation is restrictive because of the assumption that the flow is inviscid, but since it is a linear equation it can be solved for a larger number of different flow cases. Solutions to this equation represent a particular branch of fluid mechanics, sometimes referred to as perfect fluids. A detailed discussion of perfect fluids is presented in Ref. 1 and 2 but is beyond the scope of this text. As will be discussed in Chapter 7, the major effects of fluid viscosity are concentrated in a thin layer called the boundary layer near the surface boundaries of the flow. Outside of the boundary layer the assumption that the flow is inviscid is very good and permits the application of this equation.
   Euler's equation of motion can be integrated between two points on a streamline in a flow to give the Bernoulli equation. As discussed earlier, a streamline is defined as a line in the flow along which the velocity vectors are tangent at a given instant of time, Fig. 6–2. When the flow is a SSSF a streamline also represents the path of the fluid particles.
   Figure 6–3 depicts a streamline pattern described by a streamwise coordinate system, (s, n). The "del operator" in this coordinate system is defined as

$$\nabla \equiv \mathbf{i}_s \frac{\partial}{\partial s} + \mathbf{i}_n \frac{\partial}{\partial n}$$

so that eq. 6–11 can be written as ($\mathbf{V} = \mathbf{i}_s v_s + \mathbf{i}_n 0$, from the definition of a streamline)

$$-g \frac{\partial z}{\partial s} - \frac{1}{\rho} \frac{\partial P}{\partial s} = v_s \frac{\partial v_s}{\partial s} = a_s \left\{ \begin{array}{l} \text{The fluid acceleration} \\ \text{in the } s \text{ direction} \end{array} \right\} \qquad (6\text{–}14\text{a})$$

$$-g \frac{\partial z}{\partial n} - \frac{1}{\rho} \frac{\partial P}{\partial n} = a_n \left\{ \begin{array}{l} \text{The fluid acceleration} \\ \text{in the } n \text{ direction} \end{array} \right\} \qquad (6\text{–}14\text{b})$$

The fluid acceleration a_n is the familiar centripetal acceleration that the fluid particles experienced because their path, the streamline, has a radius of curvature R_c. This acceleration can be written as $a_n = -v_s^2/R_c$, and is negative since it is directed toward the center of curvature of the streamline, while the n direction is positive away from the center of curvature.

Integrating eq. 6–14a between the points 1 and 2 on the streamline gives

$$-g(z_2 - z_1) - (P_2 - P_1)/\rho = (v_{s2}^2 - v_{s1}^2)/2$$

Since $\mathbf{V} = \mathbf{i}_s v_s + \mathbf{i}_n 0$,

$$P_1 + \rho V_1^2/2 + \rho g z_1 = P_2 + \rho V_2^2/2 + \rho g z_2$$

or

$$\frac{P_1}{\rho g} + \frac{V_1^2}{2g} + z_1 = \frac{P_2}{\rho g} + \frac{V_2^2}{2g} + z_2 \tag{6-15}$$

This is Bernoulli's equation and is valid if the flow is:

1. Inviscid with no heat transfer (outside of the boundary layer; reversible and adiabatic).
2. A steady state, steady flow (time independent).
3. Incompressible ($\rho = $ const).
4. Along a streamline (points 1 and 2 on the same streamline).

Since Bernoulli's equation is derived from Euler's equation of motion, it is also true that

5. There is no work transfer between parts 1 and 2.

Along a streamline Bernoulli's equation can be written as

$$P + \rho \frac{V^2}{2} + \rho g z = \text{const} \tag{6-16}$$

Thus, if a fluid satisfies the above conditions and the values of P, $\rho V^2/2$, and $\rho g z$ are known at one point in the flow, their sum will remain constant at all other points along a streamline passing through the known point.

The component of Euler's equation normal to a streamline, eq. 6–14b, demonstrates an important characteristic of the pressure variation in a flow. Rearranging this equation gives

$$\frac{1}{\rho}\frac{\partial P}{\partial n} = -g\frac{\partial z}{\partial n} - a_n \tag{6-17}$$

$$= -g\frac{\partial z}{\partial n} + \frac{V^2}{R_c}$$

If the streamlines are straight (infinite radius of curvature, $R_c = \infty$), the only way the pressure can vary normal to a streamline is to change its elevation, $\partial z / \partial n \neq 0$. If the streamlines are horizontal, $\partial z / \partial n = 0$ and $R_c \neq \infty$, there can be a gradient in pressure normal to the streamlines depending upon the value of V^2 and the curvature of the streamline, R_c. The pressure, P, is the same at every point in a flow whose streamlines are straight ($R_c = \infty$) and horizontal ($\partial z / \partial n = 0$).

6.3.3   Static, Dynamic, and Stagnation (Total) Pressures

Consider the case in which there is flow in a horizontal plane. Then, since $z_1 = z_2$, eq. 6–13 or eq. 6–15 can be written as

$$P_1 + \rho \frac{V_1^2}{2} = P_2 + \rho \frac{V_2^2}{2} = P_T \text{ (a constant)} \qquad (6\text{–}18)$$

Each term in this form of Bernoulli's equation represents a pressure. The term P is the thermodynamic pressure and is termed the *static pressure* (or commonly just pressure). While the value of the static pressure is related to the velocity of the flow by eq. 6–18, that is, as V^2 increases P decreases or as V^2 decreases P increases, a value of static pressure would exist if there were no velocity. The term $\rho V^2 / 2$ exists only when there is a flow and is termed the *dynamic pressure*. Outside of the boundary layer Bernoulli's equation states that the sum of the static and dynamic pressures, termed the *stagnation* or *total pressure*, is a constant along a horizontal streamline. Physically the stagnation pressure represents the static pressure which will occur at a point in a fluid if the fluid at that point is brought to rest ($V = 0$) with a reversible and adiabatic process. This concept is important for the experimental measurement of fluid velocity.

Equation 6–17 shows that if the streamlines of the flow are straight, $R_c = \infty$, the only change in pressure from streamline to streamline is due to the change in elevation of the streamline, $\partial z / \partial n$. Using this fact it is possible to measure the static pressure in a flow with a wall pressure "tap" as shown in Fig. 6–4a. The pressure tap is a small hole whose axis is perpendicular to the wall. Great care must be taken to assure that the hole is perpendicular to the wall and free of machining burrs. If so, accurate measurements of static pressure can be obtained by connecting the tap to a suitable sensor, a manometer or pressure transducer. Note that the difference between the static pressure at the wall tap and another pressure is actually measured. If the other pressure is known, like the atmospheric pressure, the static pressure can be determined.

In the fluid away from the wall, or where the streamlines are curved, the static pressure is measured by a static pressure probe, Fig. 6–4b. Such a probe must be aligned parallel to a streamline and carefully calibrated in a known flow to compensate for the effect of its intrusion into the flow.

In addition to the static pressure, the Pitot-static probe shown in Fig. 6–4b

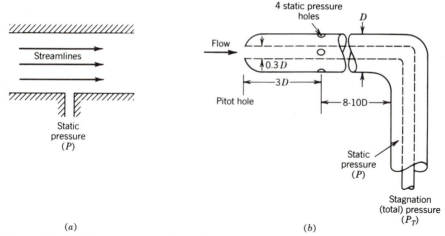

Figure 6-4 Techniques for measuring static and stagnation pressure. (*a*) Wall static pressure tap. (b) Pitot-static probe.

can also be used to measure the stagnation pressure of the flow by using the hole in the front of the probe. This hole faces into the flow and captures the flow causing the flow velocity to decelerate to zero, or stagnate. The pressure in this hole equals the stagnation pressure of the flow. When you place your hand through the open window of a moving car, the pressure on the part of your hand facing into the flow is the stagnation pressure of the flow. On the back of your hand the pressure is less than the stagnation pressure since there is a small velocity in this region. This difference in pressure creates a force that pushes your hand backward.

The Pitot-static probe can then be used to measure the velocity of the flow at a point. Taking the difference between the stagnation and static pressures, $P_T - P$, eq. 6–18 allows the magnitude of the velocity to be calculated as

fluid in manometer
$P_T - P = \rho g \Delta z$ (manometer)

$$V = \sqrt{\frac{2(P_T - P)}{\rho}} \qquad (6\text{–}19)$$

EXAMPLE 6–1

Consider the flow of 2.2 kg/s air at 20 °C through the nozzle shown in Fig. E6–1. Measurement of the pressure at station 2 indicates that the pressure is atmospheric, $P_2 = P_{atm} = 101$ kPa. The cross-sectional area at station 1 is $A_1 = 0.15\ m^2$ and at station 2, $A_2 = 0.03\ m^2$. Determine the gauge pressure at station 1 assuming the flow of the air outside of the wall boundary layers to be steady and incompressible ($\rho = 1.23\ kg/m^3$).

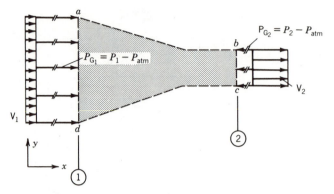

Figure E6-1 Free-body diagram of a nozzle control volume.

SOLUTION

If the flow properties at stations 1 and 2 are assumed to be uniform in the y direction, the gauge pressure, P_{G_1}, in the flow outside of the boundary layers on the wall can be found by applying Bernoulli's equation along the streamline which is coincident with the centerline of the nozzle.

From eq. 6–15 with $z_1 = z_2$, since the flow is in the x, y plane, Fig. E6–1,

$$P_1 + \rho V_1^2/2 = P_2 + \rho V_2^2/2$$

Therefore

$$P_1 = P_2 + \rho(V_2^2 - V_1^2)/2$$

From the conservation of mass

$$V_1 = \frac{\dot{m}}{\rho A_1} = \frac{2.2}{1.23(0.15)}$$

$$= 11.92 \text{ m/s}$$

$$V_2 = \frac{\dot{m}}{\rho A_2} = \frac{2.2}{1.23(0.03)}$$

$$= 59.62 \text{ m/s}$$

Therefore,

$$P_1 = \tfrac{1}{2}(1.23)[(59.62)^2 - (11.92)^2] = 103.1 \text{ kPa}$$

The gauge pressure at 1 is

$$P_{G_1} = P_1 - P_{atm} = 103.1 - 101$$

$$= 2.1 \text{ kPa}$$

COMMENT

This example demonstrates a common use of Bernoulli's equation. Note that before using Bernoulli's equation, the flow must be examined to determine a region in the flow where all of the assumptions made to arrive at Bernoulli's equation are valid. Again, the law of conservation of mass must be used to determine the averaged velocities at stations 1 and 2.

6.4 FLUIDS AT REST

6.4.1 Special Case of Bernoulli's Equation

A second special class of fluid problems to be considered is concerned with fluids that are at rest ($V = 0$). Practical examples of such problems are (1) the determination of the variation of fluid pressure and, hence, the force experience by a water storage dam as a function of the depth of water behind the dam, and (2) the use of manometers to measure the pressure at a point in a moving fluid. As we will see, this latter application takes advantage of the different densities exhibited by different fluids.

A fluid at rest is also a case in which there are no viscous shear stresses acting on the fluid. This is because there is no relative velocity or motion between fluid particles and, hence, no velocity gradients in the fluid. Such a case is by definition a steady state, steady flow (SSSF), and, if the flow is incompressible, the Bernoulli equation, eq. 6–13 or 6–15, with $V^2 = 0$ can be applied.

$$\frac{P_1}{\rho g} + \overset{=0}{\frac{V_1^2}{2g}} + z_1 = \frac{P_2}{\rho g} + \overset{=0}{\frac{V_2^2}{2g}} + z_2 \qquad (6\text{–}20)$$

$$P_1 - P_2 = \rho g(z_2 - z_1) \qquad (6\text{–}21)$$

where z is positive in the upward vertical direction. This relationship clearly shows that as one increases the depth of water in an open container (point 1 being the bottom of the container and point 2 the water surface), as ($z_2 - z_1$) becomes larger, so does the quantity ($P_1 - P_2$). Since the pressure at the surface of the water in the open container is the pressure of the atmosphere, P_{atm}, and is essentially constant, the pressure P_1 must increase.

Equation 6–21 also demonstrates that the pressure is constant in a fluid at a constant depth. Since z is the vertical direction, $z_1 = z_2$ at a constant fluid depth. Therefore, $P_1 = P_2$ at a constant fluid depth.

As discussed in Section 5.4.2, the existence of differences in pressure on the surface of an object, a control volume in Section 5.4.2, will result in a force on the object. This also applies in the special case of fluid statics. The force generated by a fluid at rest at a point on the walls of the container is the product of the difference between pressure at the *point* inside the container and the pressure outside the container and the area *normal* to this pressure difference. If the pressure varies within the container this product must be integrated over the entire container surface to determine the total force. This will be demonstrated in Example 6–4.

An additional force associated with fluids at rest occurs when an object is submerged in the fluid. This force is termed the *buoyancy force* and is determined from the familiar *Archimedes' principle*. Usually introduced in introductory physics courses, Archimedes' principle states "a submerged object experiences an upward vertical force which equals the weight of the fluid which the submerged object displaces." This buoyancy force due to Archimedes' principle less the weight of the object is the net force on the object. If the buoyancy force is greater than the weight of the object, the object will float on the surface of the fluid. Conversely, if the buoyancy force is less than the weight of the object, the object will sink. If the two forces are equal, the object will stay at any depth at which it is placed. This condition is termed neutral buoyancy.

EXAMPLE 6–2

Water is stored in a container whose cross section is shown in Fig. E6–2. In one leg of the container there is mercury (Hg) which, because it is denser than water, $\rho_{Hg} = 13.55\ \rho_w$, has settled to the bottom of the container. The pressure at the free surface is the atmospheric pressure. Determine the pressure at the points A, B, C, and D, at the distances $-z_1$, $-z_2$, and $-z_3$ below the free surface. (Note that the ratio $\rho_{Hg}/\rho_w = 13.55$ is the specific gravity of mercury. The specific gravity of some other liquids are given in Table A–11.)

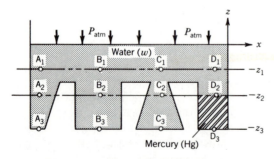

Figure E6-2 A container of liquid at rest.

SOLUTION

With the coordinate system shown in Fig. E6–2,

$$z_0 = 0$$

$$P_0 = P_{atm}$$

The pressure at $-z_1$ is then

$$P_1 = P_0 - \rho_w g(-z_1 - z_0)$$

$$= P_{atm} + \rho_w g z_1$$

Since points A_1, B_1, C_1, and D_1 are all at the same elevation $(-z_1)$ and in the same fluid, water,

$$P_{A_1} = P_{B_1} = P_{C_1} = P_{D_1} = P_{atm} + \rho_w g z_1$$

Similarly at $-z_2$,

$$P_{A_2} = P_{B_2} = P_{C_2} = P_{D_2} = P_{atm} + \rho_w g z_2$$

and at the bottom of the container, $-z_3$,

$$P_{A_3} = P_{B_3} = P_{C_3} = P_{atm} + \rho_w g z_3$$

At the point D_3, the pressure is the sum of the pressure at point D_2 and the pressure due to the mercury which has a vertical height $(z_3 - z_2)$.

$$P_{D_3} = P_{D_2} - \rho_{Hg} g[-z_3 - (-z_2)]$$

$$= P_{D_2} + \rho_{Hg} g(z_3 - z_2)$$

$$= P_{atm} + \rho_w g z_2 + 13.55 \rho_w g(z_3 - z_2)$$

$$= P_{atm} + \rho_w g(13.55 z_3 - 12.55 z_2)$$

COMMENT

This example demonstrates that the pressure in a static fluid:

1. Varies only with depth in the same fluid.
2. Increases as the depth of the fluid increases.
3. Is independent of the shape of the container.
4. Is the *same* at all points in a horizontal plane, that is, constant depth, in the *same* fluid.

5. Is a function of the density of the fluid and the pressure at the surface of the fluid.

EXAMPLE 6–3

Consider the flow of water at 10° C through the horizontal test section of a water tunnel, Fig. E6–3. A Pitot-static probe is inserted into the flow beyond the test section wall boundary layer. The probe is connected to a mercury U-tube manometer whose legs show a difference in the height of mercury of $\Delta z_{Hg} = 52$ mm. Determine the water velocity at the probe location.

SOLUTION

Since the Pitot-static probe is located outside of the flow boundary layers, Bernoulli's equation can be used to analyze the flow. Therefore the flow velocity V is given as, Eq. 6–19

$$V = \sqrt{\frac{2(P_T - P)}{\rho}}$$

To determine the quantity $(P_T - P)$, we must analyze the fluid statics problem created by the U-tube manometer. Following the procedure used in Example 6–2,

$$P_A = P_T + \rho_w g z_A \tag{a}$$

$$P_B = P + \rho_w g(z_A - \Delta z_{Hg})$$

$$P_A = P_C = P_B + \rho_{Hg} g \Delta z_{Hg}$$

$$= P + \rho_w g z_A + g(\rho_{Hg} - \rho_w)\Delta z_{Hg} \tag{b}$$

$$= P + \rho_w g z_A + \rho_w g\left(\frac{\rho_{Hg}}{\rho_w} - 1\right)\Delta z_{Hg}$$

Equating (a) and (b),

$$P_T + \rho_w g z_A = P + \rho_w g z_A + \rho_w g\left(\frac{\rho_{Hg}}{\rho_w} - 1\right)\Delta z_{Hg}$$

or

$$P_T - P = \rho_w g\left(\frac{\rho_{Hg}}{\rho_w} - 1\right)\Delta z_{Hg}$$

From Table A–9, ρ_w = 1000 kg/m³ and from Table A–11, ρ_{Hg}/ρ_w = 13.55. Since Δz_{Hg} = 52 mm,

$$V = \sqrt{\frac{2\rho_w g\left(\dfrac{\rho_{Hg}}{\rho_w} - 1\right)}{\rho_w}\Delta z_{Hg}} = \sqrt{2(9.807)(13.55 - 1)\frac{52}{1000}}$$

$$= 3.578 \text{ m/s}$$

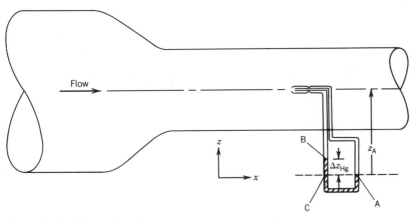

Figure E6-3   Pitot-static probe in a water tunnel test section.

COMMENT

If a horizontal reference line is drawn through the legs of the U-tube manometer we know that the pressure at any point on that line must be the same when the line is through the same fluid (see Example 6–2). We also know that the pressure at the surface of the mercury in the right leg, P_A, equals the total pressure measured by the Pitot-static probe *plus* the weight of the column of water of height z_A above this point. In the left leg of the U-tube the pressure at the surface of the mercury, P_B, is the static pressure measured by the pitot-static probe plus the weight of the column of water of height $(z_A - \Delta z_{Hg})$ above this point. The pressure at point c in the left leg, a distance z_A from the probe centerline, equals the pressure at point A in the right leg and the pressure P_B plus the weight of the column of mercury above c.

EXAMPLE 6–4

The sides of a square container are 0.3 m wide and 0.9 m high. The container is filled with water (ρ_w = 1000 kg/m³) to a depth of 0.8 m. The top of the container is sealed and the air (ρ_a = 1.225 kg/m^3) at the top of the container is at a constant pressure of 1.05 atm. Determine the force on:

(a) The bottom of the container.

(b) The top of the container.

(c) The wetted portion of the sides of the container.

The atmospheric pressure on the outside of the container is 101.3 kPa.

SOLUTION

(a) The pressure on the inside of the bottom of the container is (z = 0 at water surface and positive upward)

$$P_{bot} = P_a + \rho_w g(z_{sur} - z_{bot})$$

$$= 1.05(101,300) + 1000(9.807)[0 - (-0.8)]$$

$$= 114.2 \text{ kPa}$$

$$F_{bot} = (P_{atm} - P_{bot})A_{bot}$$

$$= (101,300 - 114,200)(0.3)(0.3)$$

$$= -1161 \text{ N (acts vertically downward)}$$

(b) The pressure on the inside of the top of the container is

$$P_{top} = P_a$$

$$= 1.05(101,300)$$

$$= 106.4 \text{ kPa}$$

$$F_{top} = (P_{top} - P_{atm})A_{top}$$

$$= (106,400 - 101,300)(0.3)(0.3)$$

$$= 459 \text{ N (acts vertically upward)}$$

(c) The pressure at any depth in the water is, where z_{sur} = 0,

$$P_z = P_a + \rho_w g(z_{sur} - z)$$

The force of the wetted portion of one side is

$$F_{wetted\ side} = \int_{A_{wetted}} (P_z - P_{atm})\,dA = -\int_0^{-0.8} (P_z - P_{atm})w\,dz$$

where w is the width of the side. Therefore,

$$F_{\text{wetted side}} = - \int_0^{-0.8} (P_a - P_{\text{atm}}) w\, dz$$

$$- \rho_w g \int_0^{-0.8} (0 - z) w\, dz$$

$$= -(106{,}400 - 101{,}300)0.3 \int_0^{-0.8} dz$$

$$+ 1000(9.807)0.3 \int_0^{-0.8} z\, dz$$

$$= -1530z \Big|_0^{-0.8} + 2942\frac{z^2}{2}\Big|_0^{-0.8}$$

$$= 2165.4\,\text{N (acts outward from center of container)}$$

COMMENT

The coordinate system used is that the vertical (z) direction is positive upward with its origin at the surface of the water. The pressure on the bottom occurs from the air pressure over the water plus a column of water of depth $\Delta z = 0.8$ m. The force on the top or bottom is the product of the area of the top or bottom and the *net* pressure, the inside pressure less the outside pressure. The pressure of the air is assumed to be constant between the water surface and the top of the container.

The force on a side of the container must be determined by integration of the pressure, which varies from the water surface to the bottom, over the side. The negative sign in front of the integral is required since the integration is in the minus z direction making $dA = -w\, dz$.

EXAMPLE 6–5

A sphere which is 1 m in diameter is filled with benzene and submerged in the ocean to a depth of 100 m. What must the weight of the sphere alone (empty) be to have it remain at 100 m?

SOLUTION

From Table A–11 the specific gravities of benzene and salt water are 0.879 and 1.025, respectively. $\rho_w = 1000$ kg/m³.

$$F_{buoy} = \rho_{salt\ water}\ g\ \text{(volume of sphere)}$$

$$= 1000(1.025)(9.807)\ \tfrac{4}{3}\ \pi\ (0.5)^3$$

$$= 5263\ N$$

The weight of the benzene, neglecting the thickness of the sphere wall, is

$$W_{benz} = \rho_{benz}\ g\ \text{(volume of sphere)}$$

$$= 1000(0.879)(9.807)\ \tfrac{4}{3}\ \pi(0.5)^3$$

$$= 4514\ N$$

When the sphere is filled with benzene and submerged to a depth of 100 m, it will remain at that depth if

$$W_{sphere} + W_{benz} = F_{buoy}$$

Therefore,

$$W_{sphere} = 5263\ N - 4514$$

$$= 749\ N$$

COMMENT

The net force on the sphere is the sum of the buoyancy force, the weight of the empty sphere, and the weight of the benzene required to fill the sphere. The buoyancy force is vertically upward and equals the weight of the salt water displaced by the submerged sphere. For the benzene-filled sphere to remain at a fixed depth it must be neutrally buoyant, that is, the weight of the sphere and the benzene must equal the buoyancy force.

*6.4.2 Fluids at Rest or Experiencing a Constant Acceleration

The behavior of a fluid experiencing only a constant acceleration **a** can be predicted by using Euler's equation of motion simplified for the case of a fluid at rest. If $\mathbf{V} = 0$, Euler's equation of motion, eq. 6–11, becomes

$$-\nabla P - \rho g \nabla z = 0 \qquad (6\text{--}22)$$

since the fluid has zero momentum. Equation 6–22 therefore represents the resultant force on a fluid due to a gravitational field ($\rho g \nabla z$) and pressure changes (∇P). It also says that the resultant force on a fluid particle at rest is zero, that is, it is in equilibrium. A fluid particle is also in equilibrium if it experiences a constant acceleration $\mathbf{a} = \mathbf{i} a_x + \mathbf{j} a_y + \mathbf{k} a_z$. Therefore,

$$-\rho g \nabla z - \nabla P = \rho \mathbf{a} (\mathbf{a} \text{ constant})$$

or,

$$-\frac{\partial P}{\partial x} = \rho a_x$$

$$(6\text{--}23)$$

$$-\frac{\partial P}{\partial y} = \rho a_y$$

$$-\rho g - \frac{\partial P}{\partial z} = \rho a_z$$

If we consider the origin of the coordinate system to be at x_1, y_1, z_1, the integration of eq. 6–23 from the origin to another point x_2, y_2, z_2 gives

$$-(P_2 - P_1) = \rho a_x (x_2 - x_1)$$

$$-(P_2 - P_1) = \rho a_y (y_2 - y_1) \qquad (6\text{--}24)$$

$$-(P_2 - P_1) = \rho (a_z + g)(z_2 - z_1)$$

If $a_x = a_y = a_z = 0$, eq. 6–24 shows that the pressure in the fluid varies only in z direction and equals

$$P_1 - P_2 = \rho g (z_2 - z_1)$$

which is identical to eq. 6–21. This tells us that the pressure in a plane $x_2 y_2$, which is at a constant depth z_2, is constant. If, however, the fluid experiences a constant acceleration a_x, a_y, and a_z in the x, y, and z direction, respectively, the pressure will vary in a plane of constant z.

EXAMPLE 6–6

At the end of the spring semester you have to take your fish aquarium home for the summer. The question is "How much water can you leave in the aquarium without spilling the water in the car?" The aquarium is 0.3 x 0.6 x 0.3 m and the maximum acceleration that your car is capable of providing in the horizontal direction is 0.3 that due to gravity.

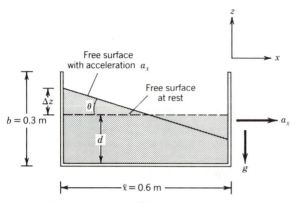

Figure E6-6   Container of liquid at constant acceleration.

SOLUTION

Assume that the road to be travelled is free of bumps and is straight. Also, sloshing of the water can be neglected so that the only accelerations experienced by the water are due to the car, a_x, and due to gravity. Figure E6–6 shows the water surface subjected to these accelerations. From eq. 6–23 ($a_y = a_z = 0$),

$$-\frac{\partial P}{\partial x} = \rho a_x$$

$$-\frac{\partial P}{\partial z} = \rho g$$

Since the surface of the water experiences a constant pressure, the atmospheric pressure P_{atm}, then

$$dP_{sur} = \frac{\partial P}{\partial x}\,dx + \frac{\partial P}{\partial z}\,dz = 0,\ \text{that is}\ \ P_{sur} = \text{const}$$

Therefore, along the surface

$$-\rho a_x\,dx - \rho g\,dz = 0$$

or

$$\frac{dz}{dx}\bigg]_{sur} = -\frac{a_x}{g} = -\frac{0.3g}{g} = -0.3$$

Since the slope of the free surface is constant, it is a plane surface. From Fig. E6–6, the height that the surface rises, Δz, when undergoing an acceleration a_x is

$$\Delta z = \frac{\ell}{2} \tan \theta = \frac{\ell}{2} \left[-\frac{dz}{dx} \right]_{sur}$$

$$= \frac{\ell}{2} \frac{a_x}{g} = \frac{\ell}{2} (0.3)$$

The maximum value of Δz, that is, when the water will spill, for the aquarium whose depth b is 0.3 m is

$$\Delta z_{max} = b - d = 0.3 - d$$

Where d is the depth of water at rest, $a_x = 0$. The allowable depth without spilling is then

$$d = 0.3 - \Delta z_{max}$$

$$= 0.3 - \frac{\ell}{2} (0.3)$$

$$= 0.21 \text{ m}$$

COMMENT

If a fluid experiences a constant acceleration it can be analyzed as a fluid at rest since there is no relative motion between the fluid and its container. The surfaces on which the pressure in the fluid is constant are surfaces parallel to the free surface of the fluid. These surfaces are also normal to the direction of the resultant acceleration on the fluid.

*6.5   ISENTROPIC FLOW OF COMPRESSIBLE FLUIDS

6.5.1   Introduction

So far we have only discussed the flow of fluids in which the density ρ is constant, an incompressible flow. There are a number of flows in which the density is a function of both the position in the flow and time, ρ $\{x, y, z, t\}$. These flows are called *compressible flows*.

The terms subsonic, transonic, and supersonic are familiar terms related to aircraft and spacecraft travel. They are also terms that refer to the velocity of the flow in terms of the *speed of sound*, c, in the fluid. The ratio of local fluid velocity to the speed of sound in the fluid is called the *Mach* number, M = V/c. This nondimensional parameter is important when discussing compressible flows. Therefore, we must define the speed of sound in a fluid before discussing compressible isentropic flows.

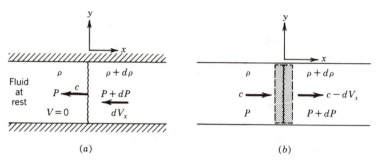

Figure 6-5 Propogation of a pressure wave in a duct. (a) Relative to duct. (b) Relative to wave front.

6.5.2 Speed of Sound

Sound is the propagation of plane, weak, pressure pulses in a fluid. The velocity of propagation, the speed of sound, can be determined by applying laws of the conservation of mass and linear momentum across a pressure wave front traveling through a stationary fluid in a duct of cross-sectional area A. Figure 6–5a depicts such a wave front.

If we consider a control volume of infinitesimal length in the flow direction which contains the wave front and moves with it, the pressure, density, and velocity at the inlet and exit of the control volume are shown in Fig. 6–5b. Note that the control volume is moving with the wave front and we view the flow *relative* to the control volume which makes it appear to be steady. Relative to the control volume, the flow enters at the wave propagation velocity c and leaves at a velocity $c - dV_x$. As the plane wave passes through the fluid, the fluid pressure, density, and velocity are changed.

The conservation of mass relation for the control volume is

$$\underbrace{\rho A_c}_{\substack{\text{mass flow} \\ \text{rate entering} \\ \text{control} \\ \text{volume}}} = \underbrace{(\rho + d\rho)A(c - dV_x)}_{\substack{\text{mass flow} \\ \text{rate leaving} \\ \text{control} \\ \text{volume}}}$$

If we assume the product of the differentials $d\rho\, dV_x$ to be small compared to $d\rho$ and dV_x, this equation reduces to

$$0 = cA\, d\rho - \rho A\, dV_x$$

or

$$dV_x = \frac{c}{\rho}\, d\rho \tag{6–25}$$

The flow is assumed to be isentropic; there are no viscous forces acting on the fluid and no heat transfer to or from it. This assumption is valid since the wave front has an infinitesimal length and there is too little time for heat transfer to occur. The flow is therefore considered to be reversible and adiabatic. The flow in Fig. 6–5 is in the horizontal or x direction and, therefore, there are no body forces in the direction of flow. With these conditions the momentum equation for the control volume can be written as

$$-(P + dP)A + PA = cA \rho [(c - dV_x) - c]$$

or

$$-A \, dP = -cA \rho \, dV_x$$

and

$$dV_x = \frac{dP}{c\rho} \tag{6-26}$$

Equating eqs. 6–25 and 6–26, we obtain

$$\frac{dP}{c\rho} = \frac{c}{\rho} \, d\rho$$

or

$$c^2 = \left[\frac{dP}{d\rho}\right]_{s=0}$$

For an ideal gas, the density and pressure in an isentropic flow are related as, eq. 3–38,

$$Pv^\gamma = \frac{P}{\rho^\gamma} = \text{const}$$

Taking the logarithm of this relation

$$\ln P - \gamma \ln \rho = \ln (\text{const})$$

and differentiating

$$\frac{dP}{P} - \gamma \frac{d\rho}{\rho} = 0$$

Therefore,

$$\frac{dP}{d\rho} = \gamma \frac{P}{\rho} = \gamma RT$$

and for an ideal gas the speed of sound becomes

$$c = \sqrt{kRT} \qquad\qquad (6\text{-}27)$$

Note that T is in degrees Kelvin.

EXAMPLE 6–7

If an airplane flies at a constant velocity of 200 m/s relative to the ground at altitudes of 2000 m (T = 275.2 K) and 11,000 m (T = 216.8 K), determine the Mach number at which the airplane is flying at each altitude.

SOLUTION

The Mach number M = V/c. Since the temperature of the atmosphere is different at 2000 m and 11,000 m, the speed of sound at these altitudes will be different.

$$\text{Speed of sound} = c = \sqrt{\delta RT}$$

From Table A–7 γ = 1.4 and R = 0.287 kJ/kg · [kj/kg · k for air. Therefore, at 2000 m

$$c = \sqrt{1.4(287)275.2}$$

$$= 332.5 \text{ m/s}$$

$$M = \frac{200}{332.5}$$

$$= 0.6015$$

and at 11,000 m

$$c = \sqrt{1.4(287)216.8}$$

$$= 295.1 \text{ m/s}$$

$$M = \frac{200}{295.1}$$

$$= 0.6777$$

COMMENT

Even though an object (an airplane) is moving at a constant velocity relative to a fixed reference (the ground), the Mach number at which it is moving is a function of the fluid temperature. This is because the speed of sound in a fluid is a function of the fluid temperature.

6.5.3 Isentropic Stagnation Properties

The solutions obtained to many problems in thermal science describe the change in a property rather than its absolute value. In order to make these changes in properties meaningful, it is necessary to have a reference state from which the change is determined. In compressible flows, the reference state is usually taken to be the local stagnation state.

The *stagnation properties* at any point in a flow are those properties that would exist in the point if the flow velocity is reduced to zero by an isentropic (reversible and adiabatic) process. In Section 6.3.3 we defined the stagnation pressure (total pressure) as the sum of the thermodynamic pressure P and the dynamic pressure $\rho V^2/2$. This definition was obtained from Bernoulli's equation which is only valid in an incompressible, inviscid, adiabatic flow. This means that the deceleration of the flow velocity to zero is a reversible (frictionless), adiabatic deceleration process.

We can write the energy equation, eq. 5–40, at a point in the flow before and after an isentropic deceleration process as

$$h + \frac{V^2}{2} = h_o \tag{6-28}$$

where the subscript o represents the stagnation condition. Since $h_o - h = c_p(T_o - T)$ for an ideal gas, eq. 6–28 provides an expression for the stagnation temperature T_o

$$T_o = T + \frac{V^2}{2c_p} \tag{6-29}$$

Recall that $c_p = R\gamma/(\gamma - 1)$ and $c = \sqrt{\gamma RT}$. Thus,

$$\frac{T_o}{T} = 1 + \frac{\gamma - 1}{2} M^2 \tag{6-30}$$

In an isentropic flow the stagnation pressure P_o is a constant throughout the flow. We also know from eq. 3–38 that

$$\frac{P_o}{\rho_o^\gamma} = \frac{P}{\rho^\gamma} = \text{const}$$

in an isentropic flow, and if the gas is ideal, $P = \rho RT$. It follows then that

$$\frac{P}{\rho T} = \frac{P_o}{\rho_o T_o} = R \quad \text{or} \quad \frac{P_o}{P} = \frac{\rho_o T_o}{\rho T} \tag{6-31}$$

and

$$\frac{P_o}{P} = \left(\frac{\rho_o}{\rho}\right)^\gamma \tag{6-32}$$

since the flow is isentropic. Substituting eq. 6–32 into eq. 6–31 gives

$$\frac{P_o}{P} = \left(\frac{T_o}{T}\right)^{\gamma/(\gamma-1)} \tag{6–33}$$

Therefore,

$$\frac{P_o}{P} = \left[1 + \frac{\gamma-1}{2}M^2\right]^{\gamma/(\gamma-1)} \tag{6–34}$$

which allows P_o to be determined knowing P, M, and γ (Table A–7) for a particular fluid.

Similarly, it follows that

$$\frac{\rho_o}{\rho} = \left[1 + \frac{\gamma+1}{2}M^2\right]^{1/(\gamma-1)} \tag{6–35}$$

Equation 6–35 can be used to demonstrate the Mach number at which the effects of fluid compressibility start to have a significant effect. The definition of an incompressible flow is one in which the density is constant. For the flow into a nozzle from a large reservoir where $V \equiv 0$, this definition means that $\rho_o \equiv 1$ everywhere in the flow. Equation 6–35 says that this can happen only if M = 0, which is a contradiction since M = 0 means $V = 0$, since c is always finite. Therefore, the existence of an incompressible flow is really an assumption.

The accepted definition of an incompressible flow is one in which $\rho/\rho_o \geqslant 0.956$. Substituting this definition into eq. 6–35 shows that for the flow of air ($\gamma = 1.4$) the flow can be considered to be incompressible if M $\leqslant$ 0.3.

We can demonstrate the validity of this definition of an incompressible flow by comparing the value of P_o/P calculated by eq. 6–34 and by eq. 6–18 which was obtained by defining ρ = const. If air ($\gamma = 1.4$) flows at M = 0.3, eq. 6–34 gives

$$\frac{P_o}{P} = \left[1 + \frac{\gamma-1}{2}M^2\right]^{\gamma/(\gamma-1)} = \left[1 + \frac{0.4}{2}(0.3)^2\right]^{1.4/4} = 1.064$$

Using eq. 6–18 for an ideal gas, $P = \rho RT$,

$$\frac{P_T}{P} = 1 + \frac{\rho}{P}\frac{V^2}{2}$$

$$= 1 + \frac{1}{2}\frac{M^2c^2}{RT}$$

$$= 1 + \frac{\gamma}{2}M^2$$

$$= 1.063$$

Thus, if M < 0.3, an error in the ratio of P_o/P (or P_T/P) of less than 0.1% is incurred by assuming the flow to be incompressible.

6.5.4   Critical Compressible Flows

A unique flow occurs when the local velocity at a point becomes sonic, M = 1.0. When M < 1.0 the flow is *subsonic*; when M > 1.0 it is *supersonic*. Supersonic and subsonic flows are significantly different. So different, in fact, that before supersonic flight was achieved, the condition at M = 1.0 was referred to as the sound barrier. Many people thought that flight at M > 1.0 was impossible. Unfortunately, a detailed study of supersonic flows is beyond the scope of this text.

The flow condition at which M = 1.0 is called the *critical flow* and represents the onset of supersonic flow. The uniqueness of this flow makes it a useful reference condition even though there may be no point in a flow where M = 1.0. It is therefore useful to determine the stagnation properties when M = 1.0. If we designate this condition by an asterisk (*), for air (γ = 1.4),

$$\frac{P_o^*}{P^*} = \left[1 + \frac{\gamma - 1}{2}\right]^{\gamma/(\gamma-1)} = 1.893$$

$$\frac{T_o^*}{T^*} = 1 + \frac{\gamma - 1}{2} = 1.200 \qquad (6\text{--}36)$$

$$\frac{\rho_o^*}{\rho^*} = \left[1 + \frac{\gamma - 1}{2}\right]^{1/(\gamma-1)} = 1.577$$

6.5.5   Effect of Flow Area Variation

To increase the flow velocity of an incompressible, subsonic flow we have seen that it is necessary to decrease the cross-sectional flow area in the direction of the flow. Such a flow geometry is called a nozzle. If the flow through the nozzle is isentropic, we have seen from Bernoulli's equation, eq. 6–18, that the stagnation pressure is constant and that as the velocity V increases, the pressure P decreases. Because subsonic flows, M < 1.0, are different from supersonic flows, M > 1.0, we must ask if an area change in a supersonic flow has the same effect as it does in a subsonic, incompressible flow. As we will discover, the answer is that it does not.

Restricting ourselves to an isentropic, one-dimensional flow with an average velocity, we can write the energy equation, eq. 6–28, along a streamline as

$$h + \frac{V^2}{2} = \text{const}$$

If we consider a differential control volume dV enclosing the streamline whose inlet and exit areas are an infinitesimal distance apart, the differential form of eq. 6–28 becomes

$$dh + \frac{d(V^2)}{2} = 0 \qquad (6\text{--}37)$$

assuming ρ = const within dV. From the relation $T\,ds = dh - dP/\rho$ (eq. 4–51), we see that $dh = dP/\rho$ for an isentropic flow ($ds = 0$). Thus,

$$\frac{dP}{\rho} = -\frac{d(V^2)}{2}$$

or

$$dP = -\rho\,V\,dV \tag{6–38}$$

Note that this is the differential form of Euler's equation of motion, eq. 6–11, without the gravitational force term included.

The conservation of mass equation, eq. 5–21, is the only relationship developed in Chapter 5 which is valid for compressible flows. If we take the natural logarithm of both sides of eq. 5–21, ρVA = const,

$$\ln \rho + \ln V + \ln A = \ln(\text{const})$$

Differentiating,

$$\frac{d\rho}{\rho} + \frac{dV}{V} + \frac{dA}{A} = 0 \tag{6–39}$$

Solving eq. 6–39 for dA/A and substituting the expression for dV from eq. 6–38 gives

$$\frac{dA}{A} = \frac{dP}{\rho V^2} - \frac{d\rho}{\rho}$$

or

$$\frac{dA}{A} = \frac{dP}{\rho V^2}\left[1 - \frac{V^2}{dP/d\rho}\right]$$

Recalling that $dP/d\rho = c^2$ for an isentropic flow,

$$\frac{dA}{A} = \frac{dP}{\rho V^2}\left[1 - \frac{V^2}{c^2}\right] = \frac{dP}{\rho V^2}[1 - M^2] \tag{6–40}$$

or from eq. 6–38

$$\frac{dA}{A} = -\frac{dV}{V}[1 - M^2] \tag{6–41}$$

Equations 6–40 and 6–41 show, respectively, that if $M < 1.0$ a decrease in cross-sectional area ($dA < 0$) results in a decrease in pressure ($dP < 0$) and an increase in velocity ($dV > 0$). This is the same trend as we observed from Bernoulli's equation for an incompressible flow through a nozzle. However, if $M > 1.0$, an increase in cross-sectional area ($dA > 0$) results in a decrease in pressure ($dP < 0$) and an increase in velocity ($dV > 0$). Thus, for a *supersonic nozzle*, and accelerating flow in which M > 1, the cross-sectional area increases in the flow direction. This area change is opposite that for a subsonic, M < 1.0, nozzle but still the velocity increases and the pressure decreases as the fluid moves in the flow direction.

The opposite effect of area change is also observed for a *diffuser* whose purpose is to decrease the flow velocity and increase the pressure. These results are summarized in Fig. 6–6.

We have examined the cases of M > 1.0 and M < 1.0. What happens when the flow is critical, M = 1.0? Equation 6–41 shows that when M = 1.0, dA/dV = 0 since A and V are both finite. This means that the cross-sectional area must be a maximum or minimum at M = 1.0. Inspection of Fig. 6–6 shows that the area must be a minimum, or a throat, when M = 1.0 since this condition connects a subsonic and supersonic nozzle.

Thus, to accelerate a flow from rest (M = 0) in a reservoir to a supersonic speed (M > 1) requires first a subsonic or converging nozzle. Under the proper values of P_o the flow will be sonic, or critical, at the throat where the area is minimum. A further increase in the flow velocity is possible only with a supersonic nozzle which has an increasing cross-sectional area. This results in what is called a *converging–diverging nozzle*.

References 1, 3, and 4 present additional discussions of compressible fluid flow.

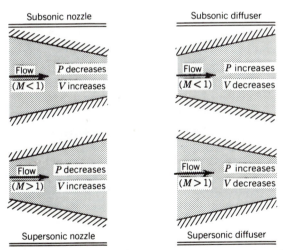

Figure 6-6   Effects of area change on velocity and pressure in a subsonic (M < 1.0) and supersonic (M > 1.0) flow.

EXAMPLE 6–8

Air is stored in a reservoir at a constant pressure and temperature. The air is discharged from the reservoir through a converging nozzle to the atmosphere at $P = 101.3$ kPa and $T = 15°$ C. What are the values of the pressure, temperature, and density at which the air must be maintained in the reservoir to produce a critical flow at the nozzle exit?

SOLUTION

From eq. 6–36, or eqs. 6–30, 6–34, and 6–35 with M = 1.0,

$$\frac{P_o^*}{P^*} = \left(\frac{P_o}{P}\right)_d = 1.893; \quad \frac{T_o^*}{T^*} = \left(\frac{T_o}{T}\right)_d = 1.200; \quad \frac{\rho_o^*}{\rho^*} = \left(\frac{\rho_o}{\rho}\right)_d = 1.577$$

The air in the reservoir is at rest and represents the stagnation conditions. Therefore, the pressure and temperature that must be maintained in the reservoir are

$$P_{res} = P_o^* = 1.893\ P^* = 1.893(101.300) = 191.8 \text{ kPa}$$

$$T_{res} = T_o^* = 1.200\ T^* = 1.200(288.2) = 345.8 \text{ K} = 72.64° \text{ C}$$

$$\rho_{res} = \rho_o^* = 1.577\ \rho^* = 1.577\left[\frac{101.3}{0.287(288.2)}\right] = 1.931 \text{ kg/m}^3$$

COMMENT

Since the flow at the exit of the converging nozzle is stated to be critical, the velocity at the exit is sonic, $M_d = 1.0$. The exit of a converging nozzle is the location of the minimum cross-sectional area or throat. The assumption that the flow is isentropic (reversible and adiabatic) means that the stagnation properties are constant throughout the flow. The isentropic flow at the exit is only critical when $P_o/P_d = 1.893$. If $P_o < 1.893P_d$ the velocity at the exit or throat will be sonic but the flow after the exit will be nonisentropic. This irreversibility will occur as shock waves.

REFERENCES

1. White, F. M., *Fluid Mechanics,* McGraw-Hill, New York, 1979.

2. Milne-Thomson, L. M., *Theoretical Hydrodynamics,* 3rd ed., McMillian, New York, 1955.

3. Fox, R. W., and McDonald, A. T., *Introduction to Fluid Mechanics,* 2nd ed., Wiley, New York, 1978.

4. Shapiro, A. H., *Compressible Fluid Flow,* Vols. I and II, Ronald Press, New York, 1958.

PROBLEMS

6-1 Consider the flow of water at 10° C through the nozzle of a water tunnel, Fig. P6-1. A Pitot tube (P) and a static pressure tap (S) are connected to a U-tube

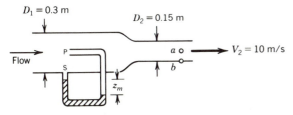

Figure P6-1 Flow through a nozzle.

manometer containing mercury (specific gravity = 13.55, Table A–11). Determine the manometer deflection z_m:
(a) When P and S are located as shown.
(b) When P is moved to location (a) and S remains in the position shown.
(c) When S is moved to location (b) with P remaining in the position shown.
(d) When P is moved to location (a) and S to position (b).
Note that the Pitot tube is not in the wall boundary layer and the static tap is not affected by the viscous effects.

6-2 A dolphin swims at a constant depth below the surface of the ocean and at a constant velocity of 8.0 m/s. If the maximum absolute pressure on the dolphin's nose is 207,870 Pa, determine the depth at which it is swimming ($P_{atm} = 101.3$ kPa, $T_w = 10°$ C).

6-3 A Pitot-static probe, Fig. 6–4b is connected to a U-tube manometer containing methyl alcohol (specific gravity 0.8). The probe is placed in an air stream of standard air ($P_{atm} = 101,300$ Pa, $T = 300$ K). If the manometer deflection is 10.3 cm, what is the velocity of the air?

6-4 Water at 303 K flows from a large tank through a converging–diverging nozzle, Fig. P6-4. The nozzle discharges to the atmosphere and has a diameter $d_d = 2.5d_t$,

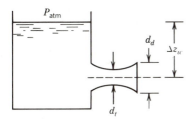

Figure P6-4 Converging–diverging nozzle.

where d_t is the nozzle throat or minimum diameter. Assuming an ideal flow, that is, no losses, determine the height of water Δz_w, at which cavitation (vaporization of the water) will start to occur in the throat.

6-5 A diver is submerged to a depth of 72 m in water having a temperature of 5° C. What is the *absolute* pressure encountered by the diver at this depth?

6-6 A glass tube is shaped as a U-tube containing both water and an unknown fluid, Fig. P6-6. The heights of the fluid levels, measured from a zero reference, are shown in centimeters. What is the specific gravity of the unknown fluid?

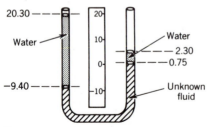

Figure P6-6 U-tube manometer.

6-7 A 0.5-m diameter cylinder shown in Fig. P6-7 is filled with two liquids, water and another with specific gravity $S = 0.88$. Determine the *net* force on the bottom of the cylinder.

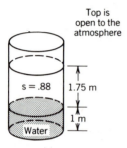

Figure P6-7 Liquid filled cylinder.

6-8 A liquid-filled U-tube, Fig. P6-8, experiences a constant acceleration a_x. Derive an expression for the acceleration a_x in terms of the distance Δz, the fluid density, and the tube geometry.

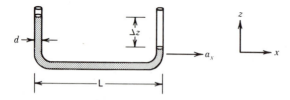

Figure P6-8 U-tube manometer.

6-9 What is the velocity in meters per second for a free-flight Mach number $M = 2.2$ in (a) air, (b) carbon dioxide, and (c) helium, all with a temperature of 260 K?

6-10 Air flows at a speed of 250 m/s, 5° C, and 71 kPa. Determine (a) the Mach number, (b) stagnation pressure, and (c) the stagnation temperature. Compare the pressure difference $(P_o - P)$ with $1/2\rho V^2$, the incompressible value of $(P_o - P)$.

6–11   Repeat Prob. 6–10 for an air flow at 100 m/s, 5° C, and 71 kPa.

6–12   Calculate the velocity at the exit of a converging nozzle which is supplied by a reservoir in which $P = 4P_{atm}$ and $T = 30°$ C. The exit diameter of the nozzle is 5 cm and it discharges to the atmosphere, $P_{atm} = 101.3$ kPa. What is the mass flow rate through the nozzle?

6–13   The reservoir pressure in Prob. 6–12 is reduced to 175.0 kPa while maintaining the same reservoir temperature. Calculate the velocity at the nozzle exit for these reservoir conditions. What is the mass flow rate through the nozzle?

6–14   A nozzle is designed to expand air from initial stagnation conditions of 1500 kPa and 450 K to an exit Mach number of 2.75. The nozzle has a minimum cross-sectional area of 6.5 cm².
(a)   Is the nozzle convergent–divergent or convergent only?
(b)   What is the mass flow through the nozzle?
(c)   Determine the exit cross-sectional area of the nozzle.

6–15   Helium gas enters a reversible adiabatic diffuser at static conditions of 100 kPa and 310 K with a Mach number 0.75. It leaves the diffuser with an exit velocity equal to one-half its entrance velocity.
(a)   Determine the exit static pressure and temperature from the diffuser.
(b)   What is the ratio of the exit-to-entrance area of the diffuser?
(c)   Sketch the process on T–s coordinates identifying both the entrance and exit static and stagnation conditions.

6–16   Carbon dioxide gas flows through a frictionless duct of constant cross-sectional area $A = 35$ cm². At a point in the duct where the stagnation temperature is 537 K, the static properties are measured as $P = 70$ kPa and $T = 340$ K. Heat is added to the CO_2 and at a point further along the duct the static pressure and temperature are $P = 128.7$ kPa and $T = 523.8$ K. Determine:
(a)   The Mach No. at the downstream location.
(b)   The stagnation temperature at the downstream location.
(c)   The heat, in kilowatts, added to the gas.

7

EXTERNAL FLOW—FLUID VISCOUS AND THERMAL EFFECTS

7.1 INTRODUCTION

In Chapter 5 we observed that the total force experienced by a control volume, or the fluid in the control volume, arises from three sources: a body force due to gravity that is proportional to the mass of the fluid, a force due to the pressure acting normal to the surface of the control volume containing the fluid, and a force parallel to the surface of the control volume due to viscous shear stresses acting on the surface. In chapters 5 and 6 two of the three contributors to the total force, gravity and pressure, were discussed in considerable detail. The difference between the sum of these contributors and the total fluid force determined from the linear momentum equation permits the contribution from the viscous shear stresses to be determined. We have already noted that the viscous forces arise in the flow near a solid boundary in a region called the fluid boundary layer. Boundary layer flows will be discussed in this and the following chapter. Methods will be presented to analyze boundary layer flows and predict the viscous forces.

In this chapter we discuss the viscous and thermal effects of *external* flows, those in which the fluid experiences only one solid boundary, Fig. 6–2, for example. Chapter 8 will discuss similar effects of an *internal* flow through a pipe or duct. As we shall see, the phenomena that occur in both cases are similar. However, the information required to describe the two flows is different.

The shear forces acting parallel to the surfaces of a fluid particle cause the fluid particle to deform from its initial shape. The definition of a fluid says that this deformation will be continuous as long as the shear force is applied. In Chapter 1 we defined a particular type of fluid, a *Newtonian fluid*, as one that exhibits a linear relationship between the rate at which the fluid deforms (strain rate) and the magnitude of the applied shear stress, τ. The constant of proportionality is the fluid's coefficient of dynamic viscosity, μ. Water and air are examples of a Newtonian fluid. The term *non-Newtonian* is used to classify all fluids that do not exhibit a linear relationship between the applied shear stress and the fluid strain rate.

As a fluid moves past a solid boundary or wall, the velocity of the fluid particles *at* the wall must equal the velocity of the wall; the relative velocity between the fluid and the wall at the surface of the wall is zero. This fact can be observed experimentally, Fig. 1–2, and is necessary to avoid a discontinuity in the flow. This condition is called the *"no slip" condition* and results in a varying magnitude of the flow velocity, or velocity gradient, as one moves away from the wall.

Figure 7–1 depicts a two-dimensional boundary layer flow and the velocity gradient, du/dy, which occurs in the direction normal to the flow. This velocity gradient occurs because of the shear stress τ_x acting on the surfaces of the particles. The action of τ_x on the bottom of the particle is to retard or decrease it's velocity. On the top surface of the particle τ_x must act to move it in the direction of flow if the fluid particle is to be in equilibrium, that is, summation of forces in any direction must be equal to zero. Therefore, the difference in velocity between the top and bottom of the fluid particle, Δu, is proportional to the distance Δy between the two surfaces

$$\Delta u \propto \tau_x \, \Delta y$$

or

$$\frac{\Delta u}{\Delta y} \propto \tau_x$$

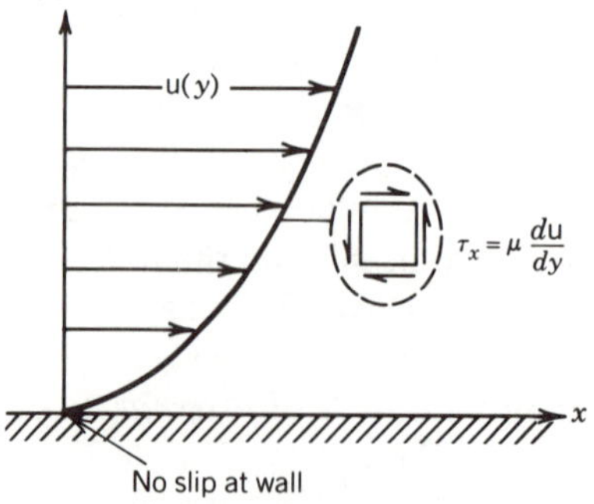

No slip at wall

Figure 7-1   Influence of viscosity on a fluid.

If the fluid is Newtonian with a coefficient of *dynamic (absolute) viscosity* μ

$$\tau_x = \mu \lim_{\Delta y \to 0} \frac{\Delta u}{\Delta y}$$

$$= \mu \frac{du}{dy} \qquad (7\text{--}1)$$

The values of μ for a number of Newtonian fluids are given in Fig. A–13 as a function of temperature. Values of the coefficient of *kinematic viscosity*, $v = \mu/\rho$, of a number of Newtonian fluids are shown in Fig. A–14.

The existence of viscous shear stresses in a fluid results in a force that resists the motion of the fluid. Energy must be added to the fluid to overcome this resistance if the flow is to be maintained. Therefore, in the design of a system with fluid flow it is important to have a minimum resistance to the flow in order to achieve maximum system efficiency, or minimum amount of required energy addition, while maintaining the flow. This requires that a method be developed to predict the magnitude of the resisting force due to viscous shear stresses in the fluid.

This chapter is concerned with the development of a means to describe the flow of a fluid past a solid surface. This will include the development of relationships required to predict the drag force on the surface.

If the temperature of the surface is different than that of the fluid, heat will be transferred. The influence of the flow on the rate of heat transfer between the surface and fluid will also be discussed in this chapter.

7.2 EXTERNAL BOUNDARY LAYERS

Consider the flow of a fluid over a thin flat plate placed in a flow having a constant velocity **U**. If an experimental study were conducted which measured the variation of velocity normal to the plate, the results would indicate the velocity distribution upstream of the plate ($x < 0$) and at the end of the plate ($x = L$) shown in Fig. 7–2. Similar effects are observed with a curved surface but, as we will discuss later, an additional effect occurs because of pressure changes in the direction of flow. We will initially consider only plane surfaces.

Since the plate in Fig. 7–2 is stationary with respect to the earth the fluid velocity at the surface of the plate is zero and increases to the free-stream velocity

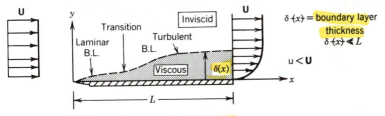

Figure 7-2 Fluid boundary layer on a flat plate.

U as one moves away from the plate. The region in which the velocity varies from 0 to 0.99U is termed the fluid boundary layer. The thickness of the fluid boundary layer normal to the plate, δ, varies in the direction of flow x, and is defined as the value of y where $u = 0.99U$. The effect of viscous shear stresses are concentrated within the boundary layer. Outside of the boundary layer, $y > δ$, the flow can be considered as inviscid ($τ_x = 0$) since $du/dy = 0$.

Experimental studies indicate that there are two boundary layer flow regimes; a *laminar flow regime* and a *turbulent flow regime*. These flow regimes can be characterized by considering the ratio of the inertia force (mass · acceleration $∝ ρL_c^3 U^2/L_c$) on a fluid particle to the viscous force (shear · stress area $∝ μL_c^2 U/L_c$) acting on the fluid particle. This ratio is dimensionless since it is the ratio of two forces and is called the *Reynolds number*

$$Re ≡ \frac{\text{inertia force}}{\text{viscous force}} ∝ \frac{ρL_c^3 U^2/L_c}{μL_c^2 U/L_c} = \frac{ρUL_c}{μ} = \frac{UL_c}{ν} \qquad (7\text{–}2)$$

The characteristic length, L_c, is associated with the particular flow; the length of the plate, the distance from the leading edge of the plate to a particular point on its surface, the boundary layer thickness, and so on. There is a critical value of Re above which the flow will be turbulent and below which it will be laminar. This critical value is termed the *transition Reynolds number* and is, as we will see, a function of several parameters including the roughness of the surface. Figure 7–3 is a holographic photograph of a boundary layer showing the transition of a laminar boundary layer to a turbulent boundary layer.

A low value of Reynolds number, Re < 1, means that the viscous forces are large compared to the inertia forces. On the other hand, large values of Re mean that while viscous forces are present, their influence is not as predominate. The magnitude of Re therefore will allow us to determine when different methods of analysis are required. For example, laminar flows occur at lower values of Reynolds number than turbulent flows and, hence, viscous forces are more dominate in a laminar flow.

The differences between a laminar and turbulent flow can best be described by experiment. For example, the introduction of dye into the laminar flow regime

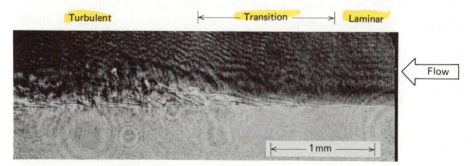

Figure 7–3   Boundary layer transition on Schiebe headform. The flow is (U = 8.0 m/s from right to left). Courtesy of Dr. J.D. van der Mulen, Netherlands Ship Model Basin.

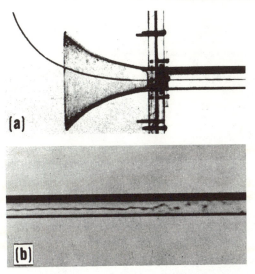

Figure 7–4 A filament of dye in a (a) laminar flow and (b) turbulent flow.[9] Used with permission.

of a liquid flowing through a pipe will show a single filament of dye, demonstrating that the fluid particles move in parallel layers or lamina, Fig. 7–4a. A similar experiment in the turbulent flow regime shows the dye particles dispersing through the flow due to the random, fluctuating motion associated with turbulence, Fig. 7–4b. For a smooth flat plate the critical value of Re_{CR} based on the distance along the plate from the plate's leading edge, is approximately 0.5 $\times$ 10^6. If the surface of the plate is roughened the value of Re_{CR} will be in the range of 8 $\times$ 10^3 − 0.5 $\times$ 10^6, depending upon the size of the roughness.

The difference between a laminar and turbulent flow can also be seen by observing the fluid velocity at a point as a function of time. Such a measurement can be made with a hot-wire anemometer. This device measures the time variation of current flowing through a thin wire ($d \approx 0.13$ mm) placed normal to the flow, and supplied with a constant power. As the velocity past the wire changes, so does the heat loss from the wire. An electrical circuit is employed to maintain a constant electrical resistance in the wire, that is, a constant temperature, and the amount of current flowing in the wire which is proportional to the fluid velocity past the wire is measured. The constant of proportionality between the current and the temperature must be determined by conducting an experiment to calibrate the wire.

A plot of the instantaneous velocity measured by a hot-wire anemometer as a function of time in a turbulent flow is shown in Fig. 7–5. The instantaneous temperature of the fluid is also time dependent as shown in Fig. 7–5. In a laminar flow, the velocity and temperature are constant with time. In a turbulent flow, u and T are composed of a time-mean component $\bar{u}$ and $\bar{T}$ plus a fluctuating component u' and T'. The fluctuating components are random with time and cause the diffusion of the dye through the fluid observed in Fig. 7–4.

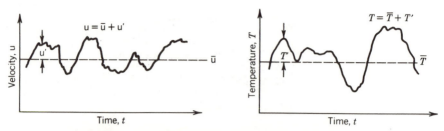

Figure 7-5   Instantaneous velocity and temperature in a turbulent flow.

If the surface of the flat plate is at a different temperature than that of the free stream of the fluid, a *thermal boundary layer* will also form on the plate. Its rate of development and thickness will be similar to that of the fluid boundary layer shown in Fig. 7–2. The dimensionless group that will indicate the relationship between fluid and thermal boundary layers is the *Prandtl number*. The Prandtl number is defined as

$$\text{Pr} \equiv \frac{\text{diffusion of momentum}}{\text{diffusion of energy}} = \frac{\nu}{\alpha} = \frac{c_p \mu}{k} \qquad (7\text{–}3)$$

Liquid metals have very low Prandtl numbers, $0.001 < \text{Pr} < 0.2$. Gases have Prandtl numbers of the order of 0.7 and liquids have Prandtl numbers in the range of $1.0 < \text{Pr} < 85,000$. The Prandtl numbers for various fluids are given in the appendix.

When the Prandtl number is equal to 1, the diffusion of momentum and energy is equal and the fluid and thermal boundary layers develop simultaneously. If the Prandtl number is greater than 1 the fluid boundary layer develops faster, while the thermal boundary develops faster for a fluid with a Prandtl number less than 1. These cases are illustrated in Fig. 7–6. The thermal boundary layer thickness, δ_T, is defined as the distance from the surface to a value of y where $(T - T_w)/(T_\infty - T_w) = 0.99$.

The presence of heat transfer will have some influence on the location of the transition from laminar to turbulent flow. It will, however, remain within the Reynolds number range previously quoted.

7.3   FLOW CHARACTERISTICS OF A BOUNDARY LAYER

In the introduction to this chapter the concept of the presence a region of flow close to solid boundaries in which the effects of viscous shear stresses are predominate was discussed. This concept was first introduced by Prandtl in 1904.

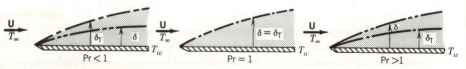

Figure 7-6   Fluid and thermal boundary layers in a flow over a flat plate.

His hypothesis of a boundary layer was arrived at by experimental observations of the flow past solid surfaces. These observations also lead Prandtl to conclude that the thickness of both a fluid and thermal boundary layer is very small compared to the distance along the surface x, that is, $\delta/x \ll 1$ and $\delta_T/x \ll 1$. Figure 7–3, which shows a laminar boundary layer undergoing transition to a turbulent boundary, demonstrates how thin a boundary layer can be. If y is the direction normal to the flow in the x direction, Prandtl observed that the velocity in the y direction is very small compared to the velocity in the x direction, $v \ll u$. He also observed that changes of the velocities u and v in the flow direction were small compared to changes normal to the flow direction, that is, $\partial /\partial x \ll \partial /\partial y$.

These observations by Prandtl lead to the conclusion that the pressure across the boundary layer, from the surface or wall to the edge of the boundary layer, is nearly constant. Therefore, at a constant value of x,

$$P_{y=0} \approx P_{y=\delta}$$

At the edge of the boundary layer the flow is inviscid ($\tau = 0$) and Bernoulli's equation is valid permitting the pressure $P_{y=\delta}$ to be found if the velocity U in the inviscid region is known.

$$P_{y=\delta} = [\text{const}] - \rho \frac{U^2}{2}$$

The variation of pressure in the flow direction is known if the variation of U in the x direction is known

$$\frac{1}{\rho} \frac{dP}{dx} = -U \frac{dU}{dx} \tag{7–4}$$

The term dP/dx is called the flow *pressure gradient*. For the case of the flow of a fluid past a flat plate, Fig. 7–2, $dU/dx = 0$ and the pressure gradient is zero. As will be discussed later, there are many flows in which $dP/dx \neq 0$.

7.4 RESISTANCE TO MOTION; DRAG ON SURFACES

In Chapter 5 we discussed the forces that can exist on a control volume. These are: (1) body forces which are due to the gravitational acceleration and are dependent upon the mass of the fluid in the control volume, (2) pressure forces which are caused by gradients, or differences, in pressure on the faces of the control volume, and (3) viscous forces which occur because of viscous shear stresses parallel to the surface of the control volume. If the direction of flow is vertically upward the body force caused by gravity, F_{grav}, will resist the motion of the fluid. Similarly, if the pressure increases in the direction of flow, F_{pres} will act to resist the fluid motion. There are situations, depending upon the flow direction, where these forces can aid the fluid motion rather than resisting it. On the other hand, the viscous force, F_{vis}, always acts to resist the fluid motion.

As a fluid moves past a solid surface a boundary layer is formed on the surface and the effects of the viscous shear stresses are concentrated in this region. Figure 7–7 depicts a boundary layer with a control volume *abcd* which includes all of the boundary layer over a length of surface *dx* in the direction of flow. Also shown in Fig. 7–7 are the pressures and shear stress, τ_w, which act on *abcd* and when multiplied by the appropriate area result in F_{pres} and F_{vis}. The force F_{grav} acts normal to the *x*, *y* plane and is therefore normal to the flow direction and does not influence the fluid motion.

The shear stresses τ_w must exist on the surface *ab* in order to satisfy the no-slip condition between the fluid and surface. The result of τ_w is to resist the motion of the fluid and make the fluid velocity zero relative to the surface at the surface. Therefore, τ_w acts in the negative *x* direction, opposite to the fluid motion through the control volume. If the control volume and surface are to be in equilibrium, the summation of forces on the surface and fluid equal zero, there must be a reaction to τ_w on the surface. This reaction force on the surface must be in the positive *x* direction, in the direction of flow, and therefore opposite to the direction of F_{vis} on the fluid. This force on the surface acts to move it in the direction of flow and is called the *viscous drag force*. The force on the surface is termed viscous drag because it results from the relative motion of fluid over the surface which generates a viscous shear stress that acts to "drag" the surface in the direction of relative fluid motion.

The drag force is therefore defined as the force on the surface of a solid object which exists because of the relative motion of a fluid over the surface. The drag force on an object is always in the direction of the relative fluid motion over the object and is equal in magnitude, but opposite in direction, to the force on the fluid which resists the movement of the fluid past the surface of the object. In this chapter we discuss two types of drag on an object; friction drag caused by viscous shear stresses and pressure drag caused by pressure gradients over the surface of the object. In both cases we will determine these drag forces by analyzing the flow through a fixed control volume and by determining the force resisting the motion of fluid through the control volume. The drag force is then defined as the reaction to this force on the fluid.

The reduction of the drag force is very important in the development of more fuel-efficient aircraft, trucks, and automobiles. The total drag times the velocity of travel is the power required to overcome the drag force and is a significant part of the total power that must be produced by the vehicle's engine. A considerable amount of research has been conducted to determine methods to reduce the drag of different moving objects. Similarly, the drag created by the wind blowing past a tall building, a large smoke stack, or television tower must be known in order to calculate the force on the object so that a safe structure can be designed. The following material will discuss the causes of drag and methods to minimize its magnitude.

The viscous shear stress in a fluid adjacent to a surface results in a viscous or *friction drag* when it is summed over the entire surface area of the solid boundary. If there is a pressure gradient in the flow direction there will also be

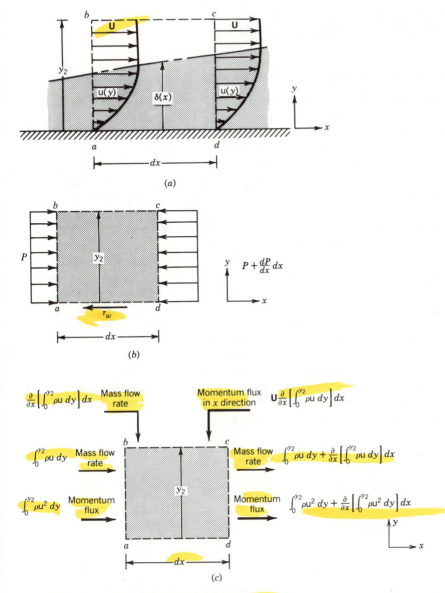

Figure 7-7 Control volume for a hydraulic boundary layer.

a varying pressure on the solid boundary. This pressure will contribute to the drag force on the surface. This drag is called the *pressure drag*. The combination of the friction and pressure drag gives the total drag on the boundary. Remember that when the term drag is discussed it is defined, by convention, to be the force *on the surface in the* direction of relative fluid motion over the surface. This is important since in many cases it is more convenient to analyze the force on the fluid, which resists its motion past the surface, and to define the drag force to be the reaction to the force on the fluid.

7.4.1 Boundary Layer Momentum Analysis

We have already developed a method that will allow us to determine the total drag on a surface boundary. This method is the linear momentum equation, eq. 5–22, for a fixed control volume. Consider the fixed control volume *abcd* shown in Fig. 7–7a which is chosen to include the boundary layer over a segment, *dx*, of the surface in the flow direction. Note that this control volume has a unit depth normal to the *x, y* plane. The total force on *abcd* in the *x* direction is equal to the change in fluid momentum through *abcd* in the *x* direction. Since the *x* direction is defined as the flow direction, the total force equals the force on the fluid which resists the motion of the fluid past the surface.

Figure 7–7b shows the pressures and shear stress that act on *abcd* and result in the total force in the *x* direction. The wall shear stress τ_w on *abcd* is the reaction to the friction drag on the surface. The pressure on *ab* is assumed to equal *P* and is uniform since $dP/dy = 0$ across the boundary layer by Prandtl's observations. On face *cd* the pressure equals that on *ab* plus the change in *P* in the *x* direction as the flow travels a distance *dx* through the control volume. The pressure on faces *bc* and *ad* are not constant but are equal in magnitude and opposite in direction and therefore cancel.

Figure 7–7c shows the mass and momentum flux through the sides of *abcd*. These are shown as integrals over the control surface since the velocity varies in the boundary layer. The existence of a mass flow through face *bc* is seen when applying the conservation of mass to *abcd*.

$$(\text{mass flow rate})_{\text{in } abcd} = (\text{mass flow rate})_{\text{out } abcd}$$

$$\underbrace{\int_0^{y_2} \rho u \, dy}_{(\text{Mass flow rate})_{ab}} + (\text{mass flow rate})_{bc} = \underbrace{\int_0^{y_2} \rho u \, dy + \frac{\partial}{\partial x}\left[\int_0^{y_2} \rho u \, dy\right] dx}_{(\text{Mass flow rate})_{cd}}$$

Therefore,

$$(\text{Mass flow rate})_{bc} = \frac{\partial}{\partial x}\left[\int_0^{y_2} \rho u \, dy\right] dx$$

Along face *bc* the fluid velocity in the *x* direction is U. The momentum flux through *bc* in the *x* direction is then

$$\mathbf{U} \cdot (\text{mass flow rate})_{bc} = \mathbf{U}\frac{\partial}{\partial x}\left[\int_0^{y_2} \rho u \, dy\right] dx$$

Applying the steady momentum equation in the x direction to control volume *abcd* gives

$$\underbrace{Py_2 - \tau_w\,dx - \left[P + \frac{dP}{dx}dx\right]y_2}_{\text{x forces on $abcd$}}$$

$$= \underbrace{\int_0^{y_2} \rho u^2\,dy + \frac{\partial}{\partial x}\left[\int_0^{y_2}\rho u^2\,dy\right]dx}_{\text{x momentum flux out of cd}} - \underbrace{\int_0^{y_2}\rho u^2\,dy}_{\substack{\text{x momentum flux}\\\text{into ab}}} - \underbrace{U\frac{\partial}{\partial x}\left[\int_0^{y_2}\rho u\,dy\right]dx}_{\text{x momentum flux into bc}} \quad (7\text{–}5)$$

Assuming an incompressible fluid, this expression can be simplified to (total derivatives are used since variations occur only in the x direction)

$$\tau_w + y_2\frac{dP}{dx} = \rho\frac{d}{dx}\int_0^{y_2}[U - u]u\,dy - \rho\frac{dU}{dx}\int_0^{y_2}u\,dy \quad (7\text{–}6)$$

Since the pressure across the boundary layer, the y direction, is constant, the force term due to the pressure gradient dP/dx can be written in terms of the free-stream velocity U using eq. 7–4

$$y_2\frac{dP}{dx} = -\rho U\frac{dU}{dx}y_2$$

$$= -\rho\frac{dU}{dx}\int_0^{y_2}U\,dy \quad (7\text{–}7)$$

Substitution of this relationship into eq. 7–5 gives an expression for the wall shear stress τ_w in terms of the fluid velocities within and outside of the boundary layer

$$\tau_w = \rho\frac{d}{dx}\int_0^{\delta(x)}[U - u]u\,dy + \rho\frac{dU}{dx}\int_0^{\delta(x)}[U - u]\,dy \quad (7\text{–}8)$$

The upper limit of integration has been changed from y_2 to δ since for $y > \delta$, $U = u$ and the contribution of each integral is zero. Note that the boundary layer thickness δ is a function of the flow direction, x.

7.4.2 Viscous Drag

Equation 7–8 relates the wall shear stress, τ_w, on the surface of a control volume coincident with a solid surface, to both the fluid velocity distribution normal to the surface and the pressure gradient in the direction of the flow. If the contribution from the pressure gradient is zero, $dP/dx = 0$ or $dU/dx = 0$ from eq.

7–4, the only force acting on the fluid in the x direction is a viscous force. The total viscous force on the fluid due to fluid viscosity, from the leading edge of the surface, $x = 0$, to a point $x = x_1$ becomes

$$D_F(x_1) = b\int_0^{x_1} \tau_w\,dx = b\rho\int_0^{x_1}\left\{\frac{d}{dx}\int_0^{\delta(x)}[U - u]u\,dy\right\}dx$$

$$= b\rho U^2\int_0^{\delta(x_1)}\frac{u}{U}\left[1 - \frac{u}{U}\right]dy \qquad (7\text{–}9)$$

where b is the width of the control volume normal to the x, y plane.

The dimensionless *average or friction drag coefficient* for a flat plate of length $L(x_1 = L)$ and width b is

$$\overline{C}_f \equiv \frac{D_F(L)}{\rho\frac{U^2}{2}bL} = \frac{2}{L}\int_0^{\delta(L)}\frac{u}{U}\left[1 - \frac{u}{U}\right]dy \qquad (7\text{–}10)$$

From this expression the friction drag for either a laminar or turbulent boundary layer can be determined, provided the velocity profile $u(y)$ is known. In turbulent boundary layers, this velocity distribution is determined experimentally, while it can be determined using the differential momentum equation for the boundary layer if the flow is laminar. From such data, the expressions for $\overline{C}_f$ and τ_w can be obtained. Table 7–1 gives the resulting expressions for both a laminar and a turbulent flow. For a laminar flow the velocity distribution can be determined from Table 7–2.

The expressions presented in Table 7–1 for both laminar and turbulent boundary layers assume that the surface of the plate is "smooth." In reality, most surfaces will have a certain degree of roughness due to their manufacture or from degradation of the surface by corrosion. The roughness of a surface is defined by the statistically mean height, h_r, of the roughness elements. For

---

Table 7–1   Summary of Boundary Layer Relationships for a Smooth Flat Plate

	Laminar	Turbulent
u/U	$= f(\eta); \eta = y\sqrt{U/\nu x}$ (Table 7–2)	$= [y/\delta]^{1/7}$
δ/x	$= 5.0[Re_x]^{-1/2}$	$= 0.371[Re_x]^{-1/5}$
$\overline{C}_f$	$= 1.328[Re_L]^{-1/2}$	$= 0.074[Re_L]^{-1/5}$ If $5\times10^5 < Re_L < 10^7$ $= 0.455[\log_{10} Re_L]^{-2.58}$ If $10^7 < Re_L < 10^9$
τ_w	$= 0.332\rho U^2[Re_x]^{-1/2}$	$= 0.0225\rho U^2[Re_\delta]^{-1/4}$ If $Re_\delta < 10^7$

Where $Re_x \equiv \rho\dfrac{Ux}{\mu}$; $Re_L \equiv \rho\dfrac{UL}{\mu}$; $Re_\delta \equiv \rho\dfrac{U\delta}{\mu}$

Table 7–2 Blasius Velocity Distribution in a Laminar Flow (ref. 1)

$y\sqrt{U/vx}$	u/U	$y\sqrt{U/vx}$	u/U
0	0	2.6	0.77246
0.2	0.06641	2.8	0.81152
0.4	0.13277	3.0	0.84605
0.6	0.19894	3.2	0.87609
0.8	0.26471	3.4	0.90177
1.0	0.32979	3.6	0.92333
1.2	0.39378	3.8	0.94112
1.4	0.45627	4.0	0.95552
1.6	0.51676	4.2	0.96696
1.8	0.57477	4.4	0.97587
2.0	0.62977	4.6	0.98269
2.2	0.68132	4.8	0.98779
2.4	0.72899	5.0	0.99155

example, if the surface of a plate were covered with sand particles, h_r would represent the mean diameter of these particles.

The effect of roughness is to decrease the value of Reynolds number at which transition from a laminar to turbulent occurs. It also increases the skin friction drag, but *only* if the boundary layer is turbulent. The influence of roughness on a laminar boundary layer is only to decrease the Reynolds number at which it undergoes transition to a turbulent boundary layer.

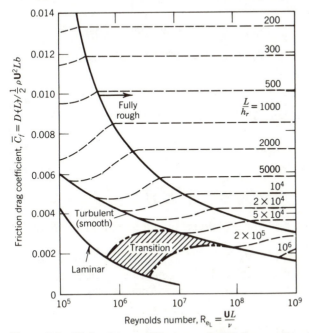

Figure 7-8 Friction drag coefficient per unit width for smooth and rough surfaces. Note: Transition Re increases for smoother surface or lower free-stream turbulence.

Figure 7–8 presents a plot of the variation of $\overline{C}_f$ with Re_L for a smooth plate (Table 7–1), and plates with different values of dimensionless roughness L/h_r. The roughness effect is based on the empirical correlations presented by White.[1] A "smooth" plate is defined as one for which $h_r < 100v/U$.

EXAMPLE 7–1

A smooth flat plate has a total length $L = 0.75$ m. The plate is to be tested in both water and air at a velocity $U = 4.5$ m/s. The temperature of both the water and air will be 20° C and the pressure equal to the atmospheric pressure. Determine:

(a) If the flow at the end of the plate will be laminar or turbulent for each fluid.

(b) The velocity of the air necessary to make the flows similar, that is, to have equal Reynolds number, Re_L.

SOLUTION

(a) If the Reynolds numbers $Re_L < 0.50 \times 10^6$ the flow will be laminar and if $Re_L > 0.50 \times 10^6$ it will be turbulent. The kinematic viscosity of the air and water are (Fig. A–14)

$$v_{air} = 15.3 \times 10^{-6} \text{ m}^2\text{/s} \quad \rho_{air} = 1.193 \text{ kg/m}^3 \text{ (Table A–8)}$$

$$v_{water} = 1.00 \times 10^{-6} \text{ m}^2\text{/s} \quad \rho_{water} = 998 \text{ kg/m}^3 \text{ (Table A–9)}$$

In water,

$$Re_L = \frac{UL}{v_{water}} = \frac{4.5(0.75)}{1.00 \times 10^{-6}} = 3.375 \times 10^6$$

In air,

$$Re_L = \frac{UL}{v_{air}} = \frac{4.5(0.75)}{15.3 \times 10^{-6}} = 0.2206 \times 10^6$$

Therefore the flow at the end of the plate is laminar in air and turbulent in water.

(b) The flow at the end of the flat plate will be similar in air and water if the Reynolds numbers are equal for both cases. Equal values of Re_L means that the ratio of inertia force to viscous force is equal in both cases.

$$\left(\frac{UL}{v}\right)_{air} = \left(\frac{UL}{v}\right)_{water}$$

Thus

$$U_{air} = U_{water}\left(\frac{L_{water}}{L_{air}}\right)\left(\frac{\nu_{air}}{\nu_{water}}\right)$$

$$= 4.5\left(\frac{0.75}{0.75}\right)\left(\frac{15.3 \times 10^{-6}}{1.00 \times 10^{-6}}\right)$$

$$= 68.85 \text{ m/s}$$

COMMENT

The effect of fluid viscosity is demonstrated by comparing the flow of water and air. The flows of two fluids are *similar* if the ratio of fluid forces, inertia to viscous forces in this case, are equal. Because of the differences in density and dynamic viscosity, or kinematic viscosity, the case of air flow requires a much larger velocity **U** than the water flow if the two flows are to be similar.

EXAMPLE 7–2

(a) Calculate the total drag per unit width due to friction (D_F) on the smooth flat plate described in Example 7–1.

(b) Estimate the thickness of the boundary layer at the end of the plate when it is tested in both air and water.

(c) Compare the values of $\overline{C}_f$ and drag due to friction experienced by the plate when tested in air and water at the same Reynolds number.

SOLUTION

(a) Consider first the case of the flat plate tested in air. Example 7–1 shows that the flow is laminar over the entire plate. Therefore, the friction drag coefficient $\overline{C}_f$ is (Table 7-1)

$$\overline{C}_f = \frac{D_F}{\rho\frac{U^2}{2}bL} = \frac{1.328}{\sqrt{Re_L}}$$

$$= \frac{1.328}{\sqrt{0.2206 \times 10^6}}$$

$$= 2.827 \times 10^{-3}$$

The total friction drag per unit width equals the drag per unit width per side times two

$$\frac{D_F}{b} = 2\overline{C}_f \, \rho \, \frac{U^2}{2} \, L$$

$$= 2(2.827 \times 10^{-3})1.193\frac{(4.5)^2}{2}(0.75)$$

$$= 51.22 \times 10^{-3} \text{ N/m}$$

The boundary layer thickness in this case is (Table 7–1),

$$\delta\{L\} = \frac{5.0L}{\sqrt{Re_L}} = \frac{5.0(0.75)}{\sqrt{0.2206 \times 10^6}}$$

$$= 7.984 \text{ mm}$$

(b) When the plate is tested in water, Example 7–1 shows that $Re_L > 0.50 \times 10^6$ and the flow is turbulent at the trailing edge of the plate. The flow near the leading edge of the plate is initially laminar and then undergoes transition to a turbulent flow. It will, however, be assumed for purposes of calculating D_F/b and $\delta\{L\}$ that the flow over the entire plate is turbulent. The validity of this assumption will be checked by also calculating the location of the transition point and summing the contributions of the laminar and turbulent boundary layers.

In water:

$$\overline{C}_f = \frac{D_F}{\rho \, \dfrac{U^2}{2} \, bL} = 0.074(Re_L)^{-1/5}$$

$$= 0.074(3.375 \times 10^6)^{-1/5}$$

$$= 3.661 \times 10^{-3}$$

The total drag per unit width is

$$\frac{D_F}{b} = 2\overline{C}_f \, \rho \, \frac{U^2}{2} \, L$$

$$= 2(3.661 \times 10^{-3})998\frac{(4.5)^2}{2}(0.75)$$

$$= 55.49 \text{ N/m}$$

and

$$\delta\{L\} = 0.371(Re_L)^{-1/5}L$$

$$= 0.371(3.375 \times 10^6)^{-1/5}(0.75)$$

$$= 13.76 \text{ mm}$$

To check the assumption that the entire flow is turbulent, the point of transition, x_{CR}, is ($Re_{CR} = 0.50 \times 10^6$)

$$x_{CR} = \frac{Re_{CR}\nu}{U-} = \frac{0.50 \times 10^6 \, (1.00 \times 10^{-6})}{4.5}$$

$$= 0.1111 \text{ m}$$

The total drag is then the sum of the drag on a plate of 0.111 m length with laminar flow and another of $(0.75 - 0.111)$ m length with turbulent flow. Thus,

$$\frac{D_F}{b} = 2\rho\frac{U^2}{2}[\{\overline{C}_fL\}_{\text{lam}} + \{\overline{C}_fL\}_{\text{turb}}]$$

$$= 2(998)\frac{(4.5)^2}{2}\left[\frac{1.328(0.1111)}{\sqrt{\dfrac{4.5(0.1111)}{1.00 \times 10^{-6}}}} + 0.074\left\{\frac{4.5(0.639)}{1.00 \times 10^{-6}}\right\}^{-1/5}(0.639)\right]$$

$$= 53.03 \text{ N/m}$$

(c) From Example 7–1 Re_L in air and water are equal if the velocity in air is increased to 68.85 m/s.

$$\frac{(D_F)_{\text{air}}}{(D_F)_{\text{water}}} = \frac{\left(2\overline{C}_f\rho\dfrac{U^2}{2}bL\right)_{\text{air}}}{\left(2\overline{C}_f\rho\dfrac{U^2}{2}bL\right)_{\text{water}}} = \frac{(\rho U^2)_{\text{air}}}{(\rho U^2)_{\text{water}}}$$

$$= \frac{1.193(68.85)^2}{998(4.5)^2}$$

$$= 0.2799$$

COMMENT

When calculating the friction force on an object, the entire surface area of the object that is "wetted" by the flow must be considered. Thus the total drag on a flat plate has contributions from both sides of the plate.

The calculation of the drag/unit width means that the result can be applied to a number of situations as long as the width of the object is much larger than its length in the flow direction. This is necessary since the derivation of $\overline{C}_f$ is based on the assumption of a two-dimensional flow.

The assumption that the flow is turbulent over the entire length gives an answer which is approximately 4% greater than the one obtained by including both the laminar and turbulent portions. This is within the accuracy of the empirical formulai used to predict the drag.

In the case of similar flows in air and water the values of $\overline{C}_f$ are equal since the values of Re_L are equal. (Note that $\overline{C}_f$ is a ratio of the friction drag force to the inertia force.) The values of friction drag are different by the ratio of the products of ρU^2 in air and water, however.

Although the flows are similar in air and water when the ratios of the forces in the two flows are equal, equal Re_L and equal $\overline{C}_f$, the friction drag force in air is 28% of that experienced in water.

7.5   THE INFLUENCE OF PRESSURE GRADIENTS

Equation 7–10 permits us to calculate the friction drag coefficient caused by flow over a flat surface. In the derivation of this equation it was assumed that the only force acting on the fluid was that due to the presence of viscous shear stresses. The force due to pressure differences on the fluid was set equal to zero by assuming the pressure gradient in the flow direction, dP/dx, to be zero. While this assumption is valid in some cases, it is not valid in a great many other flow situations. When $dP/dx \neq 0$, a pressure force will exist which not only contributes to the *total* resistance experienced by the fluid but can result in a phenomena called *flow separation*.

7.5.1   Flow Separation

The magnitude of the pressure gradient dP/dx is dependent upon the shape of the surface, which in turn influences the variation of velocity outside of the boundary layer in the flow direction, dU/dx. We have seen that it is possible to have a flow in which the velocity U is increasing in the direction of flow, $dU/dx > 0$, by shaping the walls to form a nozzle. From eq. 7–3 this means the pressure gradient dP/dx is negative. If $dP/dx < 0$, a *favorable pressure gradient* exists since the resulting pressure force on the fluid acts in the direction of flow.

Conversely, if the velocity U is decreasing in the direction of flow, as in a diffuser where $dU/dx < 0$, the pressure gradient is positive, $dP/dx > 0$, and the resulting pressure force is opposite to the flow direction. This pressure force acts to retard the flow and is termed an *adverse pressure gradient*. The existence of an adverse pressure gradient means the momentum of the fluid is decreasing and the fluid near the surface can be brought to rest at some distance from the wall, u = 0 at $y > 0$. When this occurs the flow is said to *separate*.

The occurrence of separation can also be determined by examining the shape of the u velocity profile in the boundary layer. If the velocity of the flow near the wall, $y > 0$, becomes zero and flow separation occurs, the velocity gradient normal to the flow, $\partial u/\partial y$, must be zero at this point, Fig. 7–9. Prior to flow separation, $\partial u/\partial y > 0$. After flow separation there will be a reverse flow, flow in the minus x direction, and $\partial u/\partial y < 0$ near the wall. This flow reversal can only occur if a force in addition to τ_x is applied to the fluid. This additional force is due to the pressure gradient in the flow direction and must be in the minus

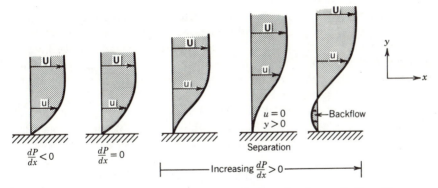

Figure 7-9 Influences of a pressure gradient on the boundary layer velocity profile.

x direction, $dP/dx > 0$. Figure 7–9 shows the effect of an increasing adverse pressure gradient on the velocity distribution in a boundary layer. A boundary layer flow over a flat plate will not experience flow separation since there is no pressure gradient, $dP/dx = 0$.

A practical example of the influence of a pressure gradient is the incompressible flow of a fluid through a rectangular converging–diverging channel, Fig. 7–10. The phenomena of separation is the same in both external and internal flows. Therefore, an internal flow case is chosen to demonstrate flow separation

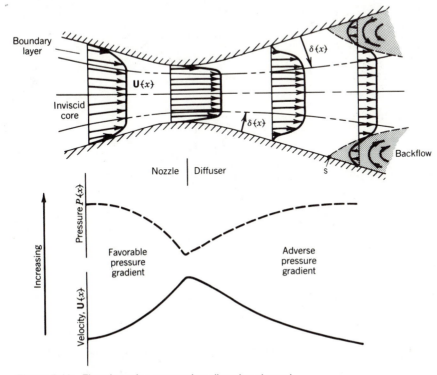

Figure 7-10 Flow through a converging–diverging channel.

since it is easier to describe the variation of velocity and pressure in the flow direction. The converging portion of this channel acts to increase the velocity in the region away from the walls, the inviscid core, and $dU/dx > 0$. Such a channel is termed a nozzle. This produces a favorable pressure gradient in the boundary layer. At the throat section, the cross-sectional area is constant as are the velocity U and the pressure.

The diverging channel in which the area increases in the flow direction is termed a diffuser since the velocity is decreasing, $dU/dx < 0$, and the pressure increasing, $dP/dx > 0$. This means that the diffuser experiences an adverse pressure gradient and can experience flow separation along its walls. This is depicted in Fig. 7–10, together with the reversed or backflow that occurs after the separation point.

The existence of flow separation in a diffusing channel represents a loss of fluid energy. Since the purpose of a diffuser is to convert dynamic pressure, $\rho U^2/2$, into static pressure, P, these losses reduce the efficiency of the diffuser. Data describing the losses in a diffuser will be presented in Chapter 8.

A second example of favorable and adverse pressure gradients over a surface

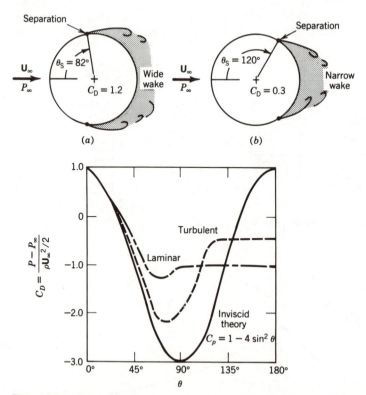

Figure 7-11 Flow and surface pressure over an infinite circular cylinder normal to the flow. (a) Laminar flow. (b) Turbulent flow.

is the flow past a circular cylinder which is infinitely wide in the direction normal to the flow, Fig. 7–11. Since the cylinder is infinitely wide the flow considered is that over a cross section of the cylinder and is a two-dimensional flow. In this case $dP/dx < 0$ over the forward portion of the cylinder and then becomes positive or adverse. The point at which $dP/dx = 0$ is located at an angle of approximately 70° from the most forward point (stagnation point) on the cylinder. The stagnation point is the location where the static pressure is a maximum since the fluid velocity at this point is zero. This is shown by the dimensionless distribution of the surface pressure coefficient C_p, Fig. 7–11. The pressure coefficient represents the ratio of the pressure force on the cylinder to the inertia force of the fluid. In the definition of C_p in Fig. 7–11, P is the surface pressure which varies over the cylinder and P_∞ and U are the pressure and velocity of the fluid, respectively, upstream of the cylinder.

When $\theta > 70°$, the point at which flow separation occurs is dependent upon whether the boundary layer is laminar or turbulent. The reason for this can be seen if a laminar and a turbulent boundary layer are nondimensionalized based on their respective boundary layer thickness δ, Fig. 7–12. This figure shows that at a constant value of y/δ the turbulent boundary layer has a larger velocity and, hence, a fluid particle at that y/δ has a greater momentum. This means that a fluid particle in a turbulent boundary layer can experience a larger adverse pressure gradient than a fluid particle in a laminar boundary layer before its velocity is reduced to zero. As a result, a laminar boundary layer will separate at a smaller value of θ, or arc length of surface, than will a turbulent boundary layer when they both experience the same adverse pressure gradient. As shown in Fig. 7–11, the laminar boundary layer separates at $\theta_s \approx 82°$, while in a turbulent boundary layer separation occurs at $\theta_s \approx 120°$ if the surface is "smooth."

The result of flow separation is a region of low-energy recirculating flow called a *wake*. As shown in Fig. 7–11, the wake associated with turbulent boundary layer over the circular cylinder is much narrower than that associated with the laminar boundary layer. A similar result is observed for the flow over a sphere. Figure 7–13 shows the wake generated by a sphere with a laminar and turbulent boundary layer. The wake associated with the turbulent boundary layer is considerably narrower.

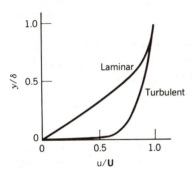

Figure 7-12 Comparison of laminar and turbulent boundary layer velocity profiles.

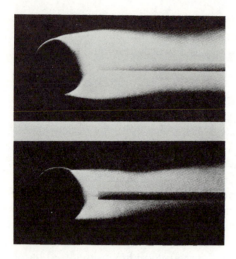

Figure 7–13 Comparison of the wakes produced by smooth and roughened spheres in an air stream at the same Reynolds number (Re ≈ 10^5), made visible by smoke injected through the tube surrounding the supporting rod. Above, the boundary layer is laminar; below, boundary layer turbulence is induced by sand grains cemented in a narrow band around the front of the sphere[7]. Used with permission.

7.5.2 Pressure Drag

If the pressure on the surface of an object is not constant (uniform), a net force will be exerted on the object. The magnitude of this force in a certain direction is determined by summing the product of the pressure and the projection of the object on a plane normal to that direction. Thus, for an infinitely wide circular cylinder of radius R, the pressure drag force per unit width is

$$D_P = \frac{\text{pressure drag}}{\text{unit width}} = \frac{1}{b} \int_{\text{AREA}} P \, dA_x$$

$$= \frac{1}{b} \int_0^{2\pi} P\{\theta\} R \cos \theta \, d\theta \qquad (7\text{–}11)$$

If $P\{\theta\}$ is constant or symmetrical about the line at $\theta = 90°$, this force is zero. In all other cases, such as shown in Fig. 7–11, there is a net force. The component of this net force in the x direction is termed the *pressure drag* and the component in the y direction, or normal to U, the *lift* on the cylinder.

For the circular cylinder shown in Fig. 7–11, the lift force is zero because of the pressure symmetry between 0–180 and 180–360°. However, the pressure drag force is nonzero.

The *total drag* (D_T) that an object can experience is then a combination of that due to the distribution of surface pressures (*pressure drag*, D_P) and the summation of the integrated shear stress (*friction drag*, D_F).

Total drag = friction drag + pressure drag

$$D_T = D_F + D_P \qquad (7\text{–}12)$$

The relative magnitudes of these contributions to the total drag are dependent upon the shape of the object. As demonstrated earlier, the drag per unit width experienced by a thin flat plate of zero thickness located parallel to the flow is completely friction drag. If the flat is situated normal to the flow, the total drag per unit width is completely pressure drag since the shear stresses on the plate act normal to the flow, Fig. 7–14.

The total drag on an object is expressed in a dimensionless form as a total drag coefficient C_D where U is the velocity approaching, or upstream of, the object

$$C_D \equiv \frac{D_T}{\rho \dfrac{U^2}{2} A}$$

The area A is arbitrary and is usually chosen as the one most convenient to define. In the literature drag coefficients will be found defined using one of three areas:

1. *Frontal area.* The cross-sectional or projected area as seen when looking in the direction of U. This definition is usually used for thick bodies such as missiles, cylinders, cars, trains, spheres, and so on.

2. *Planform area.* The area of an object projected on to a plane containing the velocity U, or as seen from above. This definition is usually used for nearly flat objects such as wings and hydrofoils.

3. *Wetted surface area.* The area exposed to, or wetted by, the fluid. This definition is used for compound surfaces like the hull of a surface ship or barge.

The pressure drag is less amenable to prediction than the friction drag. The approach usually employed to determine the pressure drag is to conduct an experiment in which the total drag is measured. The friction drag is then calculated knowing the surface area of the object and the Reynolds number of the flow, while assuming the pressure drag to be zero. The difference between the measured total drag and the calculated friction drag is the pressure drag. The total drag of various objects are documented in the literature. The most comprehensive source is that by Hoerner.[3]

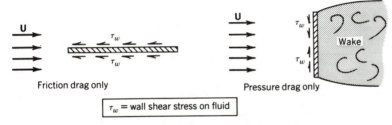

Figure 7-14   Drag on a flat plate parallel and normal to the flow.

The total drag coefficients for a number of two- and three-dimensional objects are presented in Tables 7–3 and 7–4, respectively. Care must be exercised in the application of the tables to first determine if for the object under consideration a two- or a three-dimensional drag coefficient should be used and, secondly, to determine the correct area to be used in the definition of C_D. Since the existence of a two-dimensional flow is an assumption, the drag coefficients for such cases are actually expressions of the drag per unit width of the object *normal* to the flow. The width b is measured in the z direction normal to the xy plane in which the two-dimensional flow occurs. If b is four to five times larger than the maximum dimension of the object projected onto the y axis, the assumption that the flow is two dimensional is usually valid.

The three-dimensional drag coefficients given in Table 7–4 are for objects whose major dimension is parallel to the flow direction. The area used for these drag coefficients is the area of the object projected on the yz plane. The use of these tables is demonstrated in the example problems.

Figure 7–15a presents the variation of total drag coefficient (based on frontal area) of a smooth sphere as a function of Reynolds number (based on body diameter). Figure 7–15b presents the total drag coefficient per unit width of a smooth infinite circular cylinder place normal to the flow. Note that at a Reynolds

---

Table 7–3 Drag Coefficients of Two-Dimensional Objects at Re $\approx 10^5$

Plate

$C_D = 2.0$

Square cylinder

$C_D = 2.1$

$C_D = 1.6$

Half-cylinder

$C_D = 1.2$

$C_D = 1.7$

$C_D = \dfrac{\text{drag/unit width}}{\frac{1}{2}\rho U^2 t}$

$Re = \dfrac{Ut}{\nu}$

$t =$ projected height normal to **U**

Half-tube

$C_D = 1.2$

$C_D = 2.3$

Equilateral triangle

$C_D = 1.6$

$C_D = 2.0$

Circular cylinder

	Laminar	Turbulent
$C_D =$	1.1	0.3

Elliptical cylinder

		Laminar	Turbulent
2:1	$C_D =$	0.6	0.2
4:1	$C_D =$	0.35	0.15
8:1	$C_D =$	0.25	0.1

Table 7–4   Drag of Three-Dimensional Objects Re $\approx 10^5$ (C_D Based on Frontal Area)

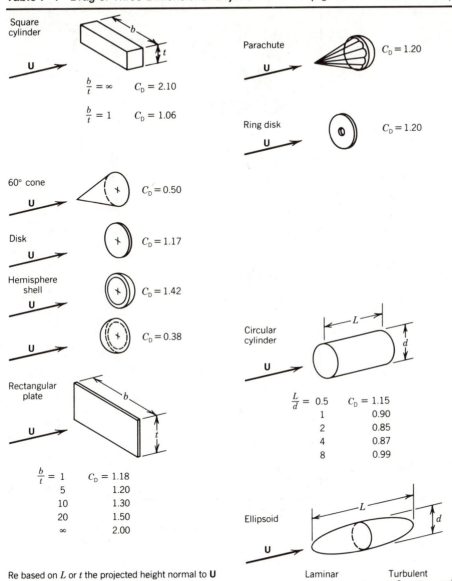

Square cylinder

U

$\frac{b}{t} = \infty$ $C_D = 2.10$

$\frac{b}{t} = 1$ $C_D = 1.06$

60° cone

U $C_D = 0.50$

Disk

U $C_D = 1.17$

Hemisphere shell

U $C_D = 1.42$

U $C_D = 0.38$

Rectangular plate

U

$\frac{b}{t} = 1$	$C_D = 1.18$
5	1.20
10	1.30
20	1.50
∞	2.00

Re based on L or t the projected height normal to **U**

Parachute $C_D = 1.20$

U

Ring disk $C_D = 1.20$

U

Circular cylinder

U

$\frac{L}{d} = 0.5$	$C_D = 1.15$
1	0.90
2	0.85
4	0.87
8	0.99

Ellipsoid

U

	Laminar	Turbulent	
		$Re_L \approx 10^5$	$Re_L \approx 10^7$
$\frac{L}{d} = 1$	$C_D = 0.47$	0.10	0.070
2	0.25	0.05	0.030
3	0.20	0.06	0.041
8	0.23	0.10	0.078

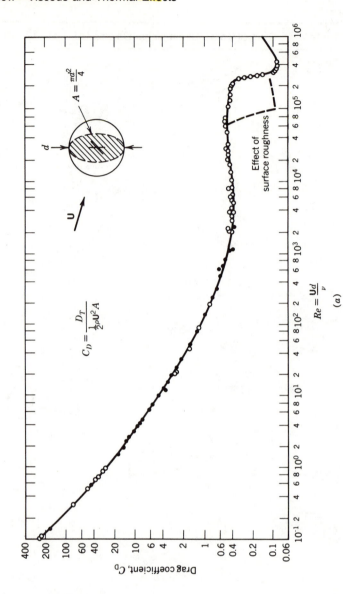

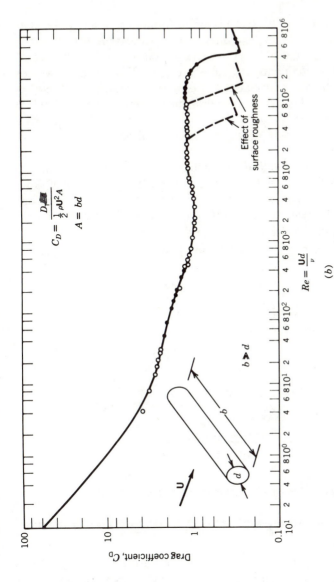

$$C_D = \frac{D}{\frac{1}{2}\rho U^2 A}$$

$$A = bd$$

$Re = \frac{Ud}{\nu}$

(b)

Figure 7-15 Drag coefficient for flow over a sphere and infinite cylinder[8]. Used with permission. (a) Drag coefficient of a smooth sphere. (b) Drag coefficient of a smooth infinite circular cylinder normal to the flow.

number of approximately 0.5×10^6 there is a significant decrease in drag coefficient for both the sphere and circular cylinder. This decrease is caused by the difference between laminar and turbulent separation on the object, the size of the resulting wake and, subsequently, the pressure drag. Thus, if the flow over the entire sphere or cylinder is turbulent the drag coefficient is lower by approximately a factor of 4. It follows that the drag on a sphere is less if its surface is roughened, thereby causing turbulent flow, Fig. 7–13. This is the reason why golf balls are dimpled. Figure 7–15 shows the influence of surface roughness on the total drag coefficient.

The pressure drag can be reduced in comparison to the friction drag by streamlining the object. That is, the length of the object in the direction of flow, L, is made large with respect to its maximum thickness, or diameter, d. Figure 7–16 shows the effect of streamlining a body of revolution. By streamlining an object, increasing its L/d, the size of the wake and hence the pressure drag is reduced. For values of $L/d > 7$–8 at least 90% of the total drag is due to friction.

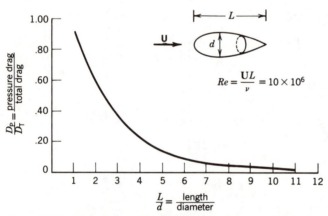

Figure 7-16   Effect of streamlining on the pressure drag of a body of revolution[3].

EXAMPLE 7–3

An advertising agent has been contracted to install a billboard which is 1.75 m high and 35 m wide. It is estimated that the maximum wind velocity which the billboard will experience is 3 m/s. In order to design the supports for this billboard, calculate the maximum drag force on the billboard. ($\nu_{air} = 14.6 \times 10^{-6}$ m²/s, $\rho_{air} = 1.23$ kg/m³.)

SOLUTION

The drag on the billboard will be assumed to be one-half that of a 3.5 $\times$ 35 m rectangle normal to the flow.

For the 3.5 × 35 m rectangle:

$$Re_L = \frac{UL}{\nu} = \frac{3(3.5)}{14.6 \times 10^{-6}}$$

$$= 719.2 \times 10^3$$

From Table 7–4 $b/h = 35/3.5 = 10$ and $C_D = 1.3$. Then for the billboard:

$$Drag = D_T = \tfrac{1}{2} \left[C_D \rho \frac{U^2}{2} A \right]$$

$$= \tfrac{1}{2} \left[1.3(1.23) \frac{(3.0)^2}{2} (3.5)35 \right]$$

$$= 440.7 \text{ N}$$

COMMENT

The billboard in this problem can be approximated as a flat plate placed normal to the flow with no flow between the bottom of the billboard and the ground. In this case, the total drag is completely due to pressure drag. The ground on which the billboard is mounted is assumed to be represented by the streamline through the stagnation point at the center of a rectangular plate placed normal to the flow. Therefore, the drag on the billboard is one-half that of a rectangular plate with twice the height of the billboard.

It can be correctly argued that this representation is not exactly the same as the actual billboard since the effect of the boundary layer over the ground is being neglected. However, the representation by one-half of a rectangular flat plate is the best representation for which we have drag data.

EXAMPLE 7–4

The piling supports for a river loading dock are constructed of circular cylinders embedded in the river bottom. The water is 6.0 m deep and flows by the pilings at maximum velocity of 1.5 m/s. If the diameter of a piling is 0.2 m, determine the maximum drag on the piling. The water temperature is 10° C and the pilings are very rough.

SOLUTION

Neglecting the effect of the water–air interface, a two-dimensional flow will be assumed over the piling.

$$\rho = 1000 \text{ kg/m}^3 \text{ (Table A–9)}$$

$$\nu = 12.96 \times 10^{-6} \text{ m}^2\text{/s (Table A–9)}$$

$$Re = \frac{Ud}{\nu} \quad \frac{1.5(0.2)}{12.96 \times 10^{-6}} = 23.15 \times 10^3$$

From Table 7–3, $C_D = 0.3$.
  The total drag per unit width is

$$\frac{D_T}{b} = \rho \frac{Ud^2}{2} C_D$$

$$= 1000 \frac{(1.5)^2}{2}(0.2)\,0.3$$

$$= 67.50 \text{ N/m}$$

The total drag on the piling is

$$D_T = \frac{D_T}{b} \cdot b$$

$$= 67.50(6.0)$$

$$= 405.0 \text{ N}$$

COMMENT

Assuming the effects of the air–water interface (free-surface) are negligible means that the circular pilings are fixed between two parallel end-plates, the river bottom and surface. Therefore, the flow over the pilings is the same at any distance between the river bottom and surface and the two-dimensional drag coefficient is found in Table 7–3. The turbulent value is used since the problem states that the surface of the piling is very rough. Hence, there is no laminar boundary layer, only a turbulent one. After determining the drag per unit width (height in this case), the total drag is determined by multiplication by the depth of water.

EXAMPLE 7–5

An airship travels at a ground speed of 38.6 m/s (75 kts) into a headwind of 6.4 m/s. The air temperature is 250 K. The airship is a 4:1 ellipsoid with a maximum diameter of 9 m. Estimate:

(a) The total drag on the airship.
(b) The pressure drag on the airship. (Note: the surface area of an ellipsoid is $4\pi ab$ where a and b are the semimajor and minor axes, respectively.)
(c) The total drag without the headwind.
(d) Compare the power that must be added to the fluid to propel the airship in (a) and (c)

SOLUTION

(a) The total drag is estimated using the total drag coefficient from Table 7–4.

$$\nu = 11.44 \times 10^{-6} \text{ m}^2/\text{s (Table A–8)}$$

$$\rho = 1.395 \text{ kg/m}^3 \text{ (Table A–8)}$$

$$\text{Re}_L = \frac{UL}{\nu} = \frac{45(36)}{11.44 \times 10^{-6}}$$

$$= 141.6 \times 10^6$$

Therefore, the flow is turbulent and $C_D = 0.041$ (Table 7–4)

$$D_T = \rho \frac{U^2}{2} A C_D$$

$$= 1.395 \frac{(45.0)^2}{2} \frac{\pi(9)^2}{4} 0.041$$

$$= 3.684 \times 10^3 \text{ N}$$

(b) The pressure drag $D_P = D_T - D_F$ (eq. 7–12). To *estimate* the friction drag we will calculate the friction drag on an *equivalent flat plate*, a flat plate with the same surface area as the airship.

$$A_{\text{sur}} = 4\pi(4.5)18$$

$$= 1017.9 \text{ m}^2$$

At $\text{Re}_L = 141.6 \times 10^6$ and assuming a smooth surface, since there is no information stated about surface roughness,

$$\overline{C}_f = \frac{D_F}{\rho \dfrac{U^2}{2} A_{\text{sur}}} = 0.455 \, [\log_{10} \text{Re}_L]^{-2.58}$$

Therefore

$$\overline{C}_f = 0.455 \, [\log_{10}(141.6 \times 10^6)]^{-2.58}$$

$$= 2.028 \times 10^{-3}$$

and

$$D_F = \rho \frac{U^2}{2} A_{sur} C_D$$

$$= 1.395 \frac{(45.0)^2}{2}(1017.9)(2.028 \times 10^{-3})$$

$$= 2.915 \times 10^3 \text{ N}$$

Hence

$$D_P = D_T - D_F$$

$$= 3.684 \times 10^3 - 2.915 \times 10^3$$

$$= 0.769 \times 10^3 \text{ N}$$

(c) Same as part (a) with $U = 38.6$ m/s.

$$D_T = \rho \frac{U^2}{2} A C_D$$

$$= 1.395 \frac{(38.6)^2}{2} \frac{\pi(9)^2}{4} 0.041$$

$$= 2.711 \times 10^3 \text{ N}$$

(d) Fluid power required $= D_T U$.
Case (a):

$$D_T U = 3.684 \times 10^3 (45.0)$$

$$= 165.8 \text{ kW}$$

Case (b):

$$D_T U = 2.711 \times 10^3 (38.6)$$

$$= 104.6 \text{ kW}$$

COMMENT

This problem demonstrates how the contribution of the pressure drag to the total drag of an object can be estimated. The total drag on the object is calculated from a correlation of experimental values of total drag measured on a number of geometrically similar objects. The friction drag is then calculated as a flat plate with the *same* surface area as the object. The difference between the total drag and the friction drag gives the pressure drag.

Note that the velocity which must be used to calculate the total drag is the velocity *relative* to the object. Thus, the existence of a headwind results in a higher airship drag when the velocity of the airship relative to the ground, its ground speed, is maintained constant. The headwind also results in a higher level of required power input to the air to achieve the same ground speed. To determine the motor power required the efficiency of the propeller must be known. The motor power required will be greater than the fluid power, since the propeller efficiency will be less than 100%.

7.6 CONVECTION HEAT TRANSFER COEFFICIENT

The concept of a *convection heat transfer coefficient, h*, was introduced in Chapter 1 and was defined in terms of the heat flux at the fluid–surface interface and a temperature difference,

$$h \equiv \frac{\dot{q}''}{\Delta T} \qquad \text{W/m}^2 \cdot \text{K} \tag{7–13}$$

The heat flux is equal to the rate of heat transfer from the surface divided by the surface area, $\dot{q}'' = \dot{Q}/A$. The heat flux is a vector acting perpendicular to the surface and is considered to be positive when the heat flows from the surface to the fluid. The temperature difference, ΔT, is the difference between the temperature of the surface, T_w, and the temperature of the fluid outside of the boundary layer, T_∞.

The heat flux and the temperature of the surface will depend on the location along the plate in an external flow; thus a *local heat transfer coefficient* must be defined

$$h_x \equiv \left. \frac{\dot{q}_x''}{(T_w - T_\infty)} \right|_x \tag{7–14}$$

where x is the coordinate tangent to the heating surface. The local heat transfer coefficient can also be expressed in terms of the temperature gradient in the fluid at the fluid–surface interface using Fourier's law,

$$\dot{q}_x'' = -k \left. \frac{\partial T}{\partial y} \right|_{y=0}$$

to obtain

$$h_x = \left. \frac{-k(\partial T/\partial y)|_{y=0}}{(T_w - T_\infty)} \right|_x \tag{7–15}$$

The total rate at which heat is transferred from an isothermal surface is most conveniently obtained using an average heat transfer coefficient

$$\dot{Q} = \bar{h}A(T_w - T_\infty) \tag{7–16}$$

The *average heat transfer coefficient*, $\bar{h}$, is obtained by integrating the local heat transfer coefficient over the complete surface length, L,

$$\bar{h} = \frac{\int_0^L h_x \, dx}{L} \tag{7–17}$$

The boundary layer development on an isothermal flat plate for a fluid with a Prandtl number of 1, $\delta = \delta_T$, together with the temperature profile in the fluid at several locations along the plate are shown in Fig. 7–17. It is seen that as one moves downstream away from the leading edge of the plate the thickness of the laminar boundary layer increases while $\partial T/\partial y|_{y=0}$ decreases. In the turbulent boundary layer region the value of the fluid temperature gradient at the

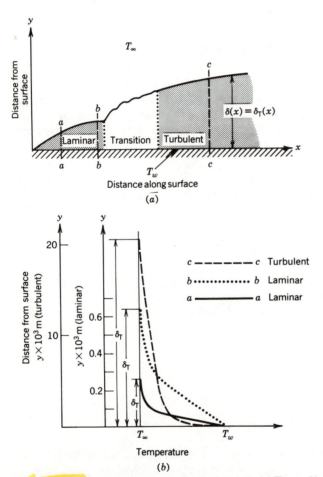

Figure 7–17   Thermal boundary layer on a flat plate. (a) Thermal boundary layer. (b) Temperature profile. Prandtl number of one.

wall also decreases as one moves downstream but its magnitude is considerably greater than that observed in the laminar flow region. The increase in the rate of heat transfer in the turbulent flow region is associated with the random fluctuations of the fluid particles which increase the mixing of the fluid, and thereby enhance the transfer of thermal energy between the surface and the fluid.

An important relationship between the convection heat transfer coefficient and the boundary layer thickness can be obtained by assuming the temperature to vary linearly across the thermal boundary layer. This assumption neglects the effect of the moving fluid on the temperature distribution; thus the thermal energy transfer across the thermal boundary layer is due entirely to conduction. While Fig. 7–18 shows that this approximation is inaccurate, it does enable us to observe a significant effect. It allows the local heat transfer coefficient to be approximated as

$$h_x = \frac{\dot{q}_x''}{(T_w - T_\infty)} \simeq \frac{-k(T_\infty - T_w)/\delta_T}{(T_w - T_\infty)}$$

or

$$h_x \simeq \frac{k}{\delta_T}$$

We can thus conclude that the local heat transfer coefficient is directly proportional to the thermal conductivity of the fluid and inversely proportional to the thickness of the thermal boundary layer.

The above expression can be rearranged to yield

$$\frac{h_x}{k} \simeq \frac{1}{\delta_T}$$

Introducing the distance from the leading edge of the plate, x, as a characteristic length we obtain

$$\frac{h_x x}{k} \simeq \frac{x}{\delta_T}$$

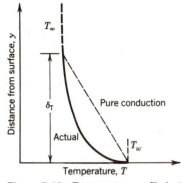

Figure 7-18 Temperature profile in thermal boundary layer.

The quantity on the left side of this expression is a dimensionless heat transfer coefficient and is called the *Nusselt* number. It is subscripted with x to indicate that it is a local quantity.

$$\mathrm{Nu}_x \equiv \frac{h_x x}{k} \tag{7-18}$$

The average Nusselt number is then

$$\overline{Nu} \equiv \frac{\overline{h} L}{k} \tag{7-19}$$

where L is the length of the plate. Correlations to determine the convection heat transfer coefficient are usually expressed in terms of the Nusselt number.

In Section 7–2 the similarities between a fluid boundary layer and a thermal boundary layer are discussed. These similarities are shown in Fig. 7–6 and indicate that at a Prandtl number equal to 1 the two boundary layers are identical. This suggests that a similarity exists between the transfer of momentum and heat. A simple relationship between the average friction drag coefficient of a flat plate, $\overline{C}_f$, and the average convection heat transfer coefficient is

$$\frac{\overline{C}_f}{2} = \frac{\overline{h}}{\rho c_p \mathrm{U}} \tag{7-20}$$

This is often referred to as Reynolds analogy. The term on the right side of eq. 7–20 is dimensionless and is called the *Stanton* number

$$\overline{St} \equiv \frac{\overline{h}}{\rho c_p \mathrm{U}} = \frac{\overline{Nu}}{\mathrm{Re}_L\, \mathrm{Pr}} \tag{7-21}$$

Reynolds analogy can thus be expressed as

$$\frac{\overline{C}_f}{2} = \overline{St} \tag{7-22}$$

The accuracy of this expression is dependent on the Prandtl number of the fluid. To increase its range of applicability, a modification can be made which yields an expression of reasonable accuracy in the Prandtl number range of 0.6 to 60.

$$\frac{\overline{C}_f}{2} = \overline{St}\mathrm{Pr}^{2/3} \tag{7-23}$$

This is known as the Chilton–Colburn analogy and is valid for both laminar flows over a flat plate and turbulent flows over surfaces of any shape.

EXAMPLE 7–6

Air at an average temperature of 26.85° C (300 K) flows over a fully rough flat plate 1 m long at a velocity of 100 m/s. Estimate the average convection heat transfer coefficient.

SOLUTION

The properties of air at 26.85° C (300 K) are obtained from Table A–8.

$$\rho = 1.1614 \text{ kg/m}^3 \qquad \nu = 15.89 \times 10^{-6} \text{m}^2/\text{s}$$

$$c_p = 1.007 \text{ kJ/kg·K} \qquad k = 26.3 \times 10^{-3} \text{W/m·K}$$

$$\text{Pr} = 0.707$$

The value of the Reynolds number at $x = L$ is

$$\text{Re}_L = \frac{UL}{\nu} = \frac{(100)1}{15.89 \times 10^{-6}} = 6.293 \times 10^6$$

Figure 7–8 may be used to find the friction drag coefficient, $\overline{C}_f = 0.00575$. The Chilton–Colburn analogy, eq. 7–23, is used to determine the Stanton number

$$\frac{\overline{C}_f}{2} = \overline{\text{St}} \text{Pr}^{2/3}$$

or

$$\overline{\text{St}} = \frac{\overline{C}_f}{2} \text{Pr}^{-2/3}$$

$$= \frac{0.00575}{2} (0.707)^{-2/3}$$

$$= 3.623 \times 10^{-3}$$

The average convection film coefficient is estimated to be

$$\overline{h} = \overline{\text{St}} \rho c_p U$$

$$= 3.623 \times 10^{-3} (1.1614)(1007)(100)$$

$$= 423.7 \text{ W/m}^2\text{·K}$$

COMMENT

The correlations usually available to determine the heat transfer coefficient for turbulent flow are only applicable for smooth surfaces. The Chilton–Colburn analogy provides us with a method that takes into account the effect of surface roughness on the convection heat transfer coefficient.

7.7 FORCED CONVECTION HEAT TRANSFER

Heat transfer from a surface to a moving stream is called *forced convection* heat transfer if the fluid's movement is created by a fan or blower. The expression for the dimensionless heat transfer coefficient, Nu_x is given as a function of the local Reynolds number and the Prandtl number

$$Nu_x = f(Re_x, Pr) \tag{7–24}$$

The exact form of the functional relationship will depend on the geometrical configuration of the surface; the characteristics of the flow, whether it is laminar or turbulent flow; and the thermal boundary conditions on the surface. The thermal boundary conditions usually cited are either a uniform surface temperature or a uniform wall heat flux. The *thermophysical properties used in the dimensionless groups are evaluated at the free-stream temperature unless stated otherwise.*

7.7.1 Flat Plate

7.7.1.A   Uniform Surface Temperature

If the entire surface of a smooth flat plate is at a uniform surface temperature and the flow is *laminar*, $Re_x < 5 \times 10^5$, the local value of the Nusselt number is

$$Nu_x = 0.332(Re_x)^{1/2}Pr^{1/3} \tag{7–25}$$

This expression can be rearranged and solved for the local convection heat transfer coefficient

$$h_x = 0.332\frac{k}{x}(Re_x)^{1/2}Pr^{1/3} \tag{7–26}$$

or

$$h_x = 0.332k\left(\frac{\rho U}{\mu}\right)^{1/2}\left(\frac{c_p\mu}{k}\right)^{1/3}x^{-1/2} \tag{7–27}$$

The total rate of heat transfer from one side of an isothermal heated plate of width b and length L is obtained by integrating the local rate of heat transfer over the entire plate length

$$\dot{Q} = b \int_0^L h_x(T_w - T_\infty)dx \tag{7-28}$$

The average heat transfer coefficient $\bar{h}$ can also be used to calculate the total rate of heat transfer from one side of a heated isothermal plate,

$$\dot{Q} = \bar{h}bL(T_w - T_\infty)$$

The average Nusselt number and heat transfer coefficient for laminar flow over a flat plate is obtained using

$$\overline{Nu} = 0.664(Re_L)^{1/2}Pr^{1/3} \tag{7-29}$$

and

$$\bar{h} = 0.664\frac{k}{L}(Re_L)^{1/2}Pr^{1/3} \tag{7-30}$$

The corresponding correlation to be used in the *turbulent* flow region, $5 \times 10^5 < Re_x < 10^7$, is

$$Nu_x = \frac{(C_{fx}/2)Re_xPr}{1 + 12.7(C_{fx}/2)^{1/2}(Pr^{2/3} - 1)} \tag{7-31}$$

where C_{fx} is the local skin friction drag coefficient

$$C_{fx} = 0.0592(Re_x)^{-1/5} \tag{7-32}$$

The average Nusselt number can be expressed as

$$\overline{Nu} = \frac{(\bar{C}_f/2)Re_LPr}{1 + 12.7(\bar{C}_f/2)^{1/2}(Pr^{2/3} - 1)} \tag{7-33}$$

where

$$\bar{C}_f = 0.074(Re_L)^{-1/5}$$

The uncertainty of the exact location of the transition from laminar to turbulent flow has been noted. An expression for the average Nusselt number for a flow that includes both laminar and turbulent flow regions is

$$\overline{Nu} = \sqrt{\overline{Nu}_{lam}^2 + \overline{Nu}_{tur}^2} \tag{7-34}$$

where $\overline{Nu}_{lam}$ is obtained from eq. 7–29 and $\overline{Nu}_{tur}$ is obtained from eq. 7–33.

When the heated plate is preceded by an unheated section as shown in Fig. 7–19 the average Nusselt number for *laminar* flow can be calculated using

$$\overline{Nu} = \frac{hL_o}{k} = 0.664(Re_L)^{1/2}(Pr)^{1/3}\left[1 - \left(1 - \frac{L_o}{L}\right)^{3/4}\right]^{2/3} \tag{7-35}$$

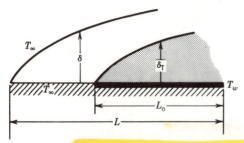

Figure 7-19 Heated section proceeded unheated section.

where L_o is the length of the heated section of the plate, as shown in Fig. 7–19. In *turbulent* flow eq. 7–33 can be used with the substitution of L_o for L in the calculation of the dimensionless groups ($\overline{Nu}$, $\overline{C}_f$, and Re_L) if $0.1 \leq L_o/L \leq 1.0$.

7.7.1.B   Uniform Heat Flux

If the entire plate surface has a constant heat flux boundary condition, the local Nusselt number is obtained from the following expression if the flow is laminar,

$$Nu_x = 0.46(Re_x)^{1/2}Pr^{1/3} \tag{7–36}$$

Heat transfer in a turbulent flow is less sensitive to thermal boundary conditions. Equation 7–33 can therefore be used for a turbulent flow without introducing an appreciable error.

7.7.2   Other Objects of Various Shapes

Gnielinski[4] has indicated that the mean Nusselt number for other objects of various shapes with a uniform surface temperature can be estimated using a relationship of the form of

$$\overline{Nu} = \overline{Nu_0} + \sqrt{\overline{Nu_{lam}^2} + \overline{Nu_{tur}^2}} \tag{7–37}$$

The appropriate values of the characteristic length, L_c, used in the calculation of the Reynolds number and the Nusselt number are given in Table 7–5. The value of $\overline{Nu}_{lam}$ is obtained using eq. 7–29 while eq. 7–33 is used to obtain $\overline{Nu}_{tur}$ as long as the Reynolds number is within the range of $1 < Re_{L_c} < 10^5$. Also

Table 7–5 Coefficients and Characteristics Length for Various Objects Forced Connection eq. 7–37[4]

Object	L_c	$\overline{Nu}_0$
Wire, cylinder, and tubes	$\dfrac{\pi}{2}d$	0.3
Spheres	d	2.0

tabulated in this table are the values of $\overline{Nu}_o$ for several different objects. Equation 7–37 can only be used when the Prandtl number is in the range $0.6 < Pr < 10^3$. When the Reynolds number is less than 1, the following expressions should be used:

Wires, cylinders and tubes:

$$\overline{Nu} = 0.75(Re_{L_c}Pr)^{1/3} \qquad (7\text{--}38)$$

Spheres:

$$\overline{Nu} = 1.01(Re_{L_c}Pr)^{1/3} \qquad (7\text{--}39)$$

where L_c is the appropriate characteristic length found in Table 7–5.

EXAMPLE 7–7

Water flows across a tube, which is 2 cm in diameter, at a velocity of 1.0 m/s. The temperature of the water is 71.85° C (345 K). A vapor is condensed inside the tube and the outer surface of the tube may be considered to be at a uniform temperature of 50° C. Determine the rate of heat transfer per unit length of the tube.

SOLUTION

The properties of water at 71.85° C (345 K) are obtained from Table A–9.

$\rho = 977$ kg/m³ $\mu = 389 \times 10^{-6}$ N·s/m²

$Pr = 2.46$ $k = 0.664$ W/m·K

The mean Nusselt number for the cylinder is obtained using eq. 7–37 and Table 7–5 if $1 < Re_L < 10^5$

$$\overline{Nu} = 0.3 + \sqrt{\overline{Nu}_{lam}^2 + \overline{Nu}_{tur}^2}$$

The characteristic length is $\pi d/2 = (\pi(0.02)/2) = 31.42 \times 10^{-3}$ m. The Reynolds number is

$$Re_{L_c} = \frac{\rho U L_c}{\mu} = \frac{977(1.0)(31.42 \times 10^{-3})}{389 \times 10^{-6}}$$

$$= 78.91 \times 10^3$$

The mean Nusselt number for laminar flow is obtained using eq. 7–29.

$$\overline{Nu}_{lam} = 0.664(Re_{L_c})^{1/2} Pr^{1/3}$$

$$= 0.664 (78.91 \times 10^3)^{1/2}(2.46)^{1/3}$$

$$= 251.8$$

For tubulent flow

$$\overline{C}_f = 0.074(Re_{L_c})^{-1/5}$$

$$= 0.0744(78.91 \times 10^3)^{-1/5}$$

$$= 7.759 \times 10^{-3}$$

and

$$\overline{Nu}_{tur} = \frac{(\overline{C}_f/2)Re_L Pr}{1 + 12.7(\overline{C}_f/2)^{1/2}(Pr^{2/3} - 1)}$$

$$= \frac{[(7.759 \times 10^{-3})/2](78.91 \times 10^3)(2.46)}{1 + 12.7[(7.759 \times 10^{-3})/2]^{1/2}[(2.46)^{2/3} - 1]}$$

$$= 456.3$$

The mean Nusselt number for the cylinder is

$$\overline{Nu} = 0.3 + \sqrt{(251.8)^2 + (456.3)^2}$$

$$= 521.5$$

The mean convection film coefficient is

$$\overline{h} = \frac{\overline{Nu}k}{L_c} = \frac{521.5(0.664)}{31.42 \times 10^{-3}}$$

$$= 11.02 \times 10^3 \text{ W/m}^2 \cdot \text{K}$$

The rate of heat transfer per unit length of the tube is

$$\dot{Q} = \overline{h}A(T_w - T_\infty)$$

$$= 11.02 \times 10^3(\pi)(0.02)(50 - 71.85)$$

$$= -15.13 \text{ kW/m}$$

COMMENT

The temperature of the surface of the tube is less than that of the surrounding fluid, therefore heat is transferred from the fluid to the tube. The rate of heat transfer will be negative as indicated by the calculation. It should also be noted that correlations used to determine the heat transfer coefficients do not depend on the direction that heat is being transferred.

7.8 NATURAL CONVECTION HEAT TRANSFER

The heat transfer from a surface to a fluid that is not moved past the surface by a fan or blower is called *natural* or *free* convection. The fluid movement in such a situation occurs only as the result of the density gradients created by the temperature field that exists near the surface. Because of the density gradient, a body force is exerted on the fluid. This body force is commonly referred to as a buoyancy force and depends upon the relative directions of the density gradients and the gravitational field.

7.8.1 Vertical Flat Plate

If one considers the heated vertical flat plate shown in Fig. 7–20a, it is observed that the cold fluid is drawn to the plate where it is heated and continues its upward movement in the manner indicated by the figure. Both thermal and hydraulic boundary layers are formed on the plate. The temperature and velocity profiles at section a–a in these boundary layers are shown in Fig. 7–20b.

A new dimensionless group is introduced in natural convection heat transfer. It is called the *Grashof* number and its local value is defined as

$$Gr_x \equiv \frac{g\beta(T_w - T_\infty)x^3}{\nu^2} \propto \frac{\text{buoyancy forces}}{\text{viscous forces}} \tag{7-40a}$$

or

$$Gr_L \equiv \frac{g\beta(T_w - T_\infty)L^3}{\nu^2} \tag{7-40b}$$

The spatial coordinate parallel to the plate in the direction of the flow is x while y is the coordinate perpendicular to the plate. The coefficient of volumetric expansion is β,

$$\beta \equiv -\frac{1}{\rho}\frac{\partial\rho}{\partial T}\bigg|_P$$

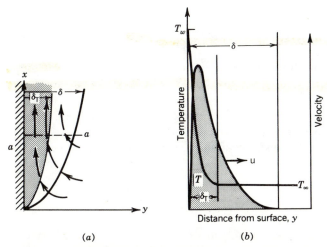

(a)   (b)

Figure 7-20 Natural convection from vertical plate. (a) Boundary layer. (b) Velocity and temperature profiles (Pr = 10).

For an ideal gas $\beta = 1/T$. The values of β for other fluids are tabulated in the appendix. The gravational acceleration, g, is 9.807 m²/s at sea level.

The Grashof number plays the same role in natural convection as the Reynolds number does in forced convection, since both are ratios of the predominate forces exerted on the fluid. The Nusselt is thus expressable as a function of the Grashof and Prandtl numbers in natural convection. Another very convenient dimensionless group used for heat transfer correlations in natural convection and to identify the transition from laminar to turbulent flow is the *Rayleigh* number. The Rayleigh number is the product of the Grashof and Prandtl numbers

$$Ra_x = Gr_x \, Pr \quad \text{or} \quad Ra_L = Gr_L \, Pr \tag{7-41}$$

Note that both the Rayleigh and Grashof numbers contain a characteristic length which is identified by a subscript. For a vertical plate, the transition from laminar to turbulent flow occurs at $Ra_x \approx 10^9$.

The following experimental correlations of the local and averaged Nusselt number are recommended by Churchill[6] for a smooth vertical plate with a uniform surface temperature.

Laminar:

$$Nu_x = 0.68 + 0.503 \, [Ra_x \psi (Pr)]^{1/4} \tag{7-42}$$

and

$$\overline{Nu} = 0.68 + 0.67 \, [Ra_L \psi (Pr)]^{1/4} \tag{7-43}$$

where

$$\psi (Pr) = \left[1 + \left(\frac{0.492}{Pr} \right)^{9/16} \right]^{-16/9} \tag{7-44}$$

Turbulent:

$$\overline{Nu} = Nu_x = 0.15 [Ra \, \psi (Pr)]^{1/3} \tag{7-45}$$

The subscript for the Rayleigh number in eq. 7–45 is either x or L. *The thermophysical properties of the fluid are evaluated at the film temperature,* $(T_w + T_\infty)/2$.

Several important observations should be made with regard to these expressions. First, since the Rayleigh number contains the temperature difference $[T_w - T_\infty]$, the convection heat transfer coefficient will be proportional to $[T_w - T_\infty]$. The expression for the heat flux is

$$\dot{q}_x'' = h_x (T_w - T_\infty) \begin{cases} \propto (T_w - T_\infty)^{5/4} \text{ laminar} \\ \propto (T_w - T_\infty)^{4/3} \text{ turbulent} \end{cases}$$

The rate of heat transfer is thus a nonlinear function of the local temperature difference. Second, the convection heat transfer coefficient in the turbulent flow region is independent of x, $\overline{h} = h_x$.

EXAMPLE 7–8

The surface of the side walls of a kitchen stove are found to be at a temperature of 37.5° C (310.7 K) when the temperature of the oven in the stove is set at 200° C (473.2 K). The stove is 0.75 m high and each of its sides are 0.7 m wide. If the temperature of the air in the kitchen is at 17.5° C (290.7 K) calculate the amount of heat lost through the sides of the stove.

SOLUTION

The properties of the air at the average temperature, $(37.5 + 17.5)/2 = 27.5°$ C (300.7 K), are obtained from Table A–8.

$\rho = 1.1614 \ \text{kg/m}^3$ $\qquad$ $\nu = 15.89 \times 10^{-6} \text{m}^2/\text{s}$

$c_p = 1.007 \ \text{kJ/kg} \cdot \text{K}$ $\qquad$ $k = 26.3 \times 10^{-3} \text{W/m} \cdot \text{K}$

$\text{Pr} = 0.707$ $\qquad$ $\beta = 1/300.7 = 3.326 \times 10^{-3} 1/\text{K}$

The Rayleigh number for the flow is

$$\text{Ra}_L = \frac{g\beta(T_w - T_\infty)L^3 \text{Pr}}{\nu^2}$$

$$= \frac{9.807(3.326 \times 10^{-3})(37.5 - 17.5)(0.75)^3(0.707)}{(15.89 \times 10^{-6})^2}$$

$$= 770.6 \times 10^6$$

The flow in laminar and the average Nusselt number can be calculated using eq. 7–43.

$$\overline{\text{Nu}} = 0.68 + 0.67 \ [\text{Ra}_L \psi + \text{Pr} \}]^{1/4}$$

$$= 0.68 + 0.67 \left\{ 770.6 \times 10^6 \left[1 + \left(\frac{0.492}{0.707}\right)^{9/16} \right]^{-16/9} \right\}^{1/4}$$

$$= 86.32$$

The value of the average heat transfer coefficient is

$$\overline{h} = \frac{\overline{\text{Nu}}k}{L} = \frac{86.3(26.3 \times 10^{-3})}{0.75}$$

$$= 3.027 \ \text{W/m}^2 \cdot \text{K}$$

The total rate of heat lost by the four sides of the stove is

$$\dot{Q} = \bar{h}A\,(T_w - T_\infty)$$
$$= 3.026(0.75)(4)(0.7)(37.5 - 17.5)$$
$$= 127.1 \text{ W}$$

If one has a uniform heat flux boundary condition on a vertical smooth flat plate, it is convenient to define a *modified Rayleigh* number as

$$Ra_x^* = Ra_x Nu_x = \frac{g\rho^2 c_p \beta \dot{q}_w'' x^4}{\mu k^2} \tag{7-46}$$

where $\dot{q}_w''$ is the uniform heat flux at the surface. The correlations recommended by Churchill[6] for this case are
Laminar:

$$Nu_x = 0.631[Ra_x^* \phi(Pr)]^{1/5} \tag{7-47}$$

and

$$\overline{Nu} = 0.726[Ra_L^* \phi(Pr)]^{1/5} \tag{7-48}$$

Turbulent:

$$\overline{Nu} = Nu_x = 0.241[Ra_L^* \phi(Pr)]^{1/4} \tag{7-49}$$

where

$$\phi(Pr) = \left[1 + \left(\frac{0.437}{Pr}\right)^{9/16}\right]^{-16/9} \tag{7-50}$$

The subscript for the Rayleigh number in turbulent flow is either x or L.

7.8.2   Other Objects of Various Shapes

Churchill[6] has also proposed a general correlation for the calculation of the natural convection heat transfer coefficient for a number of different shaped objects. The correlation is valid for both laminar and turbulent flow regions,

$$\overline{Nu} = \left[\overline{Nu}_o^{1/2} + \left(\frac{Ra_{L_c}\xi(Pr)}{300}\right)^{1/6}\right]^2 \tag{7-51}$$

where

$$\xi(Pr) = \left[1 + \left(\frac{0.5}{Pr}\right)^{9/16}\right]^{-16/9} \tag{7-52}$$

The values of L_c and $\overline{Nu}_o$ for various objects are given in Table 7–6.

Table 7–6 Parameters Used in eq. 7–51, Natural Convection.[6] Characteristic Lengths and $\overline{Nu}_o$ for Generalized Correlations

	L_c	$\overline{Nu}_o$
Inclined plate	x	0.68
Inclined disk	$9d/11$	0.56
Vertical cylinder	L	0.68
Cone	$4L/5$	0.54
Horizontal cylinder	πd	0.36π
Sphere	$\pi d/2$	π
Spheroids	$3\pi V/A$	$A^3/36V^2$
L is measured along the surface		

7.9 COMBINED NATURAL AND FORCED CONVECTION

Situations are encountered in which the fluid movement due to the action of a fan or blower is small. As a result both natural and forced convection effects may be present and must be considered when evaluating the heat transfer coefficient. This is particularly important when the buoyancy force acts in a direction parallel to the flow. When the ratio of Gr_L/Re_L^2 is approximately 1, both natural and forced convection effects must be considered. If Gr_L/Re_L^2 is greater than 1, natural convection effects dominate, while forced convection effects dominate if the ratio is less than 1.

If the buoyant force acts in the same direction as the free-stream flow field, known as *assisting flow*, the average Nusselt number for an isothermal vertical plate may be calculated using

$$\overline{Nu} = |\overline{Nu}_F^3 + \overline{Nu}_N^3|^{1/3} \tag{7-53}$$

where $\overline{Nu}_F$ is obtained using the relationships given for forced convection in Section 7.7 and $\overline{Nu}_N$ is obtained using the relationships given in Section 7.8 for natural convection. A sketch of an assisting flow situation is shown in Fig. 7–21a.

The *opposing flow* condition occurs when the buoyancy force acts in a direction counter to the free-stream flow field as shown in Fig. 7–21b. The following relationship should be used for this condition

$$\overline{Nu} = |\overline{Nu}_F^3 - \overline{Nu}_N^3|^{1/3} \tag{7-54}$$

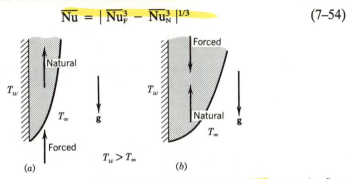

Figure 7–21 Combined natural-forced convection. $T_w > T_\infty$. (a) assisting flow (b) opposing flow.

As the values of the Nusselt numbers for forced and natural convection approach each other, the accuracy of eq. 7–54 decreases sharply.

REFERENCES

1. Schlichting, H., *Boundary Layer Theory*, 6th ed., McGraw-Hill, New York, 1968.

2. White, F. M., *Viscous Fluid Flow*, McGraw-Hill, New York, 1974.

3. Hoerner, S. F., *Fluid Dynamic Drag*, published by the author, Midland Park, NJ, 1965.

4. Gnielinski, V., Forced Convection Around Immersed Bodies, in *Heat Exchanger Design Handbook*, ed. E. V. Schlunder, Part 2, Sec. 2.5.2, Hemisphere Publishing Corporation, Washington, 1982.

5. Gebhart, B., *Heat Transfer*, 2nd ed., McGraw-Hill, New York, 1971.

6. Churchill, S. W., Free Convection Around Immersed Bodies, in *Heat Exchanger Design Handbook*, ed. E. V. Schlunder, Part 2, Sec. 2.5.9, Hemisphere Publishing Corporation, Washington, 1982.

7. Rouse, H., *Elementary Mechanics of Fluids*, Dover, New York, 1963 (originally published by J. Wiley and Sons, Inc., 1946).

8. Fox, R. W., and McDonald, A. T., *Introduction to Fluid Mechanics*, 2nd ed., Wiley, New York, 1978.

9. National Committee for Fluid Mechanics Films, *Illustrated Experiments in Fluid Mechanics*, M.I.T. Press, Cambridge, Mass., 1972.

PROBLEMS

7–1   Air at 27° C and atmospheric pressure flows past a flat plate at 15 m/s. What is the thickness of the boundary layer and the wall shear stress at a point 0.6 m from the leading edge of the plate? The transition Reynolds number is 0.3×10^6.

7–2   A boundary layer is sometimes described by its *displacement thickness*, δ^*. This quantity is defined as the distance that the wall surface would have to be moved in a direction normal to the flow so that the fluid flow rate in the boundary layer equals that of an inviscid flow past the displaced wall surface. δ^* may be expressed as

$$\delta^* = \int_0^\delta \left(1 - \frac{u}{U} \right) dy$$

Calculate δ^* for a tubulent boundary layer in terms of the local boundary layer thickness δ and the local Reynolds number. Recall that $u/U = (y/\delta)^{1/7}$ for a turbulent boundary layer.

7–3   A 1×1m wooden board weighing 105 kg slides down an inclined ramp which is covered with oil. The distance between the board and the ramp is a constant 0.5 mm. If the ramp is inclined at an angle of 22.6° from the horizontal, the velocity of the board down the ramp of 0.15 m/s. Estimate the dynamic viscosity of the oil.

7–4   Air flows past a smooth flat plate which is parallel to the flow. Determine the ratio of the friction drag over the front half of the plate, $x = 0$ to $x = L/2$, to the friction drag of the entire plate if:
(a) The flow is laminar over the entire plate.
(b) The flow is turbulent over the entire plate and $Re_L < 10^7$.

7–5   Air at standard sea level conditions and $26.85°$ C flows over a flat plate. The velocity of the air approaching the plate, the free-stream velocity, is 18 m/s. Determine the boundary layer thickness and the wall shear stress τ_w, at $x = 1$ m from the leading edge of the plate if the flow is made turbulent from the leading edge by the introduction of a boundary layer trip. If the plate has a total length of 4 m, calculate the skin friction drag per unit width on the entire length of plate for turbulent flow (both sides exposed to the free stream).

7–6   The hull of a large iron ore barge is 30 m long and 12 m wide. When fully loaded the flat bottom of the barge is 10 m below the surface of the water. When it is empty the flat bottom is submerged 2 m below the surface of the water. If the barge moves at 1.5 m/s, estimate the friction drag on the hull when it is empty and when it is loaded. The water temperature is $10°$ C and $\nu = 1.4 \times 10^{-6}$ m²/s.

7–7   Stokes law states that the total drag coefficient of a sphere is

$$C_D = \frac{24}{Re_d}$$

when $Re_d < 1$. Show that the terminal velocity, w_t, of a sphere of diameter d falling in a fluid with viscosity μ and density ρ_F is

$$w_t = \frac{gd^2(\rho_S - \rho_F)}{18\ \mu}$$

where g is the gravitational constant and ρ_S the density of the sphere. The fluid is very viscous so that $Re_d \ll 1$. How could this relation be used to determine the viscosity of a fluid?

7–8   Aunt Suzie's Pancake House is constructing a new restaurant at a very windy site in the mountains. It would be beneficial to erect a large sign to advertise the new restaurant. The sign is to be 6.5 m high, 13.0 m long and will be mounted atop two tall pylons. The structural design of the pylons requires an estimate of the force due to a 45.0 m/s wind. Consider the sign to be a smooth, thin flat plate with a completely turbulent boundary layer. Determine the wind force on the sign when the wind is directed parallel to the edge of the sign. The temperature of the air is $26.85°$ C.

7–9   Repeat Prob 7–8 when the wind is normal to the sign. Discuss the mechanisms that cause the force in each of these two cases.

7–10   A spherical balloon has a diameter of 6.4 m and is filled with helium. The pressure and temperature of the helium are the same as the atmospheric air at an altitude of 1500 m ($T = 278.4$ K and $P = 84.5$ kPa). The mass of the balloon and its payload is 65 kg. If the drag coefficient of the balloon is $C_D = 0.21$ based on the maximum cross-sectional area, determine the ascension velocity of the balloon.

7–11   What is the terminal velocity of a parachutist in standard atmospheric conditions ($T = 288.2$ K, $P = 101.3$ kPa) if the parachute is 4.5 m in diameter and the total mass falling is 420 kg?

7–12   A cold wind, $-20°$ C, flows in a nearly horizontal direction across the side of a house at a velocity of 8 m/s. The side of the house is 3 m high and the length of the side is 10 m. There are several windows in the side so that the surface is considered to be "fully rough." The surface temperature of the house is 5° C.
    (a) Estimate the convection heat transfer coefficient using the Chilton–Colburn analogy.
    (b) Calculate the rate of heat transfer from the side of the house.

7–13   Water at a velocity of 5 m/s flows over a horizontal flat isothermal plate 20 cm long. The temperature of the water is 30° C while that of the surface of the plate is 60° C. Calculate the rate of heat transfer per unit width from the upward facing surface of the plate.

7–14   Water flows over a flat electrically heated plate, 5 cm long, at a velocity of 1 m/s. The water is at a temperature of 30° C and the thermal boundary condition on the plate is that of a constant heat flux of 10 kW/m². Estimate the surface temperature of the plate at its trailing edge.

7–15   A sensing device is to be placed in a gas stream which is at a temperature of 300° C and moving at a velocity of 20 m/s. In order to function properly the sensor must be cooled. A preliminary design of the cooling system requirements may be obtained by approximating the probe as a flat plate with an uncooled section at its tip. A sketch of the unit with appropriate dimensions is given in Fig. P7–15.

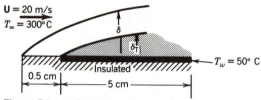

Figure P7-15   Sensing device.

Determine the average cooling, in kilowatts per meter squared, required to maintain the wall at 50° C. Consider the gas to have the same thermophysical properties as air.

7–16   A thin ribbon heating element is to be placed in an air stream which is moving at a velocity of 8 m/s and has a temperature of 15° C. The ribbon, which is a component in an air heater, is oriented parallel to the air stream as shown in Fig. P7–16. The maximum surface temperature of the ribbon for continuous operation is 150° C. Estimate the rate of heat transfer per meter of ribbon length.

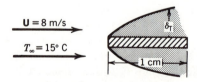

Figure P7-16   Ribbon heating element.

7–17   A bare 40-W light bulb, 10 cm in diameter, is placed outside and exposed to air which is at 14° C and is moving at a velocity of 5 m/s. It was observed that its surface temperature is approximately 36° C. It is desired to estimate the amount of heat lost by the bulb. This may be accomplished by assuming the bulb to be a sphere.

7–18   A hot air hair dryer is composed of an electrical heating element containing 0.5 mm diameter wire. The air moves over the heating element at a velocity of 35 m/s. Estimate the convection heat transfer coefficient for the transfer of heat between the wire and the air in watts per meter squared-degrees Kelvin. The thermophysical properties of the air are to be evaluated at 50° C.

7–19   An outside wall of a room is 7 m long and 3 m high. It is poorly insulated and the interior surface of the wall is at a temperature of 5° C. Estimate the rate of heat transfer to the wall due to natural convection if the temperature of the air in the house is 20° C.

7–20   A thin electrically heated vertical plate, 25 × 25 cm, is immersed in a large tank of water. The electrical energy supplied to the plate was measured and found to be 6.25 kW. Estimate the maximum surface temperature of the plate assuming it has a constant heat flux boundary condition on both vertical faces. The average temperature of the water is 5° C and the thermophysical properties used in the calculations are evaluated at 32.2° C.

7–21   A horizontal cylindrical electrical heating element 1 cm in diameter and 13 cm long is placed in 20° C air. The construction of the heating element is such that the heating element is at a uniform surface temperature of 110° C. Calculate the rate of heat transfer from the surface of the heating element.

7–22   A window 0.5 m wide and 0.4 m high has an interior surface temperature of 10° C. A small fan has been installed in order to minimize the danger of the window fogging up. The fan creates a slight upward movement of air over the window with its velocity being of the order of 1 m/s. Estimate the rate of heat loss through the window if the temperature of the air in the room is 20° C.

8

Internal Flows—Fluid Viscous and Thermal Effects

8.1 INTRODUCTION

Chapter 7 describes the effects of viscosity in an external flow, when a fluid moves past a solid surface boundary. The consequence of this motion is the formation of a boundary layer region adjacent to the surface. The effects of viscosity are concentrated in this region and result in a variation of velocity normal to the surface from a value of zero at the wall (no-slip condition) to a maximum value at the outer edge of the boundary layer. Since an external flow has no upper solid boundary, the maximum velocity equals the velocity of the flow in the free stream. The velocity variation aross the boundary layer represents a loss in fluid momentum and hence a resistance to the flow. The reaction of this resistance on the surface is a drag force.

A second category of flows in which viscosity has an important effect is termed *internal flows*. These are flows in which the fluid is completely enclosed by a solid boundary and represent the flow of a fluid in a pipe or duct, such as experienced in the movement of oil, gas, or water from one geographical point to another. The presence of viscosity creates a fluid boundary layer adjacent to the solid surface (wall) as it did in an external flow, which in turn produces a resistance to the movement of the fluid through the duct. Near the walls the same boundary layer characteristics exist as are found in external flows—no-slip condition at the wall, a velocity variation normal to the wall, and the generation of viscous shear stresses on the fluid which oppose its motion. The major difference between internal and external flows is that the internal flow is bounded and a boundary layer exists on all the enclosing surfaces. As the flow proceeds from the inlet or entrance of a duct, the thicknesses of these boundary layers increase and eventually meet at the center of the duct. Prior to this point the flow is composed of an inviscid core and a boundary layer region in which the effects of viscosity are concentrated, see Fig. 8–1. The velocity distribution across the duct changes as the fluid moves away from the duct inlet since the region affected by the viscosity is increasing in size. A *fully developed* duct flow is then

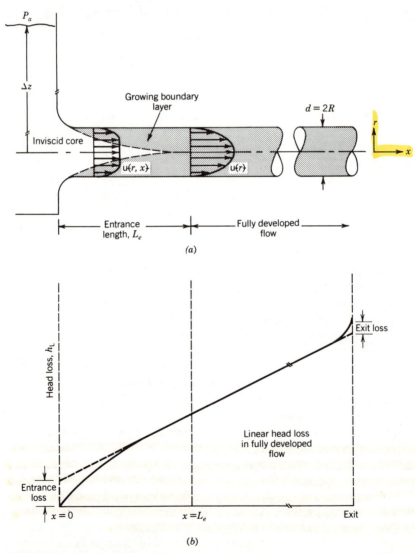

Figure 8-1 Flow from a reservoir into a constant diameter horizontal pipe. (a) Boundary layer growth. (b) Head loss in a pipe as a function of length.

defined as one in which the velocity distribution no longer changes in the flow direction. The region upstream of the point at which the flow becomes fully developed is called the entrance region or the hydrodynamic developing flow region. The entrance length, L_e, is the length of this region measured in the flow direction.

To demonstrate the internal flow of a viscous fluid, consider the flow of a liquid from a large reservoir into a circular duct of constant diameter, Fig. 8–1. The existence of a reservoir containing a liquid of depth Δz produces a flow through the system. If the duct exits to the atmosphere, the steady, incompressible, one-dimensional energy equation (first law of thermodynamics) between the surface of the reservoir and the exit of the duct, eq. 5–44, becomes

$$\frac{P_a}{\rho g} + \frac{V_i^2}{2g} + z_i = \frac{P_a}{\rho g} + \frac{V_d^2}{2g} + z_d + h_{\text{L}} \qquad (8\text{–}1)$$

where h_{L} is the total head loss in the system and V_d is the magnitude of the average velocity at the duct exit, eq. 5–14. The fluid energy available to produce a flow through this system is provided by the potential head Δz which exists between the surface of the reservoir and the exit of the duct. Since the pressure terms in eq. 8–1 cancel and V_i is very small, the available potential head must equal the velocity head of the fluid at the exit of the duct, $V_d^2/2g$, plus the total loss h_{L}.

In the system shown in Fig. 8–1, the losses occur in the transition from the reservoir to the duct (entrance loss), in the entrance length region, in the fully developed flow portion of the duct due to viscous effects alone, and at the exit as the flow leaves the duct. The exit loss is due to the expansion of the fluid from the duct of diameter d to a reservoir of infinite diameter. Figure 8–1b shows the distribution of these losses through the system. In the following sections, methods for estimating these losses will be discussed.

8.2   VISCOUS EFFECTS IN THE ENTRANCE REGION OF A DUCT

The loss term h_{L} depicted in Fig. 8–1 is composed of losses due to the friction on the walls of the duct (major losses) and losses associated with the entrance and exit of the flow to and from the duct (minor losses). The minor losses include friction effects but are predominantly caused by adverse pressure gradient effects. In the entrance of the duct the boundary layer initially forms in a manner similar to that at the leading edge of a flat plate. As shown in Fig. 8–1, the thickness of this viscous region increases as the flow proceeds along the duct, in the x direction. The rate at which the boundary layer thickness increases depends upon whether the flow is laminar or turbulent. A boundary layer is formed on each of the duct walls and eventually they meet in the center of the duct. As described above, the *entrance length* L_e is defined as the distance in the flow direction between the duct inlet and the point where the flow becomes fully developed.

LAMINAR REL 2300
TURB. RE > 2300

$$R_E = \dfrac{d_h V \rho}{\mu}$$

Within the entrance length, $0 < x < L_e$, the flow may be either laminar, turbulent, or both, depending upon the wall roughness h_r and the Reynolds number. For internal flows the Reynolds number is defined using the mean velocity V and the inside diameter of the duct d, if the duct is circular. Experiments show that the critical value of Reynolds number, the Reynolds number at which the transition from laminar to turbulent flow occurs, is approximately 2300. When the flow is turbulent and the duct is not circular, the Reynolds number is defined using the hydraulic diameter, d_h, which is defined as

$$d_h \equiv \frac{4(\text{cross-sectional area})}{\text{wetted perimeter}} = \frac{4A_c}{\mathcal{P}} \tag{8-2}$$

In both circular and noncircular ducts, the flow in the entrance length is characterized by a viscous region at the wall plus an inviscid core around the centerline of the duct.

At some axial distance from the duct entrance the inviscid core is no longer present and the viscous portion extends across the entire duct. Further velocity changes occur until the flow becomes *fully developed* and the velocity profile is invariant in the axial direction. In the fully developed flow region, $x > L_e$, the velocity profile is only a function of duct radius, or location from the centerline. Experiments show that the entrance length is different for completely laminar and turbulent flows and is a function of the Reynolds number.

$$L_e \approx 0.060(d)\,\text{Re} \quad \text{for laminar flows} \tag{8-3}$$

$$L_e \approx 4.40(d)(\text{Re})^{1/6} \quad \text{for turbulent flows}$$

The velocity profiles for fully developed laminar and turbulent internal flows are compared in Fig. 8–2.

8.3 ENERGY LOSSES IN INTERNAL FLOWS

8.3.1 Major Losses

We will first consider the loss in energy experienced by the fluid in the fully developed flow region. Figure 8–1*b* shows this loss to be linear with respect to the length of the duct and is due to the friction or shear stress on the wall of

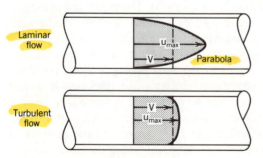

Figure 8-2 Comparison of fully developed laminar and turbulent velocity profiles.

the duct. Figure 8–3 depicts a section of a fully developed duct flow. As shown, the section has a length L and is inclined to the horizontal direction at an angle ϕ. Acting on the ends of the control volume containing the fluid are the uniform pressures P_1 and P_2. Along the cylindrical surface a shear stress τ_w acts on the fluid to retard its motion. The following analysis develops an expression for the loss due to the shear stress or friction on the control surface of this fluid for either a fully developed, incompressible, laminar, or turbulent flow in a duct of constant internal diameter.

The steady-state, steady-flow energy equation, eq. 5–44, will be used to analyze the flow shown in Fig. 8–3. The head loss occurs only from friction effects and is designated h_f to distinguish it from other loss due to flow separation. $\dot{W}_s = 0$ since there is no shaft crossing the control volume,

�# �🟊 ✫
$$\frac{P_1}{\rho g} + \frac{V_1^2}{2g} + z_1 = \frac{P_2}{\rho g} + \frac{V_2^2}{2g} + z_2 + h_f \qquad (8\text{–}4)$$

VELOCITIES ARE AVE.

From the statement of the conservation of mass for this control volume, eq. 5–17, $V_1 = V_2 = V$ since the cross-sectional areas at station 1 and 2 are equal and the density of the fluid is constant. The energy equation can be rearranged as

Pascals (not KPa)

STRAIGHT
PIPE
$$h_f = \frac{P_1 - P_2}{\rho g} + (z_1 - z_2) \qquad (8\text{–}5)$$
meters

If $z_1 = z_2$, eq. 8–5 shows that $P_1 > P_2$ since h_f is always positive. This says that the influence of friction on the walls of a duct is to produce a decrease or drop in pressure in the direction of flow.

Applying the steady-state, steady-flow, one-dimensional linear momentum equation to the control volume in the flow direction gives

$$(P_1 - P_2)\pi R^2 + \rho g \pi R^2 L \sin \phi - \tau_w\, 2\pi R L = \dot{m}(V_2 - V_1) \qquad (8\text{–}6)$$

= 0

PERIMETER

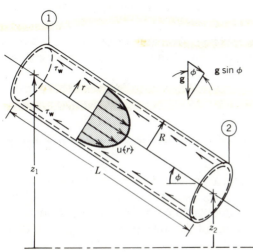

Figure 8-3 Fully developed flow in an inclined pipe.

Since $L \sin \phi = z_1 - z_2$, eq. 8–6 can be written as

$$\frac{P_1 - P_2}{\rho g} + (z_1 - z_2) = \frac{2\tau_w}{\rho g}\frac{L}{R} \tag{8–7}$$

Combining eqs. 8–5 and 8–7,

$$h_f = \frac{2\tau_w}{\rho g}\frac{L}{R} \tag{8–8}$$

In Chapter 7, the wall shear stress was seen to be a function of several variables

$$\tau_w \propto f(\mu, V, R, \rho, h_r)$$

This functional relationship can be simplified through dimensional analysis to give

$$\frac{8\tau_w}{\rho V^2} = f\left(Re, \frac{h_r}{2R}\right) \equiv f \tag{8–9}$$

The dimensionless factor f is called the Darcy–Weisbach friction factor[1].
The head loss due to τ_w, or friction on the duct, is then

$$h_f = f\frac{L}{2R}\frac{V^2}{2g} = f\frac{L}{d}\frac{V^2}{2g} \quad \text{meters} \tag{8–10}$$

Therefore to find h_f between two stations in a duct of constant diameter d, it is necessary to know the velocity V, the distance between the two stations L, and the Darcy–Weisbach friction factor f. The variation of f with Reynolds number and relative surface roughness, h_r/d, is given in Fig. 8–4 as the well-known Moody chart[2] for duct friction.

Figure 8–4 indicates that the influence of roughness is negligible if the flow is laminar. This influence of roughness is the same as that noted for laminar external flows in Chapter 7. The case of turbulent flow is another matter, however, as f depends strongly on the roughness of the duct. From studies of commercial clean ducts Moody was able to determine their average surface roughness h_r. These are listed in Table 8–1.

If the duct is noncircular and the flow is turbulent, the Moody friction chart can be used if the duct diameter d is replaced by the hydraulic diameter d_h, eq. 8–2.

EXAMPLE 8–1

Determine the pressure drop per meter of a fully developed flow in a 10-m length of cast iron duct with a square cross section of 1.152 m per side. The volume flow rate of water through the duct is 20.41 m³/s. The duct is located in a horizontal plane, that is, the flow does not change elevation. ($\nu_{water} = 1.005 \times 10^{-6}$ m²/s; $\rho = 998$ kg/m³)

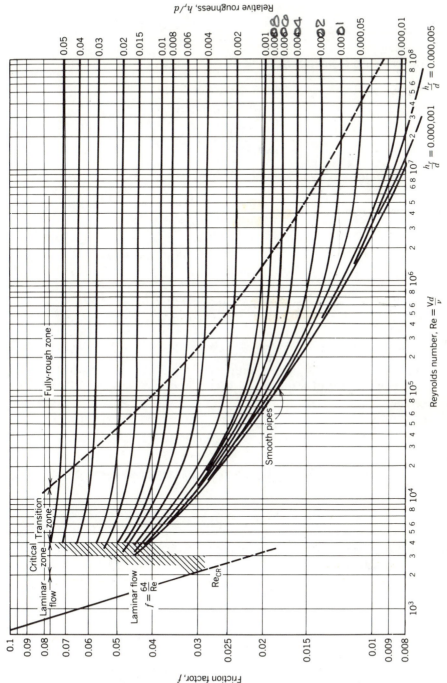

Figure 8-4   Friction factor for fully developed circular duct flows[2]. Used with permission.

SOLUTION

The hydraulic diameter d_h, eq. 8–2, is

$$d_h = \frac{4 \text{ (cross-sectional area)}}{\text{wetted perimeter}}$$

$$= \frac{4 (1.152)^2}{4 (1.152)} = 1.152 \text{ m}$$

The average velocity through the duct is

$$V = \frac{\dot{V}}{A_c} = \frac{20.41}{(1.152)^2}$$

$$= 15.38 \text{ m/s}$$

and

$$Re = \frac{V d_h}{\nu} = \frac{15.38 (1.152)}{1.005 \times 10^{-6}}$$

$$= 17.63 \times 10^6$$

Since Re > 2300 the flow is turbulent. From Table 8–1 the average roughness height of the walls of a cast iron duct is $h_r = 0.26$ mm. Hence,

$$\frac{h_r}{d_h} = \frac{0.26}{1152} = 225.7 \times 10^{-6}$$

From Fig. 8–4 the Darcy–Weisbach friction factor at Re = 17.63×10^6 and $h_r/d_h = 0.23 \times 10^{-3}$ is

$$f = 0.014$$

The pressure drop in the 10 m long duct is obtained from eq. 8–5 and 8–10 with $z_1 = z_2$:

$$P_1 - P_2 = \rho g h_f$$

$$= \rho g \left[f \frac{L}{d_h} \frac{V^2}{2g} \right]$$

$$= 998 (9.807) \left[0.014 \left(\frac{10}{1.152} \right) \frac{(15.38)^2}{2 (9.807)} \right]$$

$$= 14.34 \text{ kPa}$$

The pressure drop per meter of duct length is then

$$\frac{P_1 - P_2}{\text{meter}} = \frac{14.34 \text{ kPa}}{10 \text{ m}} = 1.434 \text{ kPa/m}$$

COMMENT

Since the duct has a square cross section and the flow is fully developed and turbulent the duct is considered as an equivalent circular duct by employing the hydraulic diameter. The calculation of the pressure drop per length of duct allows the application of this result to various duct lengths.

Table 8–1   Average Roughness of Commercial Pipes

Material (new)	h_r, mm
Riveted steel	0.9–9.0
Concrete	0.3–3.0
Wood stave	0.18–0.9
Cast iron	0.26
Galvanized iron	0.15
Asphalted cast iron	0.12
Commercial steel or wrought iron	0.046
Drawn tubing	0.0015
Glass	"Smooth"

8.3.2 Minor Losses

In addition to the loss due to friction, the system described in Fig. 8–1 experiences losses in head at the pipe inlet and exit. These losses are called minor losses, h_m, to contrast them from the major head loss, h_f, which is due only to friction. In addition to the losses in the inlet and exit of a pipe there may be other contributions to the minor loss of a system. These may be caused by valves, bends, elbows, sudden and gradual contractions and expansions, and the pressure gradients created by these devices. The summation of these minor head losses with the head loss due to friction gives the total head loss, h_L, for the system

$$h_L = h_f + \Sigma h_m \qquad (8\text{–}11)$$

The determination of the magnitude of h_m is dependent upon experimental data. These data are usually presented so that h_m is some fraction of the available velocity head $V^2/2g$. Thus

$$h_m = K \frac{V^2}{2g} \qquad (8\text{–}12)$$

where K is the dimensionless loss factor and is a function of the particular device that produces the minor loss. Substituting this expression for h_m and that for h_f

Table 8–2 Loss Factor, $K = \dfrac{h_m}{V^2/2g}$ for Open Valves, Elbows, and Tees

Nominal diameter, cm	Screwed				Flanged				
	1.3	2.5	5.0	10	2.5	5	10	20	50
Valves (fully open):									
Globe	14	8.2	6.9	5.7	13	8.5	6.0	5.8	5.5
Gate	0.30	0.24	0.16	0.11	0.80	0.35	0.16	0.07	0.03
Swing check	5.1	2.9	2.1	2.0	2.0	2.0	2.0	2.0	2.0
Angle	9.0	4.7	2.0	1.0	4.5	2.4	2.0	2.0	2.0
Elbows:									
45° regular	0.39	0.32	0.30	0.29					
45° long radius					0.21	0.20	0.19	0.16	0.14
90° regular	2.0	1.5	0.95	0.64	0.50	0.39	0.30	0.26	0.21
90° long radius	1.0	0.72	0.41	0.23	0.40	0.30	0.19	0.15	0.10
180° regular	2.0	1.5	0.95	0.64	0.41	0.35	0.30	0.25	0.20
180° long radius					0.40	0.30	0.21	0.15	0.10
Tees:									
Line flow	0.90	0.90	0.90	0.90	0.24	0.19	0.14	0.10	0.07
Branch flow	2.4	1.8	1.4	1.1	1.0	0.80	0.64	0.58	0.41

given by eq. 8–10 into eq. 8–11 gives

$$h_L = \frac{V^2}{2g}\left[f\frac{L}{d} + \Sigma K\right]$$

(8–13)

Most available data for the minor loss coefficient K are for turbulent flows only. Some of the existing values of K are presented in Tables 8–2 and 8–3 and Figs. 8–5 to 8–8.

Table 8–3 Increased Losses of Partially Open Valves

| Condition | Ratio K/K (open condition) | |
	Gate valve	Globe valve
Open	1.0	1.0
Closed, 25%	3.0–5.0	1.5–2.0
50%	12–22	2.0–3.0
75%	70–120	6.0–8.0

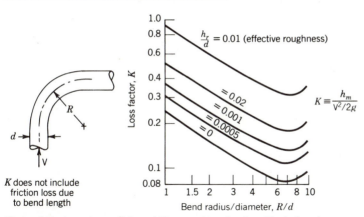

K does not include friction loss due to bend length

Figure 8–5 Loss factor K for a 90° constant radius bend including the effect of surface roughness.

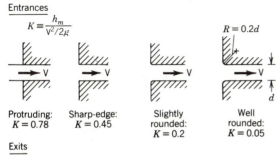

Entrances

$K \equiv \dfrac{h_m}{V^2/2g}$

Protruding: $K = 0.78$ Sharp-edge: $K = 0.45$ Slightly rounded: $K = 0.2$ Well rounded: $K = 0.05$

Exits

$K = 1.0$ for all exit shapes

Figure 8–6 Loss factor K for pipe entrances and exits.

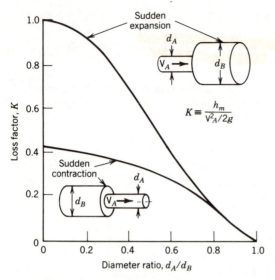

Figure 8-7   Loss factor K for sudden expansions and contractions.

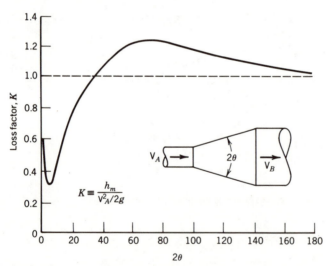

Figure 8-8   Loss factor K for a gradual conical expansion.

EXAMPLE 8–2

The ducting that connects a swimming pool and its water conditioner is shown in Fig. E8–2. The flow rate of water through this cast iron duct is 0.1 m³/s. Calculate the pressure drop through this system. ($\nu_{\text{water}} = 1 \times 10^{-6}$ m²/s; $\rho = 998$ kg/m³.)

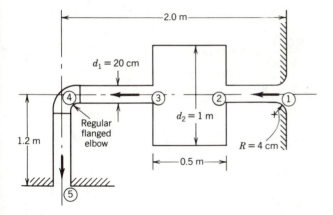

Figure E8-2 Piping system with major and minor losses.

SOLUTION

The head loss through this duct system is due to a combination of the major loss due to friction and and five minor losses. The minor losses are:

1. Well-rounded entrance ($R = 0.2d$)($K_1 = 0.05$, Fig. 8–6).

2. Sudden expansion ($d_A/d_B = 0.2$)($K_2 = 0.925$, Fig. 8–7).

3. Sudden contraction ($d_A/d_B = 0.2$)($K_3 = 0.39$, Fig. 8–7).

4. Regular flanged elbow ($K_4 = 0.26$, Table 8–2).

5. Exit ($d_A/d_B = 0$) ($K_5 = 1.0$, Fig. 8–6 or 8–7).

$$h_L = h_f + \Sigma h_m$$

$$= \left[f \frac{L}{d} \frac{V^2}{2g} \right]_{d_1} + \left[f \frac{L}{d} \frac{V^2}{2g} \right]_{d_2}$$

$$+ [K_1 + K_2 + K_3 + K_4 + K_5] \left[\frac{V^2}{2g} \right]_{d_1}$$

For the pipe with $d_1 = 0.2$ m ($L_1 = 1.5$ m $+ 1.2$ m)

$$V = \frac{\dot{V}}{A_c} = \frac{4\dot{m}}{\rho \pi d_1^2} = \frac{4(0.1)}{\pi (0.2)^2} = 3.183 \text{ m/s}$$

$$Re = \frac{Vd_1}{\nu} = \frac{15.91(0.2)}{1 \times 10^{-6}} = 636.6 \times 10^3$$

$$h_r = 0.26 \text{ mm (Table 8–1)}; \quad \frac{h_r}{d_1} = \frac{0.26}{200} = 1.3 \times 10^{-3};$$

$$f = 0.0214 \text{ (Fig. 8–4)}$$

For the pipe with $d_2 = 1$ m ($L_2 = 0.5$ m),

$$V = \frac{4\dot{m}}{\rho \pi d_2^2} = \frac{4(0.1)998}{998\pi(1)^2} = 0.1273 \text{ m/s}$$

$$Re = \frac{Vd_2}{\nu} = \frac{0.637(1)}{1 \times 10^{-6}} = 127.3 \times 10^3$$

$$h_r = 0.26 \text{ mm}; \quad \frac{h_r}{d_2} = \frac{0.26}{1000} = 0.26 \times 10^{-3}; \quad f = 0.019$$

Therefore,

$$h_L = \left[0.0214 \frac{(1.5 + 1.2)}{0.2} \frac{(3.183)^2}{2(9.807)} \right]$$

$$+ \left[0.019 \frac{(0.5)}{1} \frac{(0.1273)^2}{2(9.807)} \right]$$

$$+ [0.05 + 0.925 + 0.39 + 0.26 + 1.0]$$

$$\cdot \frac{(3.183)^2}{2(9.807)}$$

$$= 1.505 \text{ m}$$

$$P_e - P_d = \rho g h_L = 998(9.807)(1.505)$$

$$= 14.73 \text{ kPa}$$

COMMENT

The total loss in a duct system is the sum of the losses of the individual members of the system. In calculating these individual losses, it is necessary to determine the exact definition of the loss coefficient and velocity head used to present the data. Note the change in Re between the duct having $d = 0.2$ m and $d = 1$ m. This difference in Re results in different values of friction factor.

8.4   HEAT TRANSFER IN DUCTS

When heat is added to or removed from a fluid flowing in a duct, the energy content of the fluid will change as it moves through the duct. The amount of heat transferred and thus the temperature distribution in the fluid will depend upon the thermodynamic state of the fluid entering the duct, the velocity of the fluid, and the thermal boundary conditions at the wall of the duct. While certain similarities do exist between external and internal flows, there are several distinct differences that must be considered when predicting the rate of heat transfer.

8.4.1   Convection Heat Transfer Coefficient

The energy content of a fluid stream at a point in the duct is the product of the local mass velocity, ρu, and the local enthalpy, h. The total energy carried by the fluid passing through the duct at a given axial location can be obtained by integrating the local energy content of the fluid over the cross section of the duct. If the fluid does not experience a change in phase during the addition or removal of heat, a negligible error is introduced if the local enthalpy is approximated by the specific heat of the fluid times its local temperature. The total energy content of a fluid stream flowing in a circular duct of radius R is then

$$\dot{m}h = \int_0^R 2\pi\rho u c_p Tr\, dr \tag{8-14}$$

An average temperature describing the energy content of the fluid can be obtained using the mean velocity of the fluid stream, V. This temperature is called the *bulk* or *mixing cup temperature* T_b. For a circular duct this temperature is

$$T_b \equiv \frac{\dot{m}h}{\dot{m}c_p} = \frac{2\int_0^R \pi\rho u c_p Tr\, dr}{\pi R^2 \rho V c_p} \tag{8-15}$$

Note that the local velocity and the local temperature are both functions of radial location in the duct, $u(r)$ and $T(r)$. If the properties of the fluid are constant, the expression for the bulk temperature reduces to

$$T_b = \frac{2\int_0^R uTr\, dr}{R^2 V} \tag{8-16}$$

The local heat transfer coefficient for an internal flow is defined in terms of the difference between the temperature of the wall and the bulk or mixing cup temperature of the fluid at the axial location x

$$h_x \equiv \frac{\dot{q}''_x}{(T_w - T_b)} \tag{8–17}$$

8.4.2 Energy Balance on Fluid Flowing in a Duct

The conservation of energy, *energy balance*, for a control volume is used to determine the effect of the transfer of heat to or from a fluid, as it flows through a duct, on the energy content of the fluid stream. In forming the energy balance the bulk temperature of the fluid stream is used since it is directly proportional to the energy content of the fluid. An incremental control volume for a duct with a constant cross section is shown in Fig. 8–9. The flow is assumed to be steady state, steady flow with no work done on or by the fluid in the control volume. Changes in the potential and kinetic energy of the fluid are negligible. The heat added to the control volume is expressed as the product of the heat flux at the duct wall and the surface area of the duct.

$$\rho V A_c c_P T_b \big|_x + \dot{q}''_x \mathscr{P} \, \Delta x = \rho V A_c c_P T_b \big|_{x + \Delta x}$$

where $\mathscr{P}$ is the perimeter of the duct in contact with the fluid and A_c is the cross sectional area of the duct. A Taylor series expansion is used to express

$$T_b \big|_{x + \Delta x} = T_b \big|_x + \Delta x \frac{dT_b}{dx} \bigg|_x$$

The energy equation thus reduces to

$$\rho V A_c c_P T_b \big|_x + \dot{q}''_x \mathscr{P} \, \Delta x = \rho V A_c c_P T_b \big|_x + \rho V A_c c_P \frac{dT_b}{dx} \bigg|_x$$

or

$$\rho V A_c c_P \frac{dT_b}{dx} = \dot{q}''_x \mathscr{P} \tag{8–18}$$

Once the boundary conditions at the wall of the duct are defined, the bulk temperature at any axial location can be determined by integrating the energy

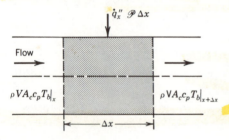

Figure 8-9 Control volume for flow in a constant cross-sectional area duct.

equation, eq. 8–18. There are two common boundary conditions used in convection heat transfer. One is a uniform wall heat flux while the other is a uniform wall temperature.

8.4.2.A Uniform Wall Heat Flux

When the heat flux at the wall of the duct is uniform, $\dot{q}''_w = $ uniform, the integration of eq. 8–18 yields

$$T_b = \frac{\dot{q}''_w \mathscr{P} x}{\rho V A_c c_P} + \text{const} \tag{8–19}$$

If the bulk temperature of the fluid stream entering the duct, $x = 0$, is T_i, the expression for the bulk temperature at any value of $x > 0$ becomes

$$T_b = \frac{\dot{q}''_w \mathscr{P} x}{\rho V A_c c_P} + T_i$$

or

$$T_b = \frac{\dot{q}''_w A}{\dot{m} c_P} + T_i \tag{8–20}$$

where A is the heat transfer surface area of the duct, the perimeter of the duct times the distance from the start of the heating, and $\dot{m}$ is the fluid mass flow rate. The bulk temperature is seen to increase linearly with the distance from the start of the heating.

The temperature of the wall, T_w, at any location may be calculated using

$$\dot{q}''_w = h_x(T_w - T_b)$$

or

$$T_w = \frac{\dot{q}''_w}{h_x} + T_b \tag{8–21}$$

The maximum wall temperature will usually occur at the exit of the heated duct where T_b is largest and h_x has its lowest value. A typical bulk fluid and wall temperature distribution in a uniformly heated duct is shown in Fig. 8–10. The variation in the local heat transfer coefficient is also shown.

8.4.2.B Uniform Wall Temperature

If the wall temperature of the duct is uniform, the local heat flux in eq. 8–18 is replaced by $h_x(T_w - T_b)$ which is obtained by rearranging eq. 8–17. The energy equation then becomes

$$\rho V A_c c_P \frac{dT_b}{dx} = h_x \mathscr{P}(T_w - T_b)$$

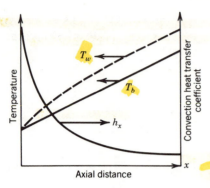

Figure 8-10 Bulk fluid and wall temperature distribution for flow in a duct with uniform wall heat flux.

Rearranging this equation we obtain

$$\frac{dT_b}{(T_w - T_b)} = \frac{h_x \mathcal{P}}{\dot{m}c_p} \, dx$$

If the convection heat transfer coefficient is uniform or an average convection heat transfer coefficient, $\bar{h}$, is used, the equation can be integrated to yield

$$\ln (T_w - T_b) = \frac{\bar{h}\mathcal{P}x}{\dot{m}c_p} + \text{const}$$

The integration constant is evaluated using the entrance condition, $x = 0$ and $T_b = T_i$. The final expression for the bulk temperature is

$$\frac{T_w - T_b}{T_w - T_i} = \exp\left(-\frac{\bar{h}A}{\dot{m}c_p}\right) \qquad (8\text{–}22)$$

A typical bulk fluid temperature distribution in a duct with a uniform wall temperature is shown in Fig. 8–11.

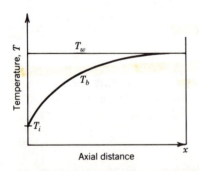

Figure 8-11 Bulk fluid and wall temperature distribution for flow in a duct with uniform wall temperature.

EXAMPLE 8–3

An uninsulated hot water pipe has an inside diameter of 2 cm. The pipe passes through a crawl space under a house where it is exposed to air at a temperature of 5° C. The temperature of the water in the pipe when it enters the crawl space is 40° C. Three (3) meters of the pipe length are exposed to the cold air before it re-enters the house. The inside surface temperature of the pipe is estimated to be nearly uniform at a temperature of 8° C. Estimate the temperature of the water at the location where the pipe re-enters the house. The mean velocity of the water is 1 m/s and the average convection heat transfer coefficient is estimated to be 4500 W/m² · K.

SOLUTION

The properties of water are evaluated at an estimated mean bulk temperature of 24° C (297.2 K) using Table A–9

$$\rho = 990 \text{ kg/m}^3 \qquad c_p = 4.181 \text{ kJ/kg·K}$$

The temperature of the water in the pipe when it re-enters the house is obtained by rearranging eq. 8–22

$$T_b = (T_i - T_w) \exp\left(-\frac{\bar{h}A}{\dot{m}c_p}\right) + T_w$$

The heat transfer surface area is $A = \pi dL = \pi(0.02)(3) = 0.1885 \text{ m}^2$ and the mass flow rate is

$$\dot{m} = \rho V A_c = 998(1)\left(\frac{\pi(0.02)^2}{4}\right) = 0.3135 \text{ kg/s}$$

The temperature of the water leaving is

$$T_b = (40 - 8) \exp\left(-\frac{4500(0.1885)}{0.3135(4181)}\right) + 8$$

$$= 24.75° \text{ C}$$

COMMENT

Based on the information available the mean bulk temperature of the water is estimated as the mean of the entering water temperature (40° C) and the pipe wall temperature (8° C). Using this estimated value the bulk temperature of the exiting water is calculated. A more accurate mean bulk temperature of the water flowing through the duct can now be calculated and is (40.0 + 24.75)/2 or 34.37° C. The thermophysical properties of the water should be determined at this temperature and the calculations repeated to see if appreciable errors have been introduced. For this example the error is negligible.

8.4.3 Entrance Region Effects

In Section 8.1 reference was made to the flow in the entrance region of a duct where the transition from a uniform velocity profile at the duct inlet to the fully developed velocity profile occurs. A similar transition occurs in the temperature profile in the fluid as shown in Figs. 8–12a and 8–12b for flow in a duct with a uniform wall temperature. Two cases will be discussed. In Fig. 8–12a the fluid enters the duct with a *uniform velocity* and *temperature*. Both velocity and temperature profiles are developing as the flow proceeds along the length of the duct. This will be referred to as the *developing* heat transfer region. Fully developed conditions are obtained when the axial velocity and the dimensionless temperature, defined as $(T_w - T)/(T_w - T_b)$, are independent of axial location. In Fig. 8–12b, the fluid enters the heating section after it has passed through an unheated section of sufficient length to allow the velocity profile to become fully developed. The section of the duct in which the dimensionless temperature profile is developing is referred to as the *"thermal"* developing heat transfer region. The convection heat transfer coefficient is a function of axial location in both the developing and the "thermal" developing regions. In the fully developed region the heat transfer coefficient is a constant. The entrance or developing flow regions are very short in turbulent flows, while they are somewhat longer in laminar flows and the entrance region affects must be considered.

8.5 LAMINAR FLOW HEAT TRANSFER COEFFICIENTS   Re < 2300

The value of the convection heat transfer coefficient for laminar flow in a duct is dependent on the geometrical cross section of the duct, the thermal boundary condition at the duct wall, and the distance from the duct entrance. The di-

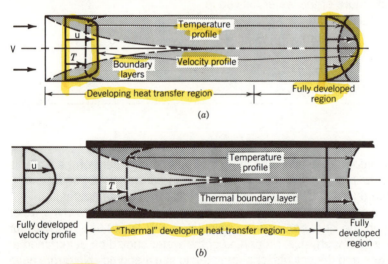

Figure 8-12   Heat transfer entry regions. (a) Developing heat transfer region. (b) "Thermal" developing heat transfer region.

mensionless heat transfer coefficient, the Nusselt number, is defined as

$$\text{Nu} \equiv \frac{hd_h}{k}$$

(8–23)

where d_h is the hydraulic diameter of the section, eq. 8–2, and k is the thermal conductivity of the fluid.

8.5.1 Circular Ducts

The Nusselt numbers for fully developed flow in a circular duct are

Uniform wall temperature: T_W Nu = 3.66

Uniform heat flux: $\dot{q}''_w$ Nu = 4.36

The Nusselt number is a function of the distance from the start of the heating section in both the developing and the "thermal" developing heat transfer regions. A typical plot of the functional relationship is shown in Fig. 8–13. The location from the start of the heating is expressed in terms of the dimensionless axial position which is defined as

WATCH DIMENSIONS' (UNITS)

$$X = \frac{x}{d_h \text{Re Pr}}$$

(8–24)

The subscript x is used to indicate the local Nusselt number. The average Nusselt number obtained by integrating the local value,

$$\overline{\text{Nu}} = \frac{1}{L} \int_o^L \text{Nu}_x \, dx$$

is also shown in the figure.

The heat transfer entrance region in a developing flow (both velocity and dimensionless temperature profiles are developing) is usually very short when compared with the total length of the heated duct. If d/L is less than 0.1 an

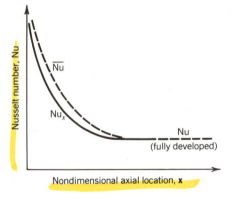

Figure 8-13   Nusselt number variation with axial location for flow within a duct.

insignificant error is introduced by using the Nusselt number correlations given for the "thermal" developing heat transfer region. If d/L is greater than 0.1 more accurate values of the heat transfer coefficient in this region can be obtained by referring to Shah and London[3].

In the "thermal" developing heat transfer region Gnielinski[4] has recommended the relationships given in Table 8–4. The *Peclet* number, Pe, is defined as the product of the Reynolds number and the Prandtl number

$$\text{Pe} = \text{RePr} \qquad \frac{d_h V \rho c_p}{k} \qquad (8\text{–}25)$$

The thermophysical properties are evaluated at the mean temperature of the fluid

$$T_m = \frac{T_{b,i} + T_{b,d}}{2}$$

The mixing cup temperature of the fluid entering the duct is $T_{b,i}$ while $T_{b,d}$ is the mixing cup temperature of the fluid leaving the duct. If the value of $\text{Pe}(d/L)$ falls outside the ranges indicated in Table 8–4 for uniform wall heat flux the value of $\text{Pe}(d/L)$ is substituted into both expressions and the larger Nusselt number is used.

Table 8–4 Nusselt Numbers for "Thermal" Developing Heat Transfer Region Circular Duct (Fully Developed Velocity Profile) Gnielinski[4], hd/k

Correlation	Remarks
Uniform duct temperature	
local	
$\text{Nu}_x = 1.077 \sqrt[3]{\text{Pe}\,\dfrac{d}{x}}$	$\text{Pe}\,\dfrac{d}{L} > 10^2$
$\text{Nu}_x = 3.66$	$\text{Pe}\,\dfrac{d}{L} < 10^2$
Mean	
$\overline{\text{Nu}} = \sqrt[3]{(3.66)^3 + (1.61)^3\,\text{Pe}\,\dfrac{d}{L}}$	
Uniform wall heat flux	
Local	
$\text{Nu}_x = 1.302 \sqrt[3]{\text{Pe}\,\dfrac{d}{x}}$	$\text{Pe}\,\dfrac{d}{L} > 10^4$
$\text{Nu}_x = 4.36$	$\text{Pe}\,\dfrac{d}{L} < 10^3$
Mean	
$\overline{\text{Nu}} = 1.953 \sqrt[3]{\text{Pe}\,\dfrac{d}{L}}$	$\text{Pe}\,\dfrac{d}{L} > 10^2$
$\overline{\text{Nu}} = 4.36$	$\text{Pe}\,\dfrac{d}{L} < 10$

EXAMPLE 8–4

Dry air with a mass flow rate of 0.987 kg/hr is to be heated by passing it through an electrically heated tube. The inside diameter of the tube is 1 cm and the heating section is 0.5 m long. An unheated section of tubing precedes the heated section so that the flow enters the heated section with a fully developed velocity profile. The maximum temperature of the air leaving the heating section is to be found under the design constraint that the maximum temperature of the tube wall cannot exceed 200° C. The temperature of the air entering the unit is 20° C.

SOLUTION

The properties of the air are evaluated at a mean temperature of 325 K using Table A–8.

$$k = 28.15 \times 10^{-3}\,\text{W/m·K} \quad c_p = 1.006\,\text{kJ/kg·K} \quad \rho = 1.075\,\text{kg/m}^3$$

$$\mu = 196.4 \times 10^{-7}\,\text{N·s/m}^2 \quad \text{Pr} = 0.7035$$

The cross-sectional area of the tube is

$$A_c = \frac{\pi d^2}{4} = \frac{\pi (0.01)^2}{4} = 78.54 \times 10^{-6}\,\text{m}^2$$

The Reynolds number is

$$\text{Re} = \frac{\rho V d}{\mu} = \frac{\dot{m}d}{A_c \mu} = \frac{(0.987/3600)(0.01)}{78.54 \times 10^{-6}\,(196.4 \times 10^{-7})}$$

$$= 1777$$

The flow is laminar since Re < 2300 and the correlations given in Table 8–4 can be used. The value of

$$\text{Pe}\,\frac{d}{L} = \text{RePr}\,\frac{d}{L} = 1777\,(0.7035)\left(\frac{0.01}{0.5}\right) = 25.0$$

The local Nusselt number is 4.36. The local convection heat transfer coefficient is

$$h_x = \frac{\text{Nu}_x k}{d} = \frac{4.36\,(28.15 \times 10^{-3})}{(0.01)} = 12.27\,\text{W/m}^2\text{·K.}$$

The wall temperature will be a maximum at the end of the heating section. The heat flux per unit area at this location, $x = 0.5$ m, is

$$\dot{q}_w'' = h_x(T_w - T_b)$$

$$= 12.27(200 - T_b)|_{L=0.5m} \tag{a}$$

The bulk temperature of the air leaving the heating section is obtained using eq. 8–20.

$$T_b|_{L=0.5m} = \frac{\dot{q}_w'' A}{\dot{m}c_p} + T_i$$

$$= \frac{\dot{q}_w''(\pi dL)}{\dot{m}c_p} + T_i$$

$$= \frac{\dot{q}_w''(\pi)(0.01)(0.5)}{(0.987/3600)(1008)} + 20$$

$$= 56.8 \times 10^{-3}\dot{q}_w'' + 20 \tag{b}$$

Equations (a) and (b) are solved simultaneously to obtain

$$\dot{q}_w'' = 1.301 \text{ kW/m}^2$$

and the bulk temperature at the exit of the duct

$$T_b|_{L=0.5m} = 56.8 \times 10^{-3}(1301) + 20$$

$$= 93.96° \text{ C}$$

COMMENT

The bulk fluid temperature leaving the duct is used to determine the mean temperature at which the thermophysical properties of the fluid are evaluated. For this problem the bulk fluid temperature at the exit is one of the items to be determined. It is therefore necessary to use a trial-and-error solution which is started by estimating the bulk fluid temperature leaving the duct.

For the conditions of this problem it is known that the bulk fluid temperature at the duct exit would be greater than 20° C but less than 200° C. The bulk fluid and wall temperature distribution in the duct will be similar to that shown in Fig. 8–10. It is estimated that a bulk fluid temperature increase of 65° C might be reasonable. Therefore, a mean fluid temperature of approximately 325 K is used for the evaluation of the thermophysical properties in the initial calculations and the results obtained are shown.

The second set of calculations would be performed with the properties evaluated at a mean temperature of

$$\frac{20 + 93.96}{2} = 56.98° \text{ C (330.1 K)}$$

The results for these calculations are $\dot{q}''_w = 1.311$ kW and $T_b|_{L=0.5m} = 94.46°$ C. The agreement is reasonable and no further calculations are required.

8.5.2 Noncircular Ducts

The Nusselt number for fully developed flow in several different noncircular ducts is tabulated in Table 8–5. More extensive results are presented by Shah and London.[3]

The Nusselt number for the "thermal" developing heat transfer region between parallel plates with a spacing of s is given in Table 8–6. A correlation suitable for the developing heat transfer region between parallel plates having a uniform wall temperature is

$$\overline{Nu} = \frac{hd_h}{k} = 7.56 + \frac{0.0312[Pe(s/L)]^{1.14}}{1 + 0.058[Pe(s/L)]^{0.64}Pr^{0.17}} \tag{8-26}$$

This equation is valid for $0.1 < Pr < 10^3$ and the complete range of $Pe(s/L)$.

8.6 TURBULENT FLOW HEAT TRANSFER

The transition from laminar to turbulent flow occurs at a Reynolds number of 2300. If the Prandtl number is greater than 0.5, the thermal boundary conditions at the wall of the duct have minimum influence on the value of the Nusselt

Table 8–5 Nusselt Numbers For Fully Developed Laminar Flow

$$Nu = \frac{hd_h}{k}$$

Configuration	Uniform Wall Temperature	Uniform Wall Heat Flux
○	3.66	4.36
□ (1/1)	2.98	3.61
▭ (1/2)	3.39	4.12
▭ (1/8)	5.60	6.49
▭ (1/∞)	7.56	8.24
△	2.35	3.00

Table 8–6 Nusselt Numbers for "Thermal"
Developing Flow Between Infinite Parallel
Plates, Gnielinski[4], hd_h/k

Correlation	Range
Uniform wall temperature	
Local	
$Nu_x = 1.958 \sqrt[3]{Pe\dfrac{s}{x}}$	$Pe\dfrac{s}{x} > 10^3$
$Nu_x = 7.56$	$Pe\dfrac{s}{x} < 10^2$
Mean	
$\overline{Nu} = 2.936 \sqrt[3]{Pe\dfrac{s}{L}}$	$Pe\dfrac{s}{L} > 10^3$
$\overline{Nu} = 7.56$	$Pe\dfrac{s}{L} < 10^2$
Uniform wall heat flux	
Local	
$Nu_x = 2.36 \sqrt[3]{Pe\dfrac{s}{L}}$	$Pe\dfrac{s}{L} > 10^4$
$Nu_x = 8.24$	$Pe\dfrac{s}{L} < 10^2$
Mean	
$\overline{Nu} = 3.55 \sqrt[3]{Pe\dfrac{s}{L}}$	$Pe\dfrac{s}{L} > 10^3$
$\overline{Nu} = 8.24$	$Pe\dfrac{s}{L} < 10^2$

number. The following correlations can be used for the determination of the Nusselt number in turbulent flow in smooth ducts.

$$0.5 < Pr < 1.5$$

$$\overline{Nu} = 0.0214(Re^{4/5} - 100)Pr^{2/5}\left[1 + \left(\frac{d_h}{L}\right)^{2/3}\right] \qquad (8\text{–}27)$$

and

$$1.5 < Pr < 500$$

$$\overline{Nu} = 0.012(Re^{0.87} - 280)Pr^{2/5}\left[1 + \left(\frac{d_h}{L}\right)^{2/3}\right] \qquad (8\text{–}28)$$

The hydraulic diameter, d_h, is used as the characteristic length in the calculation of both the Nusselt number and the Reynolds number.

The Chilton–Colburn analogy may be used to calculate the Nusselt number for fully developed turbulent flow in a rough duct. The analogy for internal flow is

$$\frac{f}{8} = \overline{St}Pr^{2/3} \tag{8-29}$$

This is rearranged to obtain

$$\overline{Nu} = \frac{f}{8}RePr^{1/3} \tag{8-30}$$

The friction factor is obtained from Fig. 8–4.

EXAMPLE 8–5

Water flows at a mean velocity of 2 m/s in a long 2 × 4 cm rectangular duct. The average bulk temperature of the water is 62° C. Estimate the mean convection heat transfer coefficient.

SOLUTION

The properties of the water at 62° C (335.2 K) are obtained from Table A–9.

$$\rho = 982 \text{ kg/m}^3 \qquad\qquad k = 0.656 \text{ W/m·K}$$

$$c_p = 4.186 \text{ kJ/kg·K} \qquad Pr = 2.89$$

$$\mu = 453 \times 10^{-6} \text{ N·s/m}^2$$

The hydraulic diameter of the duct is

$$d_h = \frac{4A_c}{\mathcal{P}} = \frac{4(0.02)(0.04)}{2(0.02 + 0.04)} = 26.67 \times 10^{-3} \text{ m}$$

The Reynolds number for the flow is

$$Re = \frac{\rho d_h V}{\mu} = \frac{982(26.67 \times 10^{-3})(2)}{453 \times 10^{-6}} = 115.6 \times 10^3$$

The mean Nusselt number may be obtained using eq. 8–28,

$$\overline{Nu} = 0.012(Re^{0.87} - 280)Pr^{2/5}\left[1 + \left(\frac{d_h}{L}\right)^{2/3}\right]$$

Since the duct is long the length correction, $[1 + (d_h/L)^{2/3}]$, is taken as 1.

$$\overline{Nu} = 0.012[(115.6 \times 10^3)^{0.87} - 280](2.89)^{2/5}$$

$$= 460.8$$

The mean convective heat transfer coefficient is

$$\bar{h} = \frac{\overline{Nu}k}{d_h} = \frac{460.8(0.656)}{26.67 \times 10^{-3}}$$

$$= 11.33 \times 10^3 \text{ W/m}^2\cdot\text{K}$$

COMMENT

How long must the duct be in order to neglect the entrance region when calculating the heat transfer coefficient? An error of 1% or less will be introduced if the duct has a L/d of approximately 1000. For the conditions of this problem the duct must be at least 26.7 m long in order to neglect entrance effects. If the duct was only 1 m long the average heat transfer coefficient would be

$$\bar{h} = 11.33 \times 10^3 \left[1 + \left(\frac{d_h}{L}\right)^{2/3}\right]$$

$$= 11.33 \times 10^3 \left[1 + \left(\frac{26.67 \times 10^{-3}}{1}\right)^{2/3}\right]$$

$$= 12.34 \times 10^3 \text{ W/m}^2\cdot\text{K}$$

8.7 HEAT EXCHANGERS

Most of the practical cycles used to convert heat to work or to pump heat utilize a working fluid. The fluid is circulated through the various components of the cycle to produce the desired effects. In two or more of the components of the cycle, heat is either added to or removed from the working fluid. This is usually accomplished by having the working fluid exchange heat with a second fluid. If there is no work done by or on the fluids as they pass through the device in which the exchange of heat takes place, the component is classified as a *heat exchanger*. To illustrate, it was previously noted that water is the usual working fluid in an electrical generating plant. The water is vaporized by the addition of heat from the hot gases formed during the combustion of a fuel. The steam is exhausted from the turbine and passes to the condenser. The vapor is condensed through the transfer of heat to a second fluid, usually water from a river, lake, or cooling tower. The condenser cooling water is at a temperature lower than the saturation temperature corresponding to the pressure of the steam leaving the turbine. Both devices are heat exchangers, although we usually refer to them as a furnace and as a condenser, respectively.

8.7.1 Classification of Heat Exchangers

Heat exchangers can be classified on the basis of their application or on the basis of the relative flow configuration of the fluid streams. Both types of classifications will be discussed in this section.

8.7.1.A Classification Based upon Application

The first method of classification to be discussed is that based upon the application of the heat exchanger.

No Phase Change.   There are two types of heat exchangers in which no phase change occurs as the fluids flow through the unit. The most common type is based upon the *shell and tube* configuration. A typical unit of this type is shown in Fig. 8–14. One of the fluids flows within the tubes while the other fluid flows around the tubes with its flow pattern being determined by the placement of the baffles. There are many different flow configurations possible for shell and tube heat exchangers. While both fluids passing through the unit are usually liquids there are exceptions and units have been constructed in which both the fluids are gases.

Plate or *compact* heat exchangers such as the one pictured in Fig. 8–15 are used primarily when it is desired to transfer heat between two gas streams or a gas stream and a liquid stream. These units have very large heat transfer surface area per unit volume. This is achieved using fins or extended surfaces such as those shown in Figs. 8–15 and 8–16.

Phase Change.   In many applications the heat exchanger is designed so that one of the fluids undergoes a phase change. Units of this type are usually of a modified shell and tube design. When vapor is formed, the heat exchanger is called a *furnace, evaporator* or a *reboiler*. The last type of unit is commonly used in the chemical processing industry. The vapor can be formed either in the tube-side or the shell-side fluid. A common type of boiler is one in which fuel and air enters the heat exchanger and a combustion process occurs which results

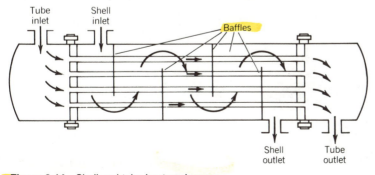

Figure 8-14   Shell and tube heat exchanger.

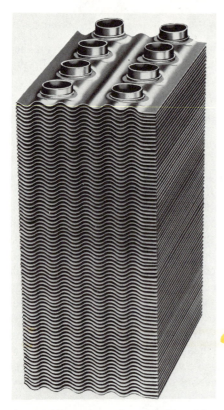

Figure 8-15 Compact heat exchanger surface. Cross section of a fin and tube heat exchanger showing corrugated aluminum fins and round copper tubes. Used by permission of The Trane Company, La Crosse, Wisconsin.

in the liberation of a considerable amount of thermal energy. The hot gases formed during the combustion process transfer heat to the liquid causing it to undergo a change in phase. A sketch of a typical unit of this type is given in Fig. 8–17. The air heaters and economizers, shown in Fig. 8–17, are also heat exchangers but no phase change takes place as the fluid streams pass through them.

When the vapor stream is condensed as it passes through the heat exchanger we call the unit a *condenser*. Once again a modified shell and tube design is used although compact heat exchangers similar to that shown in Fig. 8–15 are quite common.

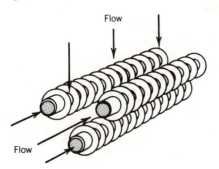

Flow

Flow

Figure 8-16 Cross flow heat exchanger with extended surfaces.

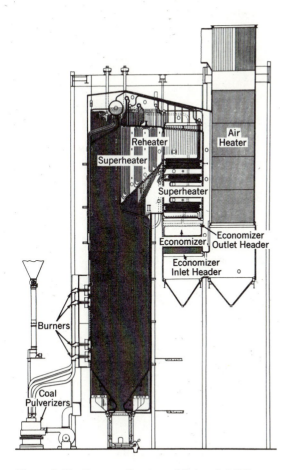

Figure 8-17 Furnace. Courtesy of Babcock & Wilcox.

Regenerators. When the two fluid streams exchange heat using the same flow passage in a periodic fashion so that at any given time only one fluid is in contact with the heat exchanger, the unit is called a *regenerator*. In the design of a regenerator special attention must be given to the ability of the heat exchange to store heat. Thus regenerators are usually very large and consist of a large mass of material. A schematic sketch of a fixed bed regenerator, typical of those used in the steel manufacturing industry is shown in Fig. 8–18. The flow of the gases is changed during the storage and heat retrieval operating periods.

8.7.1.B   Classification Based upon Flow Configuration

Five types of heat exchangers are used when classification is based upon flow configurations. These are shown in Fig. 8–19. While the configurations shown are idealizations of what truly occurs in the heat exchanger, these sketches do illustrate typical flow patterns in the units.

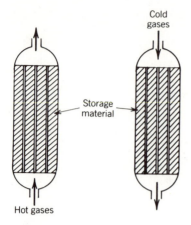

Heat storage   Heat retrieval   **Figure 8-18** Fixed bed regenerators.

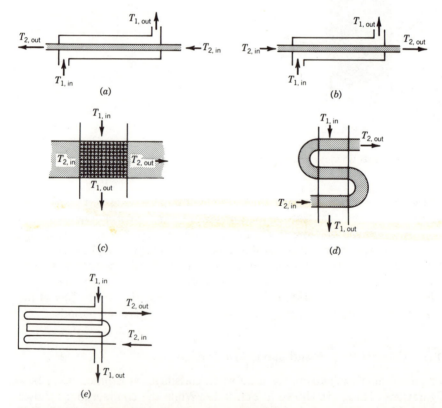

Figure 8-19 Heat exchanger classifications. (a) Counterflow. (b) Parallel flow. (c) Cross flow. (d) Cross–counterflow. (e) Multipass shell and tube flow.

Counterflow (Fig. 8–19a). The two fluid streams flow parallel to each other but in opposite directions. This is the most efficient heat exchanger flow configuration.

Parallel Flow (Fig. 8–19b). The two fluid streams flow parallel to each other, in the same direction. These units are less common since their efficiency is lower than that of the counterflow arrangement.

Cross Flow (Fig. 8–19c). The two fluid streams flow at right angles to each other. While these units are not as efficient as a counterflow design, they are usually used because of the ease with which the fluid can be routed into the exchanger. The radiator in an automobile is of this type.

Cross Counterflow (Fig. 8–19d). This flow arrangement results from a desired to design a heat exchanger which is simple to construct. As the number of passes increases, the unit's efficiency approaches that of a counterflow heat exchanger.

Multipass Shell and Tube Flow (Fig. 8–19e). The simplicity of the construction leads to the use of these types of flow arrangements for many heat exchanger applications.

8.7.2 Overall Heat Transfer Coefficient

The *overall heat transfer coefficient* in a two-fluid configuration is equal to the rate of heat transfer divided by the temperature difference between the two fluids and the surface area separating the two fluid streams.

$$U \equiv \frac{\dot{Q}}{A\Delta T} \qquad \text{W/m}^2\text{·K} \qquad (8\text{–}31)$$

It is convenient to introduce the concept of thermal resistance at this point in our discussion. The rate of heat transfer between two spatial locations is directly proportional to the temperature difference between the two locations, the driving potential for the flow of heat, and inversely proportional to the resistance to the flow of heat offered by the material and at the fluid-material interfaces, which are located between the spatial locations, the thermal resistance. This relationship can be expressed as

$$\dot{Q} = \frac{\Delta T}{R_t} \qquad (8\text{–}32)$$

The methods for calculating the rate of heat transfer at a fluid–surface interface, convection, have been presented earlier in this chapter and in Chapter 7. The expression for the thermal resistance associated with the flow of heat between a fluid and an isothermal surface can be obtained by recalling that

$$\dot{Q} = \bar{h}A(T_w - T_\infty)$$

The thermal resistance is obtained by comparing this expression with eq. 8–32 to obtain

$$R_t = \frac{1}{\bar{h}A}$$

The units for the thermal resistance are degrees Kelvin per Watt.

There are two types of thermal resistance which may be present in a heat exchanger. They are located at the fluid–surface interfaces, and across any solids that may be located between the two fluid streams. The two fluids are normally separated by a single metallic wall and the thermal resistance of the wall will be denoted by R_w. The thermal resistance at the two fluid–surface interfaces associated with convection will be denoted as R_1 and R_2. In many heat exchangers inpurities in the fluid streams will result in the buildup of a deposit on the wall of the heat exchanger. Such a surface is said to be *"fouled."* The deposits offer additional resistances to the transfer of heat, R_{f1} and R_{f2}.

The rate at which heat is transferred between two fluid streams in a heat exchanger can thus be expressed as

$$\dot{Q} = \frac{T_{b1} - T_{b2}}{R_1 + R_{f1} + R_w + R_{f2} + R_2} \tag{8–33}$$

where T_{b1} and T_{b2} are the bulk temperature of the two fluid streams. The overall heat transfer coefficient, as defined by eq. 8–31, can also be expressed in terms of the thermal resistances offered to the flow of heat between the two fluid streams

$$UA = \frac{1}{R_1 + R_{f1} + R_w + R_{f2} + R_2} \tag{8–34}$$

$$= \frac{1}{\Sigma R_t}$$

The appropriate expressions for the evaluations of the thermal resistances between two fluid streams separated by a plane wall or a duct are given in Table 8–7. The values of the fouling factor, $\mathfrak{F}$, are given in Table 8–8. The thermal resistance associated with the buildup of deposits can be obtained by dividing the appropriate fouling factor by the surface area on which the deposit is made

$$R_f = \frac{\mathfrak{F}}{A} \tag{8–35}$$

In Table 8–9, approximate values of the overall heat transfer coefficient for some typical heat exchanger are listed.

The use of the overall heat transfer coefficient is not restricted to heat exchangers. It is commonly used in the calculation of the rate of heat transfer between any two fluid streams. A typical example of such an application would be the calculation of the rate of heat transfer through the walls of a residential dwelling.

Table 8-7 Thermal Resistances in a Heat Exchanger

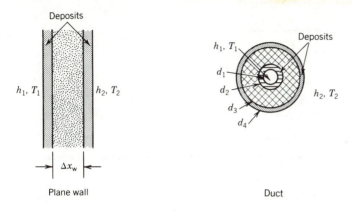

Plane wall   Duct

Description	Symbol	Plane wall	Tube
Fluid–surface	R_1	$\dfrac{1}{\bar{h}_1 A}$	$\dfrac{1}{\bar{h}_1(\pi d_1 L)}$
Fouling deposits (surface 1)	R_{f1}	$\dfrac{\mathfrak{F}_1}{A}$	$\dfrac{\mathfrak{F}_1}{\pi d_2 L}$ [a]
Wall	R_w	$\dfrac{\Delta \Pi_w}{k_w A}$	$\dfrac{\ln(d_3/d_2)}{2\pi k_w L}$
Fouling deposits (surface 2) TABLE 8-8	R_{f2}	$\dfrac{\mathfrak{F}_2}{A}$	$\dfrac{\mathfrak{F}_2}{(\pi d_3 L)}$ [a]
Fluid–surface	R_2	$\dfrac{1}{\bar{h}_2 A}$	$\dfrac{1}{[\bar{h}_2(\pi d_4 L)]}$

[a]The thickness of the scale is usually small, thus $d_2 \approx d_1$ and $d_4 \approx d_3$.

Table 8-8 Fouling Factors, $\mathfrak{F}$

	$m^2 \cdot K/W$
Seawater below 50° C	0.0001
Seawater above 50° C	0.0002
Riverwater below 50° C	0.0002–0.0001
Fuel oil	0.0009
Refrigerating liquids	0.0002
Feedwater below 50° C	0.0001
Feedwater above 50° C	0.0002
Steam (non-oil bearing)	0.00009
Industrial air	0.0004

Table 8–9 Approximate Values of Overall Heat Transfer Coefficients for Heat Exchangers, U

Application	W/m²·K
Water to water	850–1700
Gas to gas	10–40
Feedwater heater	1100–8500
Steam condenser	1100–5600
Water to oil	110–350
Steam to heavy fuel oil	56–170

EXAMPLE 8–6

The overall heat transfer coefficient, based upon the inside surface area of a tube, for a condenser in an electrical generating plant is to be determined. The condenser is of the shell and tube type with the commercial bronze tubes having an inside diameter of 3 cm and a wall thickness of 2 mm. The cooling fluid is seawater (under 50° C). The average convection heat transfer coefficient on the cooling water side is 10,000 W/m²·K while that on the outside of the tube is 50,000 W/m²·K. Assume fouling to occur on both sides of the tube but the thickness of the deposits are small.

SOLUTION

The expression for the overall heat transfer coefficient using eq. 8–34 and Table 8–7 is

$$U_1 A_1 = \cfrac{1}{\cfrac{1}{\bar{h} A_1} + \cfrac{\mathfrak{F}_1}{A_1} + \cfrac{\ln(d_2/d_1)}{2\pi k_w L} + \cfrac{\mathfrak{F}_2}{A_2} + \cfrac{1}{\bar{h}_2 A_2}}$$

Dividing through by the inside surface area of the tube, A_1, we obtain

$$U_1 = \cfrac{1}{\cfrac{1}{\bar{h}_1} + \mathfrak{F}_1 + \cfrac{d_1 \ln(d_2/d_1)}{2k_w} + \cfrac{d_1}{d_2}\mathfrak{F}_2 + \cfrac{d_1}{d_2 \bar{h}_2}}$$

The fouling factor for the cooling water, $\mathfrak{F}_1$, is 0.0001 m²·K/W while that for the steam side, $\mathfrak{F}_2$, is 0.00009 m²·K/W. These were obtained from Table 8–8. The thermal conductivity of the bronze is obtained from Table A–15, $k = 52$ W/m·K.

$$U_1 =$$

$$= \cfrac{1}{\cfrac{1}{10,000} + 0.0001 + \cfrac{(0.03)\ln(0.034/0.03)}{2(52)} + \cfrac{0.03(0.00009)}{0.034} + \cfrac{0.03}{(0.034)(50 \times 10^3)}}$$

$$= \cfrac{1}{100 \times 10^{-6} + 100 \times 10^{-6} + 36.1 \times 10^{-6} + 79.4 \times 10^{-6} + 17.65 \times 10^{-6}}$$

$$= \cfrac{1}{333.2 \times 10^{-6}}$$

$$= 3001 \, W/m^2 \cdot K$$

COMMENT

When the condenser tubes are free of deposits, the conditions at initial installation or after cleaning, the overall heat transfer coefficient is 6502 $W/m^2 \cdot K$. The presence of deposits will decrease the rate of heat transfer per unit surface area to 46.2% of that obtained with clean surfaces. More heat transfer surface area will be required to meet a specific design condition when deposits are formed, thereby increasing the cost of the heat exchanger. Great care is therefore taken to maintain the purity of the steam being condensed and to use as clean as practical cooling water.

8.7.3 Design and the Prediction of the Performance of Heat Exchangers

When designing a heat exchanger we are concerned with the determination of the required heat transfer surface area and other geometrical parameters associated with the heat exchanger. The number of tubes, tube diameter, and tube length, and the type of configuration are items required to construct the heat exchanger. For a given set of specifications, there are many different arrangements that can be used. It would be impossible to cover in this text, all the factors that must be considered in the design of a heat exchanger. The reader who is interested in obtaining a more detailed discussion on this subject is referred to the *Heat Exchanger Design Handbook*[5] and *Compact Heat Exchangers.*[6]

The prediction of the performance of a given heat exchanger is more straightforward because the geometrical configuration of the exchanger is known. The same basic relationships are used for both the design and the performance prediction of heat exchangers.

8.7.3.A   First Law Analysis of a Heat Exchanger

A first law analysis of a heat exchanger will be discussed using a control volume whose boundary is the outer surface of the heat exchanger and imaginary planes at the entrance and exit of the fluid streams. If the exchanger is perfectly insulated from the surroundings and changes in the kinetic and potential energies of the fluid streams, as they pass through the heat exchanger, are negligible, we obtain

$$\dot{m}_h h_{hi} + \dot{m}_c h_{ci} = \dot{m}_h h_{hd} + \dot{m}_c h_{cd}$$

where the subscripts *h* and *c* are used to denote the hot and cold fluid streams, respectively. The terms can be rearranged to obtain

$$\dot{m}_c(h_{cd} - h_{ci}) = -\dot{m}_h(h_{hd} - h_{hi}) \tag{8–36}$$

The enthalpy is replaced by the product of the specific heat of the fluid and the bulk fluid temperature if no change in phase takes place. The product of the mass flow rate and the specific heat is referred to as the *heat capacity* of the fluid stream

$$C \equiv \dot{m}c_p \quad \text{W/K} \tag{8–37}$$

Equation 8–36 can thus be written as

$$C_c(T_{cd} - T_{ci}) = -C_h(T_{hd} - T_{hi}) \tag{8–38}$$

The rate that heat is gained by the cold fluid is

$$\dot{Q}_c = C_c(T_{cd} - T_{ci}) \tag{8–39}$$

and the rate that heat is lost by the hot fluid is

$$\dot{Q}_h = C_h(T_{hd} - T_{hi}) \tag{8–40}$$

The statement of the conservation of energy for a heat exchanger (first law of thermodynamics) can be expressed as

$$\dot{Q}_c + \dot{Q}_h = 0 \tag{8–41}$$

8.7.3.B   NTU–Effectiveness Method

The rate at which heat is transferred between the hot and cold fluids as they pass through the heat exchanger is given by eqs. 8–39 through 8–41. The heat transfer surface area required to transfer the heat can be calculated using either the *log mean temperature difference*, (LMTD), or *number of transfer units* (NTU)–effectiveness method. Each method has its own advantages and disadvantages which are adequately described in Refs. 5 and 6. The NTU–effectiveness method will be described in this section.

The NTU is defined as

$$NTU \equiv \frac{UA}{C_{min}} \tag{8-42}$$

where C_{min} is the minimum heat capacity of the two fluid streams, $\dot{m}c_p$. The *effectiveness*, ε, is defined as the ratio of the actual rate of heat transfer by the fluid streams divided by the maximum possible rate of heat transfer

$$\varepsilon \equiv \frac{\dot{Q}_c}{\dot{Q}_{max}} = \frac{-\dot{Q}_h}{\dot{Q}_{max}} \tag{8-43}$$

The maximum possible rate of heat transfer is equal to the minimum fluid heat capacity times the maximum temperature difference across the heat exchanger. The maximum temperature difference is the difference between the hot and cold fluids entering the heat exchanger, thus

$$\dot{Q}_{max} = C_{min}(T_{hi} - T_{ci}) \tag{8-44}$$

The rate of heat transfer from the hot fluid or to the cold fluid as it passes through the heat exchanger can be expressed in terms of the effectiveness

$$\dot{Q}_c = \varepsilon C_{min}(T_{hi} - T_{ci}) \tag{8-45a}$$

or

$$\dot{Q}_h = -\varepsilon C_{min}(T_{hi} - T_{ci}) \tag{8-45b}$$

It should be noted that $\dot{Q}_c = -\dot{Q}_h = \dot{Q}$ if the heat exchanger is completely insulated.

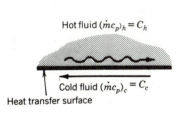

$$\varepsilon = \frac{1 - \exp[-NTU(1 - C_{min}/C_{max})]}{1 - (C_{min}/C_{max})\exp[-NTU(1 - C_{min}/C_{max})]}$$

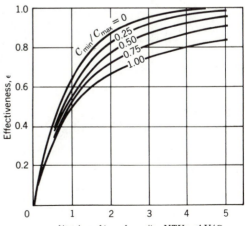

Figure 8-20 Counterflow heat exchanger.

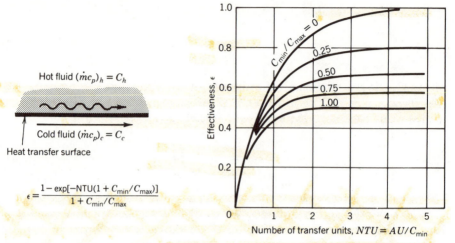

$$\epsilon = \frac{1 - \exp[-NTU(1 + C_{min}/C_{max})]}{1 + C_{min}/C_{max}}$$

Figure 8-21   Parallel flow heat exchanger.

The effectiveness of a heat exchanger is a function of the NTU, the ratio of the heat capacity of the two streams (C_{min}/C_{max}), and the geometrical configuration of the heat exchanger

$$\epsilon = f(NTU, C_{min}/C_{max}, \text{ geometry})$$

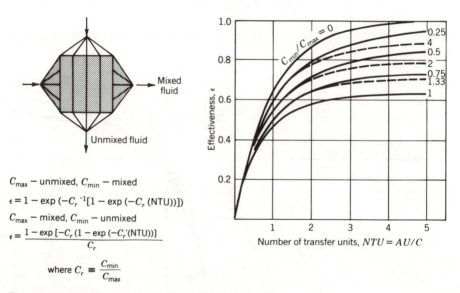

C_{max} – unmixed, C_{min} – mixed

$\epsilon = 1 - \exp(-C_r^{-1}[1 - \exp(-C_r(NTU))])$

C_{max} – mixed, C_{min} – unmixed

$\epsilon = \dfrac{1 - \exp[-C_r(1 - \exp(-C_r'(NTU))]}{C_r}$

where $C_r \equiv \dfrac{C_{min}}{C_{max}}$

Figure 8-22   Cross flow heat exchanger—one fluid mixed.

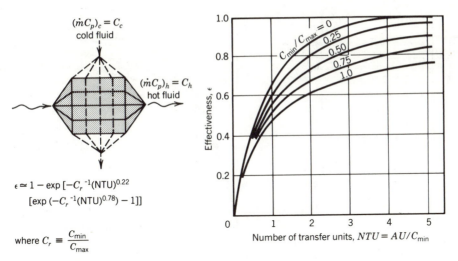

$$(\dot{m}C_p)_c = C_c$$
cold fluid

$$(\dot{m}C_p)_h = C_h$$
hot fluid

$$\epsilon \approx 1 - \exp\left[-C_r^{-1}(NTU)^{0.22}\right]$$
$$\left[\exp\left(-C_r^{-1}(NTU)^{0.78}\right) - 1\right]\right]$$

where $C_r \equiv \dfrac{C_{min}}{C_{max}}$

Effectiveness, ϵ

$C_{min}/C_{max} = 0$, 0.25, 0.50, 0.75, 1.0

Number of transfer units, $NTU = AU/C_{min}$

Figure 8-23 Cross flow heat exchanger—both fluids unmixed.

This functional relationship is given in the form of curves for several different heat exchanger configurations in Figs. 8–20 through 8–25. It is usually not practical to design or use a heat exchanger which has an effectiveness less than 0.70. It should be noted that when a change of phase takes place in one of the fluid streams, the maximum heat capacity is infinite since the specific heat of that fluid may be considered to be infinite. The ratio of C_{min}/C_{max} is equal to 0. Under these conditions the effectiveness of the heat exchanger is independent of its geometrical configuration and is that equal to the effectiveness of a counterflow heat exchanger.

$$\epsilon = (1 - e^{NTU}) \ *$$

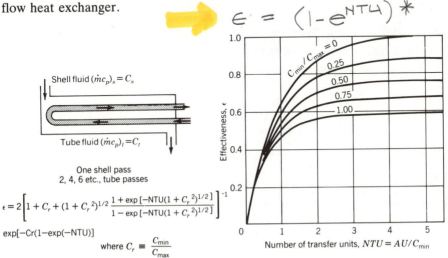

Shell fluid $(\dot{m}c_p)_s = C_s$

Tube fluid $(\dot{m}c_p)_t = C_t$

One shell pass
2, 4, 6 etc., tube passes

$$\epsilon = 2\left[1 + C_r + (1 + C_r{}^2)^{1/2}\,\frac{1 + \exp\left[-NTU(1 + C_r{}^2)^{1/2}\right]}{1 - \exp\left[-NTU(1 + C_r{}^2)^{1/2}\right]}\right]^{-1}$$

$$\exp[-Cr(1-\exp(-NTU))]$$

where $C_r \equiv \dfrac{C_{min}}{C_{max}}$

Effectiveness, ϵ

$C_{min}/C_{max} = 0$, 0.25, 0.50, 0.75, 1.00

Number of transfer units, $NTU = AU/C_{min}$

Figure 8-24 Parallel–counterflow heat exchanger.

* FOR PHASE CHANGES IN ONE FLUID

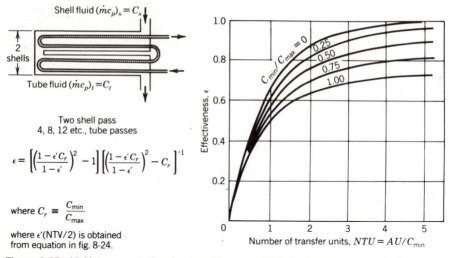

$$\epsilon = \left[\left(\frac{1 - \epsilon' C_r}{1 - \epsilon'} \right)^2 - 1 \right] \left[\left(\frac{1 - \epsilon' C_r}{1 - \epsilon'} \right)^2 - C_r \right]^{-1}$$

where $C_r \equiv \dfrac{C_{min}}{C_{max}}$

where $\epsilon'(NTV/2)$ is obtained
from equation in fig. 8-24.

Figure 8-25 Multipass counterflow heat exchanger—two-shell pass.

EXAMPLE 8–7

A heat exchanger is to be designed to cool 2 kg/s of oil from 120 to 40° C. After careful consideration a one-shell pass, six-tube pass unit was selected. Each tube pass is composed of 25 thin tubes with a diameter of 2 cm connected in parallel. The oil is to be cooled using water which enters the heat exchanger at 15° C and is discharged at 45° C. A schematic sketch of the unit is shown in Fig. E8–7. The overall heat transfer coefficient is 300 W/m²·K. Determine the:

(a) Mass flow rate of the water.

(b) Total heat transfer surface area.

(c) Length of the tubes.

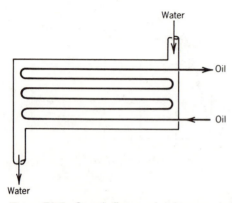

Figure E8-7 One-shell pass, six-tube-pass shell, and tube heat exchanger.

SOLUTION

The physical properties of the oil and water are obtained from Tables A–10 and A–9.

Oil   80° C (353.2 K)   Cooling water   30° C (303.2 K)

c_{po} = 2.132 kJ/kg·K c_{pw} = 4.178 kJ/kg·K

An energy balance of the two fluid streams is used to find the mass flow rate of the water

$$-\dot{m}_o c_{po}(T_{od} - T_{oi}) = \dot{m}_w c_{pw}(T_{wd} - T_{wi})$$

$$-2(2.132 \times 10^3)(40 - 120) = \dot{m}_w(4.178 \times 10^3)(45 - 15)$$

$$\dot{m}_w = 2.722 \text{ kg/s}$$

The heat capacities of the two fluid streams are

$$C_o = \dot{m}_o c_{po} = 2(2.132 \times 10^3) = 4.264 \times 10^3 \text{ W/K}$$

$$C_w = \dot{m}_w c_{pw} = 2.722(4.178 \times 10^3) = 11.37 \times 10^3 \text{ W/K}$$

and

$$\frac{C_{min}}{C_{max}} = \frac{C_o}{C_w} = \frac{4.264 \times 10^3}{11.37 \times 10^3} = 0.375$$

The total heat transferred between the two streams can be calculated using an energy balance on either the oil or water. If the oil stream is used we obtain

$$\dot{Q} = -C_o(T_{od} - T_{oi})$$

$$= -4.264 \times 10^3(40 - 120)$$

$$= 341.1 \times 10^3 \text{ W}$$

The effectiveness of the heat exchanger is

$$\varepsilon = \frac{\dot{Q}}{C_{min}(T_{hi} - T_{ci})} = \frac{341.1 \times 10^3}{(4.264 \times 10^3)(120 - 15)} = 0.7619$$

The NTU is found using Fig. 8–24 to be 2.3. The total heat transfer surface area of the heat exchanger is

$$A = \frac{NTU \, C_{min}}{U} = \frac{2.3(4.264 \times 10^3)}{300} = 32.69 \text{ m}^2$$

The heat transfer surface area per pass is

$$A_p = \frac{A}{N_p} = \frac{32.69}{6} = 5.448 \text{ m}^2$$

The heat transfer surface area per tube is

$$A_t = \frac{A_p}{N_t} = \frac{5.448}{25} = 0.2179 \text{ m}^2$$

The length of the tube is

$$L = \frac{A_t}{\pi d} = \frac{0.2179}{\pi(0.02)} = 3.468 \text{ m}$$

COMMENT

Some difficulties may be encountered in obtaining an accurate value of NTU using Fig. 8–24. The analytic expressions used to generate the NTU–ε curves, Figs. 8–20 through 8–25, are given to aid you. Since the tubes are thin $d_1 \simeq d_2$.

EXAMPLE 8–8

A cross flow compact heat exchanger has a surface area of 205 m². It is used to preheat the air entering the combustor of a gas turbine using the combustion gases leaving the gas turbine. The air enters at 15° C while the hot gases enter the unit at 200° C. Both fluids, having a mass flow rate of 2 kg/s, remain unmixed as they pass through the heat exchanger. The overall heat transfer coefficient is 35 W/m²·K. It is desired to estimate the temperature of the gases leaving the heat exchanger.

SOLUTION

The specific heat for the gases will be evaluated at the mean gas temperature of $(200 + 15)/2 = 107.5°$ C and is found using Table A–8 to be $c_p = 1.012$ kJ/kg·K. The heat capacities of the two streams are equal.

$$C_h = C_c = \dot{m}c_p = 2(1.012 \times 10^3) = 2.024 \times 10^3 \text{ W/K}$$

The NTU is

$$NTU = \frac{UA}{C_{min}} = \frac{35(205)}{2.024 \times 10^3} = 3.545$$

Figure 8–23 is used to determine the effectiveness with NTU = 3.545 and C_{min}/C_{max} = 1. The effectiveness is 0.71. Since

$$\varepsilon = \frac{\dot{Q}}{C_{min}(T_{hi} - T_{ci})}$$

the heat transfer between the fluids is

$$\dot{Q} = \varepsilon C_{min}(T_{hi} - T_{ci})$$

$$= 0.71(2.024 \times 10^3)(200 - 15)$$

$$= 265.8 \times 10^3 \text{ W}$$

The temperatures of the gases leaving the exchangers are:
   Hot gases:

$$\dot{Q} = -C_h(T_{hd} - T_{hi})$$

$$265.8 \times 10^3 = -2.024 \times 10^3(T_{hd} - 200)$$

$$T_{hd} = 68.68° \text{ C}$$

Cold gases:

$$\dot{Q} = C_c(T_{cd} - T_{ci})$$

$$265.8 \times 10^3 = 2.024 \times 10^3(T_{cd} - 15)$$

$$T_{cd} = 146.3° \text{ C}$$

REFERENCES

1. Rouse, H., and Ince, S., *History of Hydraulics,* Dover, New York, 1963 (originally published by J. Wiley and Sons, Inc., 1946).

2. Moody, L. F., Friction Factors for Pipe Flow, *ASME Transactions* **66**, 671–684 (1944).

3. Shah, R. K., and London, A. L., *Laminar Flow Forced Convection in Ducts,* Academic Press, New York, 1978.

4. Gnielinski, V., Forced Convection in Ducts, in *Heat Exchanger Design Handbook,* Ed. E. V. Schlunder, Part 2, Sec. 2.5.1, Hemisphere Publishing Corporation, Washington, 1982.

5. Schlunder, E. U., et al., *Heat Exchanger Design Handbook,* Hemisphere Publishing Corporation, Washington, 1982.

6. Kays, W. M., and London, A. L., *Compact Heat Exchangers,* 2nd ed., McGraw-Hill, New York, 1964.

PROBLEMS

8–1   What is the Reynolds number of a flow of oil at a flow rate of 0.6 m³/s in a 15-cm pipe if the dynamic viscosity of the oil is $\mu = 0.999$ N·s/m² and its specific gravity is 0.89? Is the flow laminar or turbulent? ($T = 290$ K.)

8–2   The Reynolds number of an incompressible fluid flow in a 20-cm diameter pipe is 1900. What is the Reynolds number in the 12-cm diameter pipe that is connected to the larger diameter pipe by a reducing fitting? What flow regime, laminar or turbulent, exists in the two pipes?

8–3   The velocity distribution in a laminar flow between two parallel flat plates can be expressed as

$$u = ay(s - y)$$

where a is a constant, s the distance between the plates, and y the distance measured normal to the lower plate. Determine the ratio of the average velocity to the maximum velocity between the two plates.

8–4   Water flows from reservoir A to reservoir B through 280 m of straight tubing. Both reservoirs are open to the atmosphere and the water temperature is 293 K. A flow rate of 0.009 m³/s is required through the 75-mm diameter drawn tubing. Neglecting minor losses at the exit and entrance to the reservoirs, calculate the required difference in reservoir water level to maintain the flow rate.

8–5   A pump delivers 0.01 m³ of water per second through a 10-cm pipeline of new commercial pipe, Fig. P8–5, with regular screwed elbows. If the pump discharge

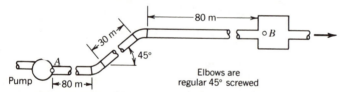

Figure P8-5   Pipeline.

(point A) pressure is 690 kPa absolute, what is the pressure at point B? The water temperature is 21.85° C.

8–6   Water at 285 K flows through a galvanized iron pipe at a rate of 0.3 m³/s. The inside diameter of the pipe is 190 mm. Determine the friction factor associated with this flow and the pressure drop per unit length of pipe.

8–7   A pump is required to move oil at 310 K from a tanker unloading terminal at sea level to a refinery storage tank located 150 m above sea level. The total length of pipe between the terminal and the storage tank is 200 m. The inside diameter of the pipe is 20 cm, it is made of wrought iron, and contains three regular 90° flanged elbows. The operating flow rate is 0.356 m³/s. Neglecting the head loss associated with the inlet and exit of the pipe, determine:
(a)   The shaft input power to the pump if its efficiency is 0.85.
(b)   If the inlet and exit to the pipe are "sudden," estimate the head loss associated with each.
(c)   Repeat part (a) including the pipe inlet and exit losses determined in part (b).

8-8   A large cistern is filled with water to a depth of 25.0 m. A well-rounded nozzle is located in the side of the cistern at a distance of 5 m above the bottom of the tank. The depth of the water is maintained constant by continuously adding water at a temperature of 285 K. Determine the velocity of the water leaving the cistern:
(a)   If the nozzle has an inside throat diameter of 1.3 cm.
(b)   When a 10-m horizontal length of *smooth* pipe is added to the nozzle of part (a).

8-9   An experimental test facility, Fig. P8–9, has been designed to measure the local heat transfer coefficient. The local heat flux is measured using a heat flux meter.

Figure P8-9   Experimental facility.

A thermocouple is used to measure the surface temperature of the duct. A thermocouple probe and a Pitot tube are used to measure the temperature and velocity distribution in the fluid in order to calculate the bulk fluid temperature. The experimental results yielded

$$\dot{q}'' = 12{,}980 \text{ kW/m}^2, \quad T_w = 52.1° \text{ C} \quad \text{and} \quad T_m = 18.25° \text{ C}$$

Calculate the local heat transfer coefficient.

8-10   Water flows through a heated duct, 3-cm inside diameter, at a mean velocity of 1 m/s. The bulk temperature of the water entering the heated duct is 18° C. Twenty (20) kW of energy are transferred to the water. Calculate the bulk temperature of the water at the point where it leaves the pipe. Changes in the kinetic and potential energies of the fluid streams can be neglected.

8-11   An electrically heated duct with uniform heat flux boundary condition is used to increase the bulk temperature of air from 20 to 80° C. The inside diameter of the duct is 3 cm and the duct is 3 m long. The mass flow rate of the air through the duct is 0.075 kg/s. Calculate the required heat flux, $\dot{q}''_w$.

8-12   Air enters a circular duct 3 cm in diameter with a mean velocity of 20 m/s. The inner surface of the duct is at a uniform temperature of 80° C while the bulk temperature of the air entering the tube is 15° C. Determine the length of the duct required to obtain a fluid exit bulk temperature of 35° C. The average convection heat transfer coefficient is 80 W/m²·K.

8-13   The surface of a circular heating duct is held at a uniform temperature of 80° C. Water passes through the duct with a mass flow rate of 2 kg/s. The duct has an internal diameter of 3 cm and a length of 5 m. The mean bulk temperature of the water entering the duct is 10° C. Estimate the bulk· temperature of the water leaving the duct when the mean heat transfer coefficient is 11,000 W/m²·K.

8-14   A small air-cooled condenser is to be designed. The air passes through a number of small circular ducts which have essentially a uniform wall temperature. The ducts are 5 mm in diameter and 4 cm long. Estimate the average convection heat transfer coefficient for the air if the Reynolds number of the air flow is 1500. The thermophysical properties of the air are evaluated at a temperature of 27° C.

8–15   A compact heat exchanger is used to heat oil. The engine oil flows through small circular ducts 1 mm in diameter and 10 cm long. The average bulk temperature of the oil is 37° C and the mass flow rate of the oil through each duct is 0.025 kg/s. Determine the average heat transfer coefficient assuming the wall of the duct to be at a uniform temperature.

8–16   A section of a compact heat exchanger is composed of triangular cross sectioned ducts as shown in the sketch. The gas flows through the ducts with a mean velocity

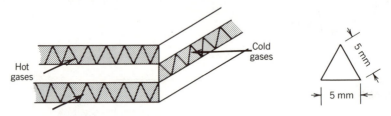

Figure P8-16   Compact heat exchanger.

of 10 m/s. Consider the gas to have thermophysical properties similar to that of air at 77° C. Estimate the average convection film coefficient assuming fully developed flow and uniform duct wall temperature.

8–17   Air is to be used to cool a solid material undergoing a heat generating reaction. Holes 1 cm in diameter are drilled through the material. The thickness of the plate is 8 cm and the thermal boundary condition at the surface of the holes is considered to be a constant heat flux. The air enters the holes with a uniform velocity profile, 1.5 m/s, and a temperature of 20° C. Estimate the rate heat is removed by the air per hole if the maximum temperature of the material must not exceed 200° C.

8–18   Hot air flows through a rectangular shaped duct, 7.5 by 30 cm. The air enters the duct with a bulk temperature of 60° C and a velocity of 60 m/s. The duct is 16 m long and the temperature of the air surrounding it may be so low that the duct walls can be considered to be at a uniform temperature of 4° C. It was decided that if the minimum temperature of the air leaving the duct was lower than 57° C, the duct should be insulated. Do you recommend that the duct be insulated?

8–19   Air at an average temperature of 300 K flows through a rough concrete duct 10 cm in diameter at a mean velocity of 2 m/s. The average roughness of the duct is 2 mm. Estimate the value of the convection heat transfer coefficient. Compare your estimated value with that obtained using the correlations for a smooth duct.

8–20   A condenser tube is 6 m long and 2 cm in diameter. Cooling water enters the tube at a velocity of 2.5 m/s. The tube wall temperature is considered to be uniform. The average temperature of the cooling water is 11.85° C. Determine the average convection heat transfer coefficient.

8–21   Water enters a cross flow heat exchanger at a temperature of 97° C and a flow rate of 3 kg/min. The water is cooled by air entering the unit with a volumetric flow rate of 5.66 m³/min and at 30° C. The water passes unmixed through the heat exchanger while the air stream is mixed. If the water leaves the unit at 73° C, determine the temperature of the air leaving and the total heat transfer surface area required. The overall heat transfer coefficient based upon the water side surface area is 25 W/m²·K.

8–22 For a flat plate heat exchanger show that the relationship between the overall heat transfer coefficient for a clean heat exchanger, U_c, and that for a fouled heat exchanger, U_f, is

$$\frac{1}{U_f} = \Sigma \mathfrak{F} + \frac{1}{U_c}$$

where $\mathfrak{F}$ are the fouling factors.

8–23 Hot water, 1000 kg/hr, is cooled from 95 to 55° C by passing it through a tube in a heat exchanger. Cold water, 2000 kg/hr, enters the exchanger at 30° C. The

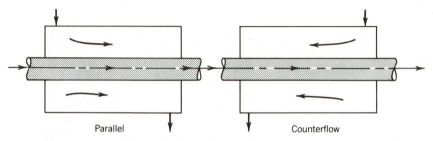

Parallel   Counterflow

Figure P8-23   Heat exchanger configurations.

overall heat transfer coefficient is 1700 W/m²·K. Calculate the heat transfer surface area required for:
(a) Parallel flow operation.
(b) Counterflow operation.

8–24 Hot air enters a cross flow heat exchanger, both fluids unmixed, at a temperature of 100° C and a mass flow rate of 3 kg/min. Cold air enters the unit at a rate of 5.66 m³/min and an entrance temperature of 30° C. The overall heat transfer coefficient, U_o, is 25 W/m²·K and the outside heat transfer surface area is 10 m². Determine the temperature of the fluids leaving the heat exchanger.

8–25 A one-shell pass, two-tube pass heat exchanger is to be designed to condense 4000 kg/hr of saturated steam. The temperature of the steam entering the unit is 20° C. Subcooling and superheating effects are neglected. The flow rate of the cooling water is 6 × 10⁵ kg/hr and the inlet temperature of the water is 15° C. The overall heat transfer coefficient based upon the steam side surface area is 3000 W/m²·K. The mean velocity of the water in the tubes is 2 m/s. The cross section area of each tube is 700 × 10⁻⁶ m² and the steam-side surface area per meter of tube length is 0.10 m²/m. Determine:
(a) Number of tubes per tube pass.
(b) Total surface area required.
(c) Temperature of cooling water out.
(d) The length of the tubes per pass.

9

Conduction Heat Transfer

9.1 INTRODUCTION

An isolated solid body is in thermal equilibrium if its temperature is identical at every spatial location in the body. If the temperature in a solid is not uniform, heat will be transferred from the high-temperature regions to the low-temperature regions. This process, called *heat conduction,* will continue until a uniform temperature field exists throughout the entire body. The basic laws of thermodynamics that govern heat conduction will now be reviewed.

Several different statements of the second law of thermodynamics have already been noted. One of these is directly applicable to the study of heat conduction. It implies that a temperature difference is required for heat to be transferred. If such a temperature difference does exist within a body, the heat will flow from the region of highest temperature to the region of lowest temperature since no work is done.

A description of the physical phenomena associated with the heat conduction process on the microscopic level is dependent upon the molecular structure of the material. In a gas, the kinetic energy of the molecules is related to the temperature of the gas. The gas molecules in a high-temperature region possess a greater kinetic energy than those molecules in a lower temperature region. Since all the molecules are in continuous random motion, collisions will occur between high-temperature and low-temperature molecules. As a result of these collisions some of the kinetic energy of the high-temperature molecules will be transferred to the low-temperature molecules. In an isolated system, this process will continue until a state of thermal equilibrium is reached, at which time any random sample of molecules will possess the same average kinetic energy.

The thermal conduction process becomes much more complicated for liquids and solids. Other microscopic energy transport mechanisms, particularly those associated with lattice vibration and free-electron transport, must be considered. Their contributions to the overall heat transfer processes may, for certain materials, become quite significant. A discussion of these mechanisms as major factors in the heat conduction process has been presented by Jakob[1] and Gebhard.[2]

Although the microscopic approach is helpful in obtaining an understanding of the phenomena involved in a particular process, a macroscopic approach is usually used to perform engineering calculations. This is certainly true in conduction heat transfer.

Nonuniform temperature fields are present in nearly all engineering applications. Several items in which the internal temperature distribution and the flow of heat are important are shown in Fig. 9–1. The nonuniform temperature field may be created by sources of energy involving nuclear, chemical, or electrical resistance heat generation; friction between moving parts; or the flow of energy between a fluid and a solid surface. The temperature distribution in the system is governed in part by conduction. In Chapters 7 and 8 we have observed that the conduction or diffusion of energy within the fluid influences the temperature distribution in the fluid and the magnitude of the convection heat transfer coefficient. This chapter will concentrate on the conduction process in solids.

The basic law that governs the conduction of heat, expressed in terms of macroscopic quantities, was proposed by Fourier in 1811. The temperature distribution in a material is considered to be a function of spatial location and time, $T(x, y, z, t)$. Fourier postulated that the rate of heat transfer per unit surface area is proportional to the temperature gradient normal to the surface. This may be expressed mathematically as

$$\dot{q}'' = \frac{\dot{Q}}{A} \propto \frac{\partial T}{\partial \eta} \tag{9–1}$$

where η is the coordinate perpendicular to the surface through which heat is being transferred as shown in Fig. 9–2. An equality sign is obtained through the

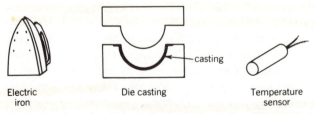

Electric   Die casting   Temperature
iron   sensor

Figure 9-1   Items in which conduction heat transfer is important.

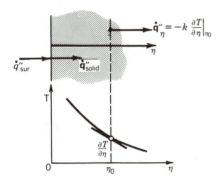

Figure 9-2   Heat flux.

introduction of the *thermal conductivity* of the material, *k. Fourier's law* becomes

$$\dot{q}'' = -k\frac{\partial T}{\partial \eta}$$
(9-2)

In order to understand the significance of this expression several observations will be made. The term on the left side of the equation, $\dot{q}''$, denotes the rate of heat transfer per unit area and is commonly referred to as the *heat flux*. It is a vector quantity. The heat flux is normal to the surface through which the heat is transferred and is positive if directed toward a negative temperature gradient.

The thermal conductivity, *k*, is a thermophysical property of the material through which the heat flows. It is directly related to the microscopic mechanism involved in the transfer of heat within the material. Since the mechanism is well understood only for gases at low temperature, the values of the thermal conductivity for most substances must be obtained experimentally. The internal structure of the material can be greatly influenced by the thermodynamic state of the material. The thermal conductivity of most materials, therefore, exhibits a temperature dependence. The influence of pressure on the thermal conductivity of solids and liquids is only significant when operating at extremely high pressures. The thermal conductivity of gases exhibits a pressure dependence, however, at both high pressures and very low pressures, where the mean-free molecular path lengths are very large. Values of the thermal conductivity for solids are listed in Tables A–15 and A–16. The thermal conductivity for air, water, and oil is listed in Tables A–8 through A–10.

9.2   HEAT CONDUCTION EQUATION AND BOUNDARY CONDITIONS

The development of the mathematical model that governs heat conduction in a solid is based upon the first law of thermodynamics. Consider the solid block shown in Fig. 9–3 with no internal heat generation. A thermodynamic system is defined which contains the complete block where the system boundaries are the exposed surfaces of the block. A statement of the first law of thermodynamics

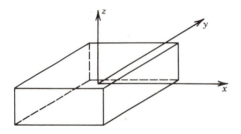

Figure 9-3   Heat conduction in a solid.

for the system reduces to

$$\dot{Q} = \frac{dU}{dt} \tag{9-3}$$

This expression indicates that the rate of change of the internal energy of the system, dU/dt, is a function of the rate of heat transfer across the boundaries of the system.

There are many factors that influence the rate at which heat is transferred across the boundaries of the system. These include the manner in which the heat passes from the surroundings to the system boundaries and the rate at which the heat passes from the boundaries to the interior of the solid. The first law requires that at the system boundaries the rate of heat transfer from the surroundings be equal to the rate of heat transfer into the solid. This is shown schematically in Fig. 9–2. Since the rate of heat transfer into the solid can be expressed by Fourier's law, eq. 9–2, we obtain

$$\dot{q}''_{sur} = -k\frac{\partial T}{\partial \eta}\bigg|_{\eta=0}$$

In Chapter 1 the different modes of heat transfer between the surroundings and a surface were described. The different types of thermal boundary conditions which may be present at the system boundaries are summarized in Table 9–1. If the surroundings are a gas, heat is transferred to the boundaries by both convection and radiation. Usually one of these modes dominates so the heat transfer associated with the other mode can be neglected. However, both convection and radiation must be considered when a natural convection boundary condition is present. If the surroundings are a solid or a liquid, Fourier's law is needed to equate the rate of heat transfer in the system and surroundings at the system boundary. It is customary to use the convection heat transfer coefficient for evaluation of the heat transfer from a fluid to the boundaries of a solid.

In many applications the surroundings behave as an infinite heat sink/source where the boundaries of the system will always be at the temperature of the heat sink/source. Another type of boundary conditions is that encountered when the surroundings supply a uniform heat flux to the system boundaries. A special case of this type of boundary condition exists when the heat flux is zero representing an adiabatic or perfectly insulated surface.

It should be noted that $\dot{q}''_{rad}$, rate of heat transfer by radiation, introduces a nonlinearity into the boundary conditions since the radiation heat transfer is

Table 9–1   Types of Boundary Conditions

Classification		
I. Heat source/sink	$T = T_s \big|_{\eta=0}$	T_s
II. Constant heat flux	$-k\dfrac{\partial T}{\partial \eta} = \dot{q}''_0 \big|_{\eta=0}$	$\dot{q}''_0 \rightarrow \eta$
III. Adiabatic	$-k\dfrac{\partial T}{\partial \eta} = 0$	$\dfrac{\partial T}{\partial \eta}=0$
IV. Two solids	$-k_1\dfrac{\partial T}{\partial \eta} = -k_2\dfrac{\partial T}{\partial \eta}$	① ②
V. Convection	$-k\dfrac{\partial T}{\partial \eta} = h(T_\infty - T)\big|_{\eta=\infty}$	T_∞, h
VI. Radiation	$-k\dfrac{\partial T}{\partial \eta} = \dot{q}''_{rad}\big|_{\eta=0}$	$\dot{q}''_{rad}$
VII. Combined radiation–convection	$-k\dfrac{\partial T}{\partial \eta} = h(T_\infty - T) + \dot{q}''_{rad}$	T_∞, h $\dot{q}''_{rad}$

proportional to the absolute temperature raised to the fourth power. A nonlinear convection boundary condition can also be present if the convection heat transfer coefficient is a function of the difference between the temperature of the surroundings and the system boundary. This latter condition is present in natural convection, condensation, and boiling. The presence of the nonlinearity at the boundary greatly complicates the mathematical model for the conduction process. Consequently nonlinear boundary conditions will be excluded from future discussions.

From Table 9–1 it can be seen that the rate of heat transfer across the boundary of the system is proportional to the temperature gradient in the system normal to the system boundary. Intuition indicates that the temperature distribution in the solid and thus the rate of heat transfer at the boundary is dependent on the boundary conditions. An interrelationship must then be established which relates the internal transfer of heat within the system and the boundary conditions. This is obtained by using a modified form of the first law of thermodynamics for the solid which is called the *heat conduction equation*.

Consider the infinitesimal system or element of the solid shown in Fig. 9–4. The material in the element is isotropic and homogeneous. A conversion of some form of energy, electrical, chemical, or nuclear, into thermal energy may take place within the system. This conversion will be referred to as *internal heat generation, $\dot{Q}^*$*. The rate of heat transfer to the element is the sum of the rate of heat transfer across the boundaries of the element and the rate at which thermal energy is generated internally. The first law statement, in the absence of work being done by the element, becomes

$$\dot{Q} + \dot{Q}^* = \frac{\partial U}{\partial t} \tag{9–4}$$

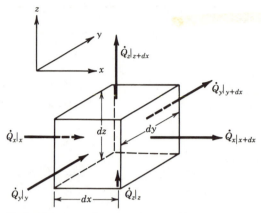

Figure 9-4 Incremental element of solid.

Fourier's law is used to evaluate the net rate of heat transfer by conduction across the six plane surfaces of the element. The heat generated within the element is expressed in terms of $\dot{q}^*$, the specific internal heat generation per unit volume. The rate of change of the internal energy of the material in the element is equal to the product of the mass of the material, its specific heat, and the rate of increase of the element's temperature.

The differential form of the generalized heat conduction equation, when the thermal conductivity of the material is assumed to be constant, is obtained by taking the limit as the volume of the element approaches zero, limit dx dy dz → 0. The final expression is

$$k\left[\frac{\partial^2 T}{\partial x^2} + \frac{\partial^2 T}{\partial y^2} + \frac{\partial^2 T}{\partial z^2}\right] + \dot{q}^* = \rho c \frac{\partial T}{\partial t} \qquad (9\text{–}5)$$

or

$$k \nabla^2 T + \dot{q}^* = \rho c \frac{\partial T}{\partial t} \qquad (9\text{–}6)$$

The expression is referred to as the *heat conduction equation* in the *cartesian coordinate* system. The subscript for the specific heat has been deleted since the heat conduction equation is only valid for a solid for which $c_p \simeq c_v$. A more detailed discussion of the derivation of this equation may be obtained in Ref. 1 to 3 and 5 to 11.

The temperature distribution within the solid and the rate of heat transfer across the boundaries of the solid can be determined by integrating the heat conduction equation. The appropriate boundary and initial conditions are then used to evaluate the integration constants.

The generalized *heat conduction equation* in the *cylindrical coordinate* system may be developed in a similar manner. Expressed in differential form, the heat conduction equation is

$$\maltese \maltese \quad k\left[\frac{\partial^2 T}{\partial r^2} + \frac{1}{r}\frac{\partial T}{\partial r} + \frac{1}{r^2}\frac{\partial^2 T}{\partial \theta^2} + \frac{\partial^2 T}{\partial z^2}\right] + \dot{q}^* = \rho c\frac{\partial T}{\partial t} \qquad (9\text{--}7)$$

9.3 STEADY-STATE HEAT CONDUCTION

When the boundary conditions are independent of time the temperature distribution within the solid is only a function of the spatial coordinates. The rate of change of internal energy is equal to zero and the heat conduction equations reduce to the following forms:

Cartesian coordinates:

$$\frac{\partial^2 T}{\partial x^2} + \frac{\partial^2 T}{\partial y^2} + \frac{\partial^2 T}{\partial z^2} + \frac{\dot{q}^*}{k} = 0 \qquad (9\text{--}8)$$

Cylindrical coordinates:

$$\frac{\partial^2 T}{\partial r^2} + \frac{1}{r}\frac{\partial T}{\partial r} + \frac{1}{r^2}\frac{\partial^2 T}{\partial \theta^2} + \frac{\partial^2 T}{\partial z^2} + \frac{\dot{q}^*}{k} = 0 \qquad (9\text{--}9)$$

9.3.1 One-Dimensional Steady-State Conduction

Heat transfer across an infinite slab like that shown in Fig. 9–5 is an example of one-dimensional heat conduction. In the absence of internal heat generation the steady-state heat conduction equation reduces to

$$\frac{d^2 T}{dx^2} = 0 \qquad (9\text{--}10)$$

Integration of the differential equation yields the following expression for the temperature distribution in the infinite slab

$$\frac{dt}{dx} = \mathfrak{A} \qquad (9\text{--}11)$$

$$T = \mathfrak{A}x + \mathfrak{B} \qquad (9\text{--}12)$$

The values of the integration constants, $\mathfrak{A}$ and $\mathfrak{B}$, are determined from the appropriate boundary conditions. For the case shown in Fig. 9–5 in which the two sides of the infinite slab are at uniform temperatures, the boundary conditions are

$$x = 0 \qquad T = T_2$$

$$x = L \qquad T = T_1$$

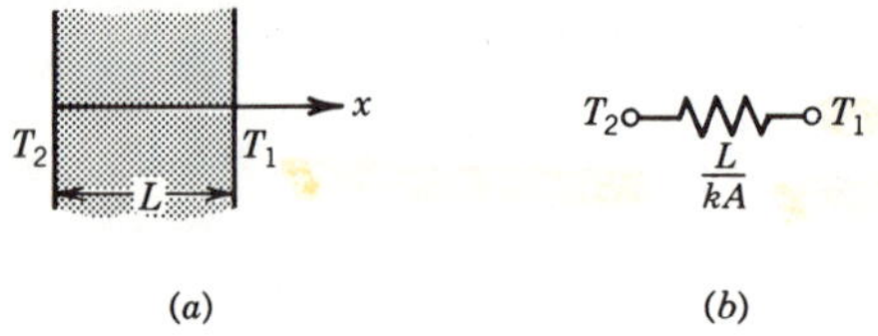

Figure 9-5 One-dimensional conduction heat transfer. (a) One-dimensional slab. (b) Equivalent thermal network.

The integration constants become

$$\mathfrak{A} = \frac{T_1 - T_2}{L}$$

and

$$\mathfrak{B} = T_2$$

When these expressions are substituted into eq. 9–12, the following expression is obtained for the temperature distribution in the slab

$$T = (T_1 - T_2)\frac{x}{L} + T_2 \tag{9–13}$$

The heat passing through the slab at $x = 0$ is obtained by Fourier's law

$$\dot{Q}|_{x=0} = \dot{q}''A|_{x=0} = -kA\frac{dT}{dx}\bigg|_{x=0} = kA\frac{(T_2 - T_1)}{L} \tag{9–14}$$

Since the temperature variation across the slab is linear the heat flux is independent of x.

The similarity between the steady-state transfer of heat and the flow of current is apparent if eq. 9–14 and Ohm's law are compared.

Equation 9–14	Ohm's Law
$\dot{Q} = \dfrac{kA}{L}\Delta T$	$i = \dfrac{\Delta e}{R}$

The analogy is summarized in Table 9–2. The importance of the analogy is that passive electrical circuit analysis techniques can be used in the solution of one-dimensional steady-state heat transfer problems. The equivalent thermal network is shown Fig. 9–5b.

If a convection boundary condition is present at $x = 0$, see Fig. 9–6a, the

Table 9-2 Analogy Between the Flow of Heat and Current for a One-Dimensional Section

	Electrical	Heat
Flow quantities	i	$\dot{Q}$
Potential driving force	Δe	ΔT
Resistance to flow	R	$R_t = \dfrac{L}{kA} + \dfrac{1}{hA}$

SURFACE
INTERNAL

boundary conditions are

$$x = 0 \qquad h(T_\infty - T) = -k\frac{dT}{dx}$$

$$x = L \qquad T = T_2$$

The integration constants in eqs. 9–11 and 9–12 are obtained through use of the boundary conditions

$$x = 0 \qquad h(T_\infty - \mathscr{B}) = -k\mathscr{A}$$

and

$$x = L \qquad T_2 = \mathscr{A}L + \mathscr{B}$$

These two expressions are solved simultaneously to find the integration constants

$$\mathscr{A} = \frac{T_2 - T_\infty}{L + k/h} \quad \text{and} \quad \mathscr{B} = T_2 + \left[\frac{T_\infty - T_2}{L + k/h}\right]L$$

The expression for the temperature distribution in the slab is

$$T = \left[\frac{T_\infty - T_2}{1 + k/hl}\right]\left(1 - \frac{x}{L}\right) + T_2 \tag{9–15}$$

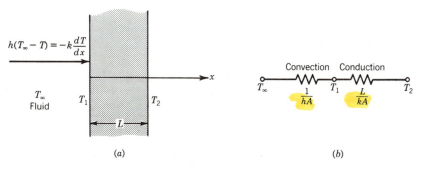

$$h(T_\infty - T) = -k\frac{dT}{dx}$$

T_∞
Fluid

T_1

T_2

L

x

(a)

Convection   Conduction

T_∞ $\dfrac{1}{hA}$ T_1 $\dfrac{L}{kA}$ T_2

(b)

Figure 9-6 Semi-infinite slab with convection at boundaries. (a) Semi-infinite slab. (b) Equivalent network for slab.

and for the rate of heat transfer across the slab

$$\dot{Q} = \frac{T_\infty - T_2}{(L/kA + 1/hA)} \tag{9-16}$$

(INDEPENDENT OF X)

The rate of heat transfer could likewise be evaluated at the convection boundary using

$$\dot{Q} = hA(T_\infty - T_1)$$

$$= \frac{T_\infty - T_1}{(1/hA)}$$

where T_1 is the temperature of the surface.

The rate of heat transfer may also be obtained from the equivalent thermal network analysis. The resistance offered to the transfer of heat by the convection process is $1/hA$. The equivalent thermal network for the slab with the convection boundary condition is shown in Fig. 9–6b. The total resistance offered by the system is the sum of the resistance offered by the solid and the convection boundary

$$\sum R_t = \frac{L}{kA} + \frac{1}{hA} \tag{9-17}$$

The total rate of heat transfer across the slab is

$$\dot{Q} = \frac{T_\infty - T_2}{\sum R_t} \tag{9-18}$$

$$= \frac{T_\infty - T_2}{L/kA + 1/hA}$$

which is identical to eq. 9–16.

The simplicity of the thermal network analysis is apparent and should be used to determine the rate of heat transfer whenever the heat flow in the solid is one dimensional. The use of the thermal resistance concept is strickly speaking only valid for one-dimensional steady-state heat transfer. It can, however, be used to obtain *approximate* solutions for composite sections in which the heat flow is two dimensional.

A typical composite section is shown in Fig. 9–7. An approximation of the rate of heat transfer across the section is obtained using eq. 9–18. The equivalent thermal network is also shown.

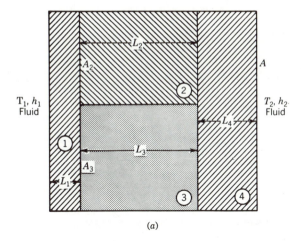

(a)

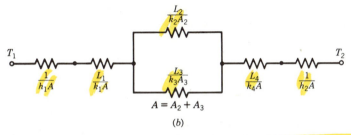

$$A = A_2 + A_3$$

(b)

Figure 9-7 Composite section. (a) Composite section. (b) Equivalent network.

EXAMPLE 9–1

A heating tape is attached to one face of a large 3 cm thick aluminum alloy 2024-T6 plate. The other face of the plate is exposed to the surroundings which are at a temperature of 20° C. The outside of the heating tape is completely insulated. The rate that heat must be supplied in order to maintain the surface of the plate exposed to the air at a temperature of 80° C under steady-state conditions is to be determined. The temperature of the surface to which the heating tape is attached is also required. The heat transfer coefficient between the plate surface and the surrounding air is 5 W/m² · K.

SOLUTION

The thermal conductivity of the aluminum alloy plate, $k = 181.8$ W/m·K at 80° C, is obtained from Table A–15. A schematic sketch of the plate is shown in Fig. E9–1a. The heat transfer is assumed to be one dimensional and the equivalent thermal network for the plate is shown in Fig. E9–1b. For steady-state operation of the heating element, all

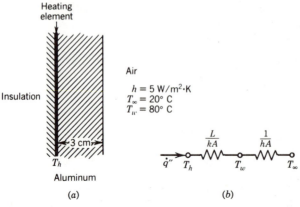

Figure E9-1   Heated wall. (a) Aluminum wall. (b) Equivalent network.

the heat supplied by the element will pass through the wall to the surroundings. The heat flux can be determined from the convection boundary condition at the surface of the plate

$$\dot{q}'' = \frac{\dot{Q}}{A}$$

$$= h(T_w - T_\infty)$$

$$= 5(80 - 20)$$

$$= 300 \text{ W/m}^2$$

The temperature of the surface to which the heating tape is attached can be obtained by using the complete thermal network

$$\dot{q}'' = \frac{T_h - T_\infty}{A\Sigma R_t} = \frac{T_h - T_\infty}{A[L/kA + 1/hA]}$$

Rearrangement gives

$$T_h = T_\infty + \dot{q}'' \left(\frac{L}{k} + \frac{1}{h}\right)$$

$$= 20 + 300 \left(\frac{0.03}{181.8} + \frac{1}{5}\right)$$

$$= 80.05° \text{ C}$$

COMMENT

An alternative method for calculating the temperature of the surface to which the heating tape is attached is to use the temperature drop across the internal resistance

$$\dot{q}'' = \frac{T_h - T_{sur}}{L/k}$$
$$= \frac{T_h - 80}{(0.03/181.8)}$$
$$T_h = 80.05° \text{ C}$$

The resistance to the transfer of heat at the convection boundary is much larger than the internal resistance. Therefore, the temperature within the plate is nearly uniform.

The thermal network analysis can also be used for the calculation of the steady-state one-dimensional rate of heat transfer in a hollow or composite cylinder. The appropriate differential equation for the hollow cylinder shown in Fig. 9–8, for steady-state heat transfer in the absence of heat generation, is

$$\frac{d^2T}{dr^2} + \frac{1}{r}\frac{dT}{dr} = 0 \tag{9-19}$$

The boundary conditions are

$$r = r_i \qquad T = T_i$$
$$r = r_o \qquad T = T_o$$

The temperature distribution in the cylinder is obtained by integrating eq. 9–19 and using the boundary conditions to determine the integration constants

$$T = T_i - \frac{T_i - T_o}{\ln(r_o/r_i)} \ln(r/r_i) \tag{9-20}$$

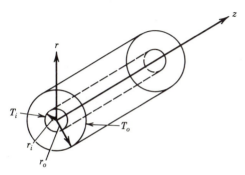

Figure 9-8 Hollow cylinder.

The expression for the total rate of heat transfer is

$$\dot{Q} = \frac{T_i - T_o}{\dfrac{\ln(r_o/r_i)}{2\pi k L}}$$

(9–21)

The equivalent resistance offered by the cylinder to the transfer of heat is

$$R_t = \frac{\ln(r_o/r_i)}{2\pi k L}$$

(9–22)

EXAMPLE 9–2

Steam at a temperature of 150° C flows through a plain carbon steel pipe 6-cm inside diameter and 4 mm thick. The pipe is insulated with a 2-cm thick layer of high-temperature glass wool blanket. The air surrounding the insulated pipe is at 18° C. Determine the rate of heat transfer per meter of pipe length.

SOLUTION

A sketch of the insulated pipe is shown in Fig. E9–2. The important parameters needed to calculate the rate of heat loss are:

$$d_1 = 0.03 \text{ m} \qquad d_2 = 0.068 \text{ m} \qquad d_3 = 0.108 \text{ m}$$

$$k_p = 60.5 \text{ W/m·K} \qquad k_{gw} = 0.076 \text{ W/m·K}$$

$$h_s = 100 \text{ W/m}^2\text{·K} \qquad h_\infty = 3 \text{ W/m}^2\text{·K}$$

$$T_s = 150° \text{ C} \qquad T_\infty = 18° \text{ C}$$

The values of the thermal conductives were obtained from Tables A–15 and A–16. The equivalent thermal network is shown in Fig. E9–2b. The sum of the resistance per unit pipe length is

$$\Sigma R_t = \frac{1}{h_s A_1} + \frac{\ln(d_2/d_1)}{2\pi k_p L} + \frac{\ln(d_3/d_2)}{2\pi k_{gw} L} + \frac{1}{h_\infty A_3}$$

$$= \frac{1}{100(0.06\pi)(1)} + \frac{\ln(0.068/0.06)}{2\pi(60.5)(1)} + \frac{\ln(0.108/0.068)}{2\pi(0.076)(1)}$$

$$+ \frac{1}{3(0.108\pi)(1)}$$

$$= 2.005 \text{ m·K/W}$$

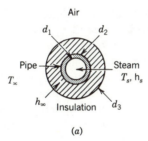

(a)

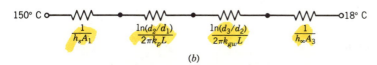

(b)

Figure E9-2   Heat loss from an insulated pipe. (a) Insulated pipe. (b) Equivalent network.

The rate of heat transfer is

$$\dot{Q} = \frac{T_s - T_\infty}{\Sigma R_t} = \frac{150 - 18}{2.005}$$

$$= 65.84 \text{ W/m}$$

COMMENT

The temperature of the outer surface of the insulation, $d = d_3$, can be calculated using the convection boundary condition.

$$\dot{Q} = hA_3(T_3 - T_\infty)$$

or

$$T_3 = \frac{\dot{Q}}{hA_3} + T_\infty$$

$$= 82.7° \text{ C}$$

This temperature is quite high and it may be necessary to decrease it by increasing the thickness of the insulation on the pipe.

9.3.2   Two-Dimensional Steady-State Conduction

In many design situations the assumption of one-dimensional heat conduction will introduce significant errors in the calculations. To obtain a more accurate estimate of the heat transfer and the temperature distribution within such solids, the solution must start with the two- or three-dimensional form of the heat conduction equation and the appropriate boundary conditions.

To further illustrate this point, consider steady-state heat transfer in the corner section shown in Fig. 9–9. A one-dimensional analysis might be reasonable to determine the rate of heat transfer in the immediate vicinity of the insulated surfaces but it would introduce large errors if used near the corner. An accurate temperature distribution in the entire section, which is required to determine the total heat transfer, can only be obtained by solving the following differential equation with the noted boundary conditions.

Differential equation:

$$\frac{\partial^2 T}{\partial x^2} + \frac{\partial^2 T}{\partial y^2} = 0 \qquad (9\text{--}23)$$

Boundary conditions:

$$x = 0 \qquad 0 \leqslant y \leqslant L_y \qquad T = T_i$$

$$x = l_x \qquad l_y \leqslant y < L_y \qquad -k\frac{\partial T}{\partial x} = h(T - T_\infty)$$

$$x = L_x \qquad 0 < y < l_y \qquad \frac{\partial T}{\partial x} = 0$$

$$y = 0 \qquad 0 \leqslant x \leqslant L_x \qquad T = T_i$$

$$y = l_y \qquad l_x \leqslant x < L_x \qquad -k\frac{\partial T}{\partial y} = h(T - T_\infty)$$

$$y = L_y \qquad 0 < x < l_x \qquad \frac{\partial T}{\partial y} = 0$$

The differential equation must be integrated and the integration constants are determined by satisfying the boundary conditions. The complexity of the mathematics involved is easily imagined and no analytical closed-form solution is available for this problem.

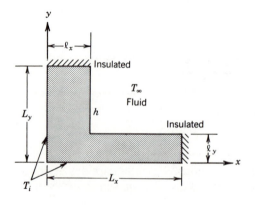

Figure 9-9 Two-dimensional heat transfer.

Exact closed-form solutions for the rate of heat transfer and temperature distribution in two- and three-dimensional solid bodies can be obtained only in a limited number of cases. We are usually thus forced to obtain the desired information by using one of the approximate techniques which have been developed.

9.3.2.A Electrical Analogy

In Section 9.3.1 the analogy between heat transfer and the flow of electrical energy for a one-dimensional configuration was presented. This analogy can be used to obtain accurate results in two-dimensional configurations by replacing the thermal resistances that are associated with the solid with a continuous electrical resistance conducting material. The material is shaped into the same geometrical configuration as the solid in which the temperature distribution is required. The appropriate electrical boundary conditions are applied to the conducting material and a probe, connected to a null voltmeter and a potentiometer, is used to locate lines of constant voltage in the material. The method is based upon the premises that it is easier to measure electrical quantities than thermal quantities and that scaling laws can be used to change the physical size of the equivalent electrical analogy.

There are several limitations on the type of two-dimensional heat conduction problems which can be easily simulated. The problems usually involve a thermal configuration that consists of a single homogeneous material. The easiest boundary conditions to simulate involve uniform temperature and adiabatic, or insulated, surfaces. The uniform temperature analogy is obtained by applying a constant voltage to the surface which is proportional to the temperature. If two or more different temperatures are present, voltages proportional to the different surface temperatures are used. An insulated surface has no electrical connection. The technique is illustrated in Fig. 9–10 where a schematic sketch of a two-dimensional solid with two different isothermal boundaries is shown. Its equivalent electrical analogy is also shown. Lines of constant potential are usually obtained by using a calibrated potentiometer and a null voltage meter.

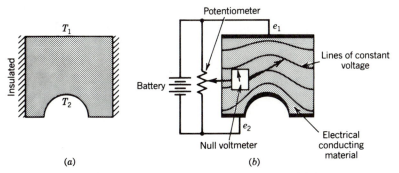

Figure 9-10 Electrical analogy for two-dimensional heat conduction problem. (a) Thermal system. (b) Electrical analog.

9.3.2.B Shape Factor

If the thermal configuration contains only two isothermal surfaces, T_1 and T_2, and a single homogeneous material, the rate of heat transfer can be calculated from the following expression,

$$\dot{Q} = Sk(T_2 - T_1) \tag{9–24}$$

where S is the shape factor and has units of length, m.

The shape factor for the one-dimensional infinite slab shown in Fig 9–5 can be found by comparing eqs. 9–14 and 9–24. It is equal to the area divided by the thickness of the wall, $S = A/L$. The shape factors for a limited number of two-dimensional configurations are tabulated in Table 9–3.

EXAMPLE 9–3

The brick chimney shown in Fig. E9–3 is 5 m high. An estimate of the total rate of heat transfer through the chimney wall is required when the inside surface is at a uniform temperature of a 100° C and the outside surface is held at a uniform temperature of 20° C.

SOLUTION

The thermal conductivity of the brick obtained from Table A–16.2 is 0.72 W/m · K. The solution technique takes into consideration the fact that planes of symmetry exist across which no heat is transferred. The rate of heat transfer through one of the corner sections shown in Fig. E9–3 is found by using the shape factor method. The total rate of heat transfer through the chimney will be equal to four times the heat transfer through the corner section. The shape factors for the three elements of the corner section are obtained from Table 9–3.

Element

1. $S_1 = \dfrac{A}{L} = \dfrac{0.3(5)}{0.1} = 15 \text{ m}$

2. $S_2 = 0.54D = 0.54(5) = 2.7 \text{ m}$

3. $S_3 = \dfrac{A_3}{L} = \dfrac{0.2(5)}{0.1} = 10 \text{ m}$

The sum of the shape factors is

$$\sum S = S_1 + S_2 + S_3 = 27.7 \text{ m}$$

Table 9-3 Shape Factors[11]

System	Schematic	Restrictions	Shape factor
Isothermal sphere buried in a semiinfinite medium		$z > D/2$	$\dfrac{2\pi D}{1 - D/4z}$
Horizontal isothermal cylinder of length L buried in a semiinfinite medium		$L \gg D$ $L \gg D$ $z > 3D/2$	$\dfrac{2\pi L}{\cosh^{-1}(2z/D)}$ $\dfrac{2\pi L}{\ln(4z/D)}$
Vertical cylinder in a semiinfinite medium		$L \gg D$	$\dfrac{2\pi L}{\ln(4L/D)}$

Table 9–3   Shape Factors[11] *(continued)*

System	Schematic	Restrictions	Shape factor
Conduction between two cylinders of length L in infinite medium	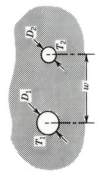	$L >> D_1 D_2$ $L >> w$	$$\dfrac{2\pi L}{\cosh^{-1}\left(\dfrac{4w^2 - D_1^2 - D_2^2}{2D_1 D_2}\right)}$$
Horizontal circular cylinder of length L midway between parallel planes of equal length and infinite width	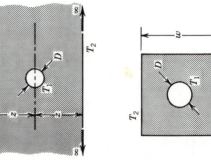	$z > D/2$	$$\dfrac{2\pi L}{\ln(8z/D)}$$
Circular cylinder of length L in a square solid or equal length	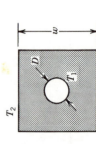	$w > D$	$$\dfrac{2\pi L}{\ln(1.08w/D)}$$

Description		Restrictions	Shape factor
Plane wall	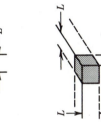	One-dimensional conduction	$\dfrac{A}{L}$
Conduction through the edge of adjoining walls	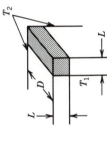	$D > L/5$	$0.54D$
Conduction through corner of three walls with a temperature difference of ΔT_{1-2} across the walls	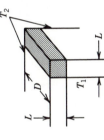	$L \ll$ and width of walls	$0.15L$
Disk of diameter D and T_1 on a semiinfinite medium of thermal conductivity k and T_2	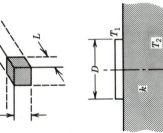	None	$2D$

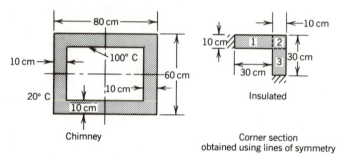

Figure E9-3   Chimney section.

The rate of heat transfer from the corner section shown in Fig. E9–3 is

$$\dot{Q}_c = \sum Sk(T_2 - T_1)$$

$$= 27.7(0.72)(100 - 20)$$

$$= 1.595 \text{ kW}$$

The total rate of heat loss through the chimney is estimated to be

$$\dot{Q} = 4\dot{Q}_c$$

$$= 4(1.595)$$

$$= 6.38 \text{ kW}$$

COMMENT

The importance of the use of the symmetry in the analysis of heat conduction problems can not be overemphasized. The solution presented is an example of the use of symmetry to simplify the calculation of the rate of heat transfer in the chimney.

9.3.2.C   Numerical Methods

The most powerful approximation methods currently available for the solution of multidimensional heat conduction problems utilizes either finite elements or finite difference techniques. A detailed description of the application of finite element techniques to heat conduction problems is presented by Myers[3] and Patankar[4] describes the use of finite difference techniques for the solution of these types of problems. Less detailed descriptions of finite difference techinques are also available in numerous heat transfer books[5,6,7]. A brief description of the use of finite difference techniques for a two-dimensional steady-state heat conduction problem will be presented to illustrate the procedure. The reader is asked to consult the stated references if more detailed information is required.

The following steps are taken for the finite difference solution of a two-dimensional steady-state heat conduction problem.

1. Determine the complete mathematical model of the problem. This includes the identification of the appropriate differential equation and the boundary conditions.

2. Subdivide the configuration using a series of orthoginal grid lines. The grid lines should match the boundaries of the section whenever possible. Each intersection of two grid lines or a grid line and a boundary, has a volume of material associated with it. These intersections are called nodes and are identified by numbers.

3. The differential equation is approximated by an algebraic finite difference equation at each node. The resulting equations are of an implicit form and contain the temperatures of the node under consideration and the temperature of each of the *immediate* surrounding nodes which are connected to it by grid or boundary lines.

4. A set of linear algebraic equations, one for each node in the configuration, is formed. If there are N nodes in the grid pattern, the set will consist of N equations which collectively contain the temperatures at the N nodes, the N unknowns. The set of equations is solved simultaneously by either direct or iterative methods to obtain the temperatures at the nodal locations.

5. The rate of heat transfer across any surface can be determined by using the temperature at the nodes and the appropiate boundary equation. The heat transfer across a convection boundary is found using

$$\dot{Q} = \Sigma\, hA(T_\infty - T_{sur})$$

where the T_{sur}'s are the temperatures of the surface nodes.

The temperature distribution in the two-dimensional steady-state configuration with internal heat generation shown in Fig. 9–11 can be obtained by using finite difference techniques and the procedure outlined previously.

1. The mathematical model for this problem consists of the differential heat conduction equation

$$\frac{\partial^2 T}{\partial x^2} + \frac{\partial^2 T}{\partial y^2} + \dot{q}^* = 0$$

and the boundary conditions. The heat conduction equation can also be expressed as

$$\left\{\begin{array}{l}\text{Net rate of}\\\text{heat transfer}\\\text{across the surface}\\\text{of a control volume}\end{array}\right\} + \left\{\begin{array}{l}\text{net rate of}\\\text{heat generated}\\\text{within the control}\\\text{volume}\end{array}\right\} = 0$$

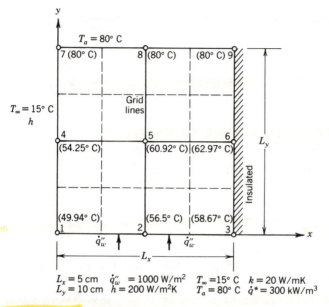

Figure 9-11 Spatial subdivisions forming nodal volumes.

or $\dot{Q} + \dot{Q}^* = 0$. The four different boundary conditions are indicated in the figure.

2. A very coarse grid pattern has been selected for this illustration. The grid spacing in the x direction is $\Delta x = L_x/2$, and in the y direction, $\Delta y = L_y/2$. A finer grid pattern, larger number of nodes, would give a more accurate result but would also require longer computational time since the number of equations to be solved simultaneously is proportional to the square of the number of nodes. The nodes are identified and the volumes associated

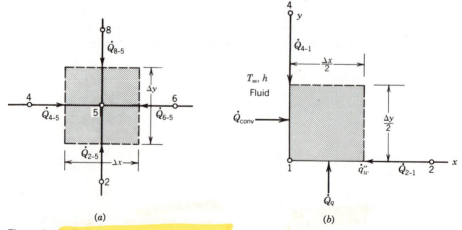

Figure 9-12 Energy balance at selected nodes. (a) Interior node. (b) Corner node.

with each node are defined by the dashed lines. The thickness of the element in a direction perpendicular to the page is considered to be unity.

3. Since the problem is steady state with internal heat generation, the first law indicates that the sum of the energy entering the volume associated with a node from the surrounding nodes plus the amount of heat generated within the volume must be equal to zero. This expression will now be used to form the finite difference approximation of the heat conduction equation at each node. The discussion begins with the interior node.

Node 5: The volume associated with this node is shown in Fig. 9–12a. The first law statement can be expressed as

$$\dot{Q}_{8-5} + \dot{Q}_{6-5} + \dot{Q}_{2-5} + \dot{Q}_{4-5} + \dot{Q}^* = 0$$

If one-dimensional heat flow is assumed to exist between adjacent nodes, Fourier's law can be used to obtain

$$k\,\Delta x\,\frac{(T_8 - T_5)}{\Delta y} + k\,\Delta y\,\frac{(T_6 - T_5)}{\Delta x} + k\,\Delta x\,\frac{(T_2 - T_5)}{\Delta y}$$
$$+ k\,\Delta y\,\frac{(T_4 - T_5)}{\Delta x} + \dot{q}^*\,\Delta x\,\Delta y = 0$$

Note that heat from all the surrounding nodes is considered to flow into node 5 when the finite difference approximation for the energy equation is formed. The expression is divided by $k\,\Delta x\,\Delta y$ and rearranged so that all the unknowns are grouped on the left side of the equation

$$\frac{T_8}{\Delta y^2} + \frac{T_6}{\Delta x^2} + \frac{T_2}{\Delta y^2} + \frac{T_4}{\Delta x^2} - 2\left[\frac{\Delta x^2 + \Delta y^2}{\Delta x^2\,\Delta y^2}\right]T_5 = -\frac{\dot{q}^*}{k}$$

Node 1: An enlarged view of node 1 and its surroundings is shown in Fig. 9–12b. Heat enters the volume of material associated with node 1 via the conduction process from nodes 4 and 2, a convection process from T_∞, and a constant heat flux process. The energy balance for node 1 is

$$\dot{Q}_{4-1} + \dot{Q}_{2-1} + \dot{Q}_q + \dot{Q}_{\text{conv}} + \dot{Q}_1^* = 0$$

or

$$k\,\frac{\Delta x}{2}\,\frac{(T_4 - T_1)}{\Delta y} + k\,\frac{\Delta y}{2}\,\frac{(T_2 - T_1)}{\Delta x} + \dot{q}_w''\,\frac{\Delta x}{2}$$
$$+ h\,\frac{\Delta y}{2}\,(T_\infty - T_1) + \dot{q}^*\,\frac{\Delta x}{2}\,\frac{\Delta y}{2} = 0$$

Rearrangement gives

$$2\,\frac{T_4}{\Delta y^2} + 2\,\frac{T_2}{\Delta x^2} - \left[2\left(\frac{\Delta x^2 + \Delta y^2}{\Delta x^2\,\Delta y^2}\right) + 2\,\frac{h}{k\,\Delta x}\right]T_1 = -\left[\frac{\dot{q}^*}{k} + \frac{2hT_\infty}{k\,\Delta x} + \frac{2\dot{q}_w''}{k\,\Delta y}\right]$$

Other Nodes: Similar procedures are used to form the finite difference approximations at nodes 4, 2, 3, and 6

$$\frac{T_7}{\Delta y^2} + \frac{2T_5}{\Delta x^2} + \frac{T_1}{\Delta y^2} - \left[2\left(\frac{\Delta x^2 + \Delta y^2}{\Delta x^2 \, \Delta y^2}\right) + 2\frac{h}{k \, \Delta x} \right] T_4 = -\left[\frac{\dot{q}^*}{k} + \frac{2hT_\infty}{k \, \Delta x} \right]$$

$$2\frac{T_5}{\Delta y^2} + \frac{T_3}{\Delta x^2} + \frac{T_1}{\Delta x^2} - \left[2\left(\frac{\Delta x^2 + \Delta y^2}{\Delta x^2 \, \Delta y^2}\right) \right] T_2 \qquad = -\left[\frac{\dot{q}^*}{k} + \frac{2\dot{q}''_w}{k \, \Delta y} \right]$$

$$2\frac{T_6}{\Delta y^2} + 2\frac{T_2}{\Delta x^2} - \left[2\left(\frac{\Delta x^2 + \Delta y^2}{\Delta x^2 \, \Delta y^2}\right) \right] T_3 \qquad = -\left[\frac{\dot{q}^*}{k} + \frac{2\dot{q}''_w}{k \, \Delta y} \right]$$

$$\frac{T_9}{\Delta y^2} + 2\frac{T_5}{\Delta x^2} + \frac{T_3}{\Delta y^2} - \left[2\left(\frac{\Delta x^2 + \Delta y^2}{\Delta x^2 \, \Delta y^2}\right) \right] T_6 \qquad = -\left[\frac{\dot{q}^*}{k} \right]$$

The finite difference expression for the nodes on the uniform temperature boundary is not obtained from the first law. The temperature of the node will always be T_a regardless of the amount of heat that flows in and out of the cell via conduction or the amount of heat internally generated. The nodal equations become

$$T_7 = T_a, \quad T_8 = T_a, \quad \text{and} \quad T_9 = T_a$$

In summary, the complete set of finite difference equations for the nodal pattern is given in Table 9–4.

4. The values of the system parameters, h, L_x, L_y, k, T_∞, $\dot{q}''_w$, $\dot{q}^*$, and T_a, are substituted into the finite difference expressions and the resulting set of equations, Table 9–5, is solved simultaneously to obtain the temperatures at the nodes in the material.

$$\begin{array}{lll} T_1 = 49.94°\text{ C} & T_4 = 54.25°\text{ C} & T_7 = 80°\text{ C} \\ T_2 = 56.50°\text{ C} & T_5 = 60.92°\text{ C} & T_8 = 80°\text{ C} \\ T_3 = 58.67°\text{ C} & T_6 = 62.97°\text{ C} & T_9 = 80°\text{ C} \end{array}$$

The temperatures are also shown in Fig. 9–11.

5. Once the temperature distribution has been determined it is possible to perform an energy balance on the section. The energy generated in the solid and that passing across the surfaces of the material are equated. For the configuration shown in Fig. 9–11 the energy balance yields

$$\left\{ \begin{array}{c} \text{Rate of} \\ \text{energy} \\ \text{generation} \end{array} \right\} = \sum \begin{array}{c} \text{rate of energy} \\ \text{crossing surfaces} \\ \text{in outward direction} \end{array}$$

The rate of energy generation per unit thickness is equal to the volume of the material times $\dot{q}^*$.

$$L_x L_y \dot{q}^* = (0.05)(0.10)(30,000)$$
$$= 150 \text{ W/m}$$

Table 9–4 Finite Difference Equations for Section Shown in Fig. 9–11

T1	T2	T3	T4	T5	T6	T7	T8	T9			RHS
$-2\left[\dfrac{\Delta x^2+\Delta y^2}{\Delta x^2\Delta y^2}+\dfrac{h}{k\Delta x}\right]$	$\dfrac{2}{\Delta x^2}$		$\dfrac{2}{\Delta y^2}$						T_1		$\dfrac{\dot{q}^*}{k}+\dfrac{2hT_\infty}{k\Delta x}+\dfrac{2\dot{q}_w''}{k\Delta y}$
$\dfrac{1}{\Delta x^2}$	$-2\left[\dfrac{\Delta x^2+\Delta y^2}{\Delta x^2\Delta y^2}\right]$	$\dfrac{1}{\Delta x^2}$		$\dfrac{2}{\Delta y^2}$					T_2		$\dfrac{\dot{q}^*}{k}+\dfrac{2\dot{q}_w''}{k\Delta y}$
	$\dfrac{2}{\Delta x^2}$	$-2\left[\dfrac{\Delta x^2+\Delta y^2}{\Delta x^2\Delta y^2}\right]$			$\dfrac{2}{\Delta y^2}$				T_3		$\dfrac{\dot{q}^*}{k}+\dfrac{2\dot{q}_w''}{k\Delta y}$
$\dfrac{1}{\Delta y^2}$			$-2\left[\dfrac{\Delta x^2+\Delta y^2}{\Delta x^2\Delta y^2}+\dfrac{h}{k\Delta x}\right]$	$\dfrac{2}{\Delta x^2}$		$\dfrac{1}{\Delta y^2}$			T_4	=	$\dfrac{\dot{q}^*}{k}+\dfrac{2hT_\infty}{k\Delta x}$
	$\dfrac{1}{\Delta y^2}$		$\dfrac{1}{\Delta x^2}$	$-2\left[\dfrac{\Delta x^2+\Delta y^2}{\Delta x^2\Delta y^2}\right]$	$\dfrac{1}{\Delta x^2}$		$\dfrac{1}{\Delta y^2}$		T_5		$\dfrac{\dot{q}^*}{k}$
		$\dfrac{1}{\Delta y^2}$		$\dfrac{2}{\Delta x^2}$	$-2\left[\dfrac{\Delta x^2+\Delta y^2}{\Delta x^2\Delta y^2}\right]$			$\dfrac{1}{\Delta y^2}$	T_6		$\dfrac{\dot{q}^*}{k}$
			$\dfrac{1}{\Delta y^2}$			-1			T_7		T_a
				$\dfrac{1}{\Delta y^2}$			-1		T_8		T_a
					$\dfrac{1}{\Delta y^2}$			-1	T_9		T_a

Table 9–5 Set of Finite Difference Equations for Section Shown in Fig 9–11

T1	T2	T3	T4	T5	T6	T7	T8	T9			RHS
−4800	3200		800						T_1		15,500
1600	−4000	1600		800					T_2		3,500
	3200	−4000			800				T_3		3,500
400			−4800	3200		400			T_4	=	13,500
	400		1600	−4000	1600		400		T_5		1,500
		400		3200	−4000			400	T_6		1,500
			400			−1			T_7		80
				400			−1		T_8		80
					400			−1	T_9		80

The rate of heat transfer per unit thickness across the surface exposed to the constant heat flux is

$$\dot{q}''_w L_x = -1000(0.05)$$
$$= -50 \text{ W/m}$$

The rate of heat transfer to the surrounding fluid, T_∞, is found by using

$$\Sigma h A \ \Delta T = h \frac{\Delta y}{2} (T_1 - T_\infty) + h \ \Delta y (T_2 - T_\infty) + h \frac{\Delta}{2} (T_5 - T_\infty)$$
$$= 200(0.025)(49.94 - 15) + 200(0.05)(54.25 - 15)$$
$$+ 200(0.025)(80 - 15)$$
$$= 892.2 \text{ W/m}$$

The total amount of heat exchanged with the reservoir at 80° C is obtained from the energy balance

$$\dot{Q} = 150 + 50 - 892.2$$
$$= -692.2 \text{ W/m (from reservoir)}$$

9.4   TRANSIENT HEAT CONDUCTION

If the thermal boundary conditions are time dependent, the complete heat conduction equation is required to describe the conduction process. In the cartesian coordinate system, the transient heat conduction equation with constant thermal conductivity is

$$k\left[\frac{\partial^2 T}{\partial x^2} + \frac{\partial^2 T}{\partial y^2} + \frac{\partial^2 T}{\partial z^2}\right] + \dot{q}^* = \rho c \frac{\partial T}{\partial t}$$

The complete mathematical model for a transient heat conduction problem includes the initial temperature distribution within the body as well as the thermal boundary conditions at the surfaces of the body.

The mathematical techniques used to obtain the transient temperature distribution within a body are quite involved for even a simple geometrical configuration. Numerical techniques, similar to those described in Section 9.3.2.C for steady-state heat conduction, currently are being used for the solution of multidimensional transient heat conduction problems. Many extensive computer programs are available which have the capability of solving problems with complicated geometries composed of different materials, variable thermophysical properties, and nonlinear boundary conditions. A complete description of these techiques is beyond the scope of this book. As a result, emphasis will not be placed on solution techniques. Instead, some of the available solutions, algebraic expressions as well as graphical charts, will be presented with emphasis placed on their use for the solution of engineering problems.

9.4.1 Lumped Parameter Analysis

A body initially at a uniform temperature, T_0, suddenly experiences a change in its thermal environment. The rate at which this change is sensed in the interior of the body will be dependent upon the resistance to the transfer of heat offered at its boundaries and the resistance to heat transfer offered internally within the material. If the thermal resistance offered at the boundaries is much greater than the internal thermal resistance, the temperature distribution within the body will be nearly uniform. The limiting case occurs when the thermal conductivity of the body is infinite, zero internal resistance to heat transfer. The temperature of the body is uniform and the time-dependent temperature can be determined by using a *lumped parameter* analysis.

The first law statement for the irregular shaped body shown in Fig. 9–13 is

$$\frac{dU}{dt} = \dot{Q} \tag{9–25}$$

The body is initially at a uniform temperature T_0. At $t > 0$, the temperature of the surrounding fluid is changed to T_∞. If the internal thermal resistance is neglected, the expression for the first law becomes

$$\rho c V \frac{dT}{dt} = hA(T_\infty - T) \tag{9–26}$$

The initial condition is

$$t = 0$$

$$T = T_0$$

The temperature distribution in the body is obtained by integrating eq. 9–26,

$$\frac{dT}{(T - T_\infty)} = -\frac{hA}{\rho c V} dt$$

$$\ln(T - T_\infty) = -\frac{hA}{\rho c V} t + \mathfrak{A}$$

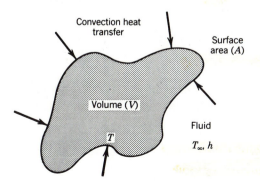

Convection heat transfer

Surface area (A)

Volume (V)

T

Fluid

T_∞, h

Figure 9-13 Transient heat transfer lumped parameter analysis.

where $\mathfrak{A}$ is the integration constant. The initial condition is used to evaluate the integration constant

$$\mathfrak{A} = \ln(T_0 - T_\infty)$$

The final expression for the temperature of the body is

$$\frac{(T - T_\infty)}{(T_0 - T_\infty)} = \exp\left(-\frac{hAt}{\rho cV}\right) \tag{9-27}$$

This expression can be transformed into a dimensionless form by introducing the following dimensionless groups.

Temperature:

$$T = \frac{(T - T_\infty)}{(T_0 - T_\infty)} \tag{9-28}$$

Length:

$$L_c \equiv \frac{V}{A} \tag{9-29}$$

The *thermal diffusivity* is introduced

$$\alpha \equiv \frac{k}{\rho c_p} = \frac{k}{\rho c}\bigg|_{\text{solid}} \quad m^2/s \tag{9-30}$$

The exponent in eq. 9–27 can be expressed in terms of two dimensionless groups frequently used in heat transfer. These are the *Biot* number

$$\text{Bi} \equiv \frac{hL_c}{k} \tag{9-31}$$

and the *Fourier* number

$$\text{Fo} \equiv \frac{\alpha t}{L_c^2} \tag{9-32}$$

Equation 9–27 in dimensionless form then becomes

$$T = \exp(-\text{BiFo}) \tag{9-33}$$

Physical significance can be attached to the Biot number. It represents a ratio between the internal resistance to the transfer of heat and the resistance to the transfer of heat offered at the boundaries

$$\text{Bi} \propto \frac{\text{internal resistance}}{\text{boundary resistance}}$$

The Biot number can be used to determine if significant errors are introduced in the calculation of the transient response of a body using the lumped parameter analysis. It has been shown that reasonable accuracy can be obtained using a lumped parameter analysis if the Bi ≤ 0.1.

EXAMPLE 9–4

A solid-steel sphere, AISI 1010, 1 cm in diameter initially at 15° C is placed in a hot air stream, $T_\infty = 60°$ C. Estimate the temperature of this sphere as a function of time after being placed in the hot air stream. The average convection heat transfer coefficient is 20 W/m² · K.

SOLUTION

The thermophysical properties of the steel sphere are obtained from Table A–15.

$$k = 63.9 \text{ W/m} \cdot \text{K}$$

$$\rho = 7832 \text{ kg/m}^3$$

$$c = 434 \text{ J/kg} \cdot \text{K}$$

The thermal diffusivity is

$$\alpha = \frac{k}{\rho c} = \frac{63.9}{(7832)(434)} = 18.80 \times 10^{-6} \text{ m}^2/\text{s}$$

The characteristic length is

$$L_c = \frac{V}{A} = \frac{(\pi/6)d^3}{\pi d^2} = \frac{d}{6} = \frac{0.01}{6} = 1.667 \times 10^{-3} \text{ m}$$

The Biot number is

$$\text{Bi} = \frac{hL_c}{k} = \frac{20(1.667 \times 10^{-3})}{63.9} = 521.8 \times 10^{-6}$$

and indicates that the lumped parameter analysis can be used without introducing an appreciable error in the calculations. The Fourier number is

$$\text{Fo} = \frac{\alpha t}{L_c^2} = \frac{18.80 \times 10^{-6}t}{(1.667 \times 10^{-3})^2} = 6.765t$$

Substitutions of these relationships into eq. 9–33 yields

$$T = \exp(-\text{BiFo})$$

$$T = T_\infty + (T_0 - T_\infty) \exp[-521.8 \times 10^{-6}(6.765t)]$$

$$= 60 + (15 - 60) \exp(-3.530 \times 10^{-3}t)$$

$$= 60 - 45 \exp(-3.530 \times 10^{-3}t)$$

The response of the steel sphere to the hot air stream is shown in Fig. E9–4.

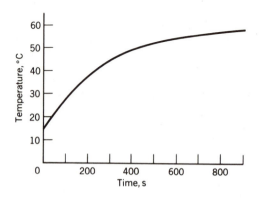

Figure E9-4   Temperature response of solid steel sphere.

COMMENT

The rate at which heat is transferred to the sphere can be calculated knowing the temperature of the sphere

$$\dot{Q} = hA(T_\infty - T)$$

$$= hA[45 \exp(-3.530 \times 10^{-3}t)]$$

$$= 20\pi(0.01)^2 (45) \exp(-3.530 \times 10^{-3}t)$$

$$= 0.2827 \exp(-3.530 \times 10^{-3}t)$$

9.4.2   One-Dimensional Heat Flow

If the transient flow of heat within a body is considered to be one dimensional, $T(x, t)$, and internal heat generation is absent, the heat conduction equation in the cartesian coordinate system reduces to

$$\frac{\partial^2 T}{\partial x^2} = \frac{1}{\alpha} \frac{\partial T}{\partial t} \tag{9-34}$$

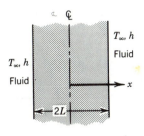

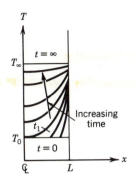

Figure 9-14   Transient heat conduction.

The corresponding equation in the cylindrical coordinates is

$$\frac{\partial^2 T}{\partial r^2} + \frac{1}{r}\frac{\partial T}{\partial r} = \frac{1}{\alpha}\frac{\partial T}{\partial t} \tag{9–35}$$

Three specific configurations will be considered in which the heat transfer is one dimensional.

9.4.2.A   Semi-infinite Solid

Although an engineer seldom encounters a *semi-infinite solid* in the strict physical sense, there are many situations in which a body behaves, in a thermal sense, as a semi-infinite solid. For instance, consider a step change in the fluid temperature surrounding the slab shown in Fig. 9–14 to occur at $t = 0$. As a result of the thermal disturbance a temperature wave moves from the surface into the material. As long as the temperature wave does not strike another boundary or encounter a temperature wave originating from another boundary, it will behave as if it were being transmitted into a semi-infinite solid. Thus, if $0 < t < t_1$, where t_1 is the time required for the wave to strike the other boundary, the semi-infinite solid solution is acceptable for the slab. For $t \ge t_1$ the procedures outlined in Section 9.4.2.B must be used.

The semi-infinite solid shown in Fig. 9–15 is initially at a uniform temperature, T_0. The surface of the body at $x = 0$ experiences a step change in its boundary

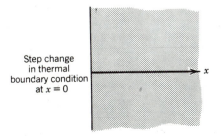

Step change
in thermal
boundary condition
at $x = 0$

Figure 9-15   Semi-infinite solid.

condition. The temperature distribution within the solid may be determined from the following expressions:

A. Step change in temperature:

$$x = 0 \quad T = T_1 \quad \text{for} \quad t > 0$$

Table 9–6 Gauss Error Function

X	erf(X)	X	erf(X)	X	erf(X)
0.00	0.00000	0.76	0.71754	1.52	0.96841
0.02	0.02256	0.78	0.73001	1.54	0.97059
0.04	0.04511	0.80	0.74210	1.56	0.97263
0.06	0.06762	0.82	0.75381	1.58	0.97455
0.08	0.09008	0.84	0.76514	1.60	0.97635
0.10	0.11246	0.86	0.77610	1.62	0.97804
0.12	0.13476	0.88	0.78669	1.64	0.97962
0.14	0.15695	0.90	0.79691	1.66	0.98110
0.16	0.17901	0.92	0.80677	1.68	0.98249
0.18	0.20094	0.94	0.81627	1.70	0.98379
0.20	0.22270	0.96	0.82542	1.72	0.98500
0.22	0.24430	0.98	0.83423	1.74	0.98613
0.24	0.26570	1.00	0.84270	1.76	0.98719
0.26	0.28690	1.02	0.85084	1.78	0.98817
0.28	0.30788	1.04	0.85865	1.80	0.98909
0.30	0.32863	1.06	0.86614	1.82	0.98994
0.32	0.34913	1.08	0.87333	1.84	0.99074
0.34	0.36936	1.10	0.88020	1.86	0.99147
0.36	0.38933	1.12	0.88079	1.88	0.99216
0.38	0.40901	1.14	0.89308	1.90	0.99279
0.40	0.42839	1.16	0.89910	1.92	0.99338
0.42	0.44749	1.18	0.90484	1.94	0.99392
0.44	0.46622	1.20	0.90131	1.96	0.99443
0.46	0.48466	1.22	0.91553	1.98	0.99489
0.48	0.50275	1.24	0.92050	2.00	0.995322
0.50	0.52050	1.26	0.92524	2.10	0.997020
0.52	0.53790	1.28	0.92973	2.20	0.998137
0.54	0.55494	1.30	0.93401	2.30	0.998857
0.56	0.57162	1.32	0.93806	2.40	0.999311
0.58	0.58792	1.34	0.94191	2.50	0.999593
0.60	0.60386	1.36	0.94556	2.60	0.999764
0.62	0.61941	1.38	0.94902	2.70	0.999866
0.64	0.63459	1.40	0.95228	2.80	0.999925
0.66	0.64938	1.42	0.95538	2.90	0.999959
0.68	0.66278	1.44	0.95830	3.00	0.999978
0.70	0.67780	1.46	0.96105	3.20	0.999994
0.72	0.69143	1.48	0.96365	3.40	0.999998
0.74	0.70468	1.50	0.96610	3.60	1.000000

The temperature distribution is

$$T = T_1 + (T_0 - T_1)\operatorname{erf}(x/2\sqrt{\alpha t}) \tag{9-36}$$

where $\operatorname{erf}(x/2\sqrt{\alpha t})$ is the Gauss error function tabulated in Table 9–6.

B. Step change in heat flux:

$$x = 0 \qquad \dot{q}''_w = -k\frac{\partial T}{\partial x} \quad \text{for} \quad t > 0$$

The temperature distribution is

$$T = T_0 + \frac{2\dot{q}''_w\sqrt{\alpha t/\pi}}{k}\exp\left(-\frac{x^2}{4\alpha t}\right) - \frac{\dot{q}''_w x}{k}[1 - \operatorname{erf}(x/2\sqrt{\alpha t})] \tag{9-37}$$

C. Step change in fluid temperature:

$$x = 0 \qquad h(T_\infty - T) = -k\frac{\partial T}{\partial x} \quad \text{for} \quad t > 0$$

The temperature distribution is

$$T = T_0 + (T_\infty - T_0)\left\{1 - \operatorname{erf}(x/(2\sqrt{\alpha t}) - \left[\exp\left(\frac{hx}{k}\right.\right.\right.$$
$$\left.\left.\left. + \frac{h^2\alpha t}{k^2}\right)\right]\cdot\left[1 - \operatorname{erf}\left(x/2\sqrt{\alpha t} + \frac{h\sqrt{\alpha t}}{k}\right)\right]\right\} \tag{9-38}$$

EXAMPLE 9–5

A very thick oak wall is initially at a uniform temperature of 20° C. It is suddenly exposed to a hot gas stream, $T_\infty = 200°$ C. Estimate the surface temperature of the wood 10 s after the gas comes in contact with the wood. The convection heat transfer coefficient is 100 W/m²·K.

SOLUTION

The thermophysical properties of the wood are obtained from Table A-16.1

$$k = 0.19 \text{ W/m·K}$$

$$\rho = 545 \text{ kg/m}^3$$

$$c = 2385 \text{ J/kg·K}$$

The thermal diffusivity is

$$\alpha = \frac{k}{\rho c} = \frac{0.19}{545(2385)} = 146.2 \times 10^{-9} \text{ m}^2/\text{s}$$

The surface temperature, $x = 0$ at $t = 10$ s, can be estimated assuming the wall to behave as a semi-infinite slab using eq. 9–38.

$$T_{sur} = T_0 + (T_\infty - T_0)\left\{1 - \text{erf}(x/2\sqrt{\alpha t}) - \left[\exp\left(\frac{hx}{k} + \frac{h^2\alpha t}{k^2}\right)\right]\right.$$

$$\left. \cdot \left[1 - \text{erf}\left(x/2\sqrt{\alpha t} + \frac{h\sqrt{\alpha t}}{k}\right)\right]\right\}$$

$$= 20 + (200 - 20)\left\{1 - 0\left[\exp\left(\frac{(100)^2(146.2 \times 10^{-9})(10)}{(0.19)^2}\right)\right]\right.$$

$$\left. \cdot \left[1 - \text{erf}\left(\frac{100\sqrt{146.2 \times 10^{-9}(10)}}{0.19}\right)\right]\right\}$$

$$= 20 + (180)[1 - 1.499[1 - 0.630]]$$

$$= 20 + (180)(0.4454)$$

$$= 100.2° \text{ C}$$

COMMENT

The temperature at a location 3cm from the surface of the wall 10 s after the gas comes in contact with the wood can be determined using eq. 9–38.

$$T_{3cm} = 20 + (180)\left\{1 - \text{erf}\left(\frac{0.03}{2\sqrt{146.2 \times 10^{-9}(10)}}\right)\right.$$

$$- \left[\exp\left(\frac{100(0.03)}{0.19} + \frac{(100)^2(146.2 \times 10^{-9})(10)}{(0.19)^2}\right)\right]$$

$$\left. \cdot \left[1 - \text{erf}\left(\frac{0.03}{2\sqrt{146.2 \times 10^{-9}(10)}} + \frac{100\sqrt{146.2 \times 10^{-9}(10)}}{0.19}\right)\right]\right\}$$

$$\approx 20.0°\text{C}$$

Since the temperature is the same as the initial temperature, the thermal wave has not penetrated to a depth of 3 cm from the surface. The semi-infinite solid solution may be used for determining the temperature distribution in the wall 10 s after it is exposed to the hot gases without introducing an appreciable error if the thickness of the oak is 3 cm or greater.

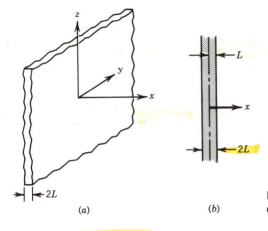

(a)   (b)

Figure 9-16 One-dimensional transient conduction. (a) Finite body. (b) Infinite slab.

9.4.2.B Infinite Slab

For the heat transfer in a body to be one dimensional, the dimensions of the body perpendicular to the direction of heat flow must be very large. An example of such a geometrical configuration is shown in Fig. 9–16a. If the thermal boundary conditions on the yz planes at $x = -L$ and $x = L$ are uniform, the boundary conditions on the other surfaces will have little influence on the heat transfer and the temperature distribution in the body. The heat transfer may be considered to be one dimensional, in the x direction, and the thermal analysis of the body may be obtained by considering it to behave as an *infinite slab* as shown in Fig. 9–16b.

The transient response of an *infinite slab* to a change in boundary conditions is obtained by solving the one-dimensional transient heat conduction equation with the initial temperature distribution in the slab and the thermal boundary conditions. For the slab shown in Fig. 9–17, which is initially at a uniform temperature of T_0, the mathematical model is:

Transient heat conduction equation (eq. 9–34):

$$\frac{\partial^2 T}{\partial x} = \frac{1}{\alpha}\frac{\partial T}{\partial t}$$

Initial condition:

$$t = 0 \qquad -L \leq x \leq L \qquad T = T_0 \qquad\qquad (9\text{–}39)$$

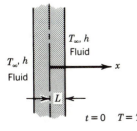

T_∞, h
Fluid

T_∞, h
Fluid

L

$t = 0 \quad T = T_0$

Figure 9-17 Infinite slab—convection boundary condition.

Boundary conditions:

$$x = 0 \qquad \frac{\partial T}{\partial x} = 0 \qquad \text{(symmetry)}$$

$$t > 0 \tag{9-40}$$

$$x = L \qquad h(T - T_\infty) = -k\frac{\partial T}{\partial x}$$

A solution procedure such as the method of separation of variables could be used to solve the differential equation, eq. 9–34. The initial and boundary conditions are used to determine the integration constants. A procedure that employs the method of separation of variables is described in detail by Myers[3].

An analysis of the mathematical model for the infinite slab, described by the differential equation and the initial and boundary conditions, indicates that the temperature distribution in the slab is a function of the following nine variables $T\{x, \rho, T_0, L, T_\infty, k, h, c, t\}$. In an attempt to reduce the number of variables, several dimensionless groups will be formed. A dimensionless length, temperature, and time (Fourier number) will be defined

$$X = \frac{x}{L}, \qquad \mathbf{T} = \frac{T - T_\infty}{T_0 - T_\infty} \qquad \text{and} \qquad \text{Fo} = \frac{\alpha t}{L^2}$$

The differential equation, eq. 9–34, may be expressed in terms of these variables utilizing

$$\frac{\partial T}{\partial x} = \frac{\partial T}{\partial \mathbf{T}} \frac{d\mathbf{T}}{dx} \frac{\partial \mathbf{T}}{\partial X} = \frac{(T_0 - T_\infty)}{L} \frac{\partial \mathbf{T}}{\partial X}$$

$$\frac{\partial^2 T}{\partial x^2} = \frac{\partial(\partial T/\partial x)}{\partial x} = \frac{dX}{dx} \frac{\partial\{[(T_0 - T_\infty)/L](\partial \mathbf{T}/\partial X)\}}{\partial X} = \frac{(T_0 - T_\infty)}{L^2} \frac{\partial^2 \mathbf{T}}{\partial X^2}$$

$$\frac{\partial T}{\partial t} = \frac{\partial T}{\partial \mathbf{T}} \frac{\partial \mathbf{T}}{\partial t} = (T_0 - T_\infty) \frac{\partial \mathbf{T}}{\partial t}$$

$$\frac{\partial T}{\partial t} = \frac{\partial \text{Fo}}{\partial t} \frac{\partial \mathbf{T}}{\partial \text{Fo}} = \frac{\alpha}{L^2} \frac{\partial \mathbf{T}}{\partial \text{Fo}}$$

The differential equation in dimensionless form becomes

$$\frac{\partial^2 \mathbf{T}}{\partial X^2} = \frac{\partial \mathbf{T}}{\partial \text{Fo}} \tag{9-41}$$

The initial condition in dimensionless form is

$$\text{Fo} = 0 \qquad 1 \leqslant X \leqslant 1 \qquad \mathbf{T} = 1 \tag{9-42}$$

The boundary condition at $x = 0$ is

$$\frac{\partial T}{\partial X} = 0 \qquad (9\text{-}43)$$

The boundary condition at $x = L$, expressed in terms of the previously introduced dimensionless groups, is

$$X = 1 \qquad h(T_0 - T_\infty)T = -k\frac{(T_0 - T_\infty)}{L}\frac{\partial T}{\partial X}$$

or

$$\frac{hL}{k}T = -\frac{\partial T}{\partial X}$$

The Biot number is introduced

$$Bi \equiv \frac{hL}{k}$$

and the boundary condition becomes

$$X = 1 \qquad BiT = -\frac{\partial T}{\partial X} \qquad (9\text{-}44)$$

The dimensionless differential equation, eq. 9-41, and initial and boundary conditions, eqs. 9-42 through 9-44, indicate that the dimensionless temperature can be expressed as a function of only three dimensionless groups (X, Fo, Bi). Indeed, this is a significant reduction in the number of variables that are required to present the solution of a one-dimensional transient temperature distribution in an infinite slab. It again emphasizes the importance of the correct mathematical model and the introduction of dimensionless groups to simplify heat transfer calculations.

The temperature in an infinite slab, Fig. 9-17, at the adiabatic surface, the plane of symmetry, and the surface in contact with the fluid can be obtained from Figs. 9-18 and 9-19. The characteristic length used in the definition of the dimensionless length is the distance from the adiabatic plane to the surface in contact with the fluid, L. The dimensionless heat loss by a slab at an initial temperature of T_0 is

$$\frac{Q_{loss}}{Q_0} = \frac{Q_{loss}}{\rho c V(T_0 - T_\infty)} \qquad (9\text{-}45)$$

The value of Q_{loss}/Q_0 is a function of $Bi^2 Fo$ and may be determined by using Fig. 9-20.

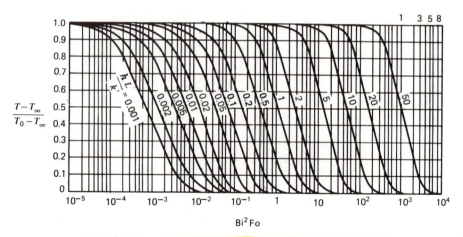

Figure 9-18 Temperature of adiabatic surface—infinite slab, **X** = 0. Used with permission.[12]

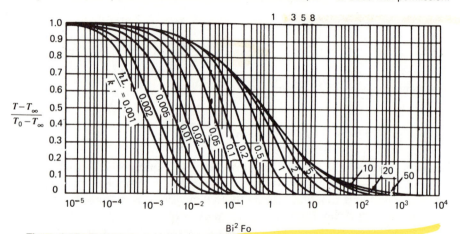

Figure 9-19 Temperature of surface—infinite slab, **X** = 1. Used with permission.[12]

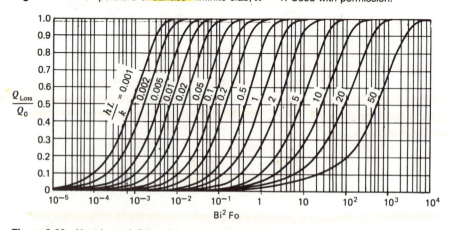

Figure 9-20 Heat loss—infinite slab. Used with permission.[12]

EXAMPLE 9–6

A large solid brick wall 15 cm thick reaches a uniform temperature of 0° C during a winter evening. At 9:00 A.M. the surrounding air warms to a temperature of 15° C. The air remains at that temperature until 3:00 P.M. Estimate the temperature at the centerline and the surface of the brick wall at noon. The average temperature of the brick and the amount of heat that has been transferred from the air to the brick are to be determined. The convection heat transfer coefficient may be considered to be constant at 50 W/m²·K.

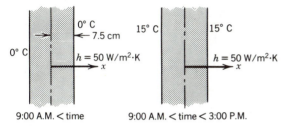

9:00 A.M. < time   9:00 A.M. < time < 3:00 P.M.

Figure E9-6   Brick wall.

SOLUTION

The wall is large so that the heat transfer will be assumed to be one dimensional. It is also assumed that the wall experiences an instantaneous change in its thermal boundary conditions at 9:00 A.M. The thermophysical properties of the brick are found in Table A–16.2.

$$k = 0.72 \text{ W/m·K} \quad \text{and} \quad \alpha = 449.1 \times 10^{-9} \text{m}^2/\text{s}$$

The Biot number is

$$\text{Bi} = \frac{hL}{k} = \frac{50(0.075)}{0.72}$$

$$= 5.208$$

The Fourier number at noon, $t = 3[3600] = 10800$ s, is

$$\text{Fo} = \frac{\alpha t}{L^2} = \frac{449.1 \times 10^{-9}(10800)}{(0.075)^2}$$

$$= 0.8623$$

The $\text{Bi}^2\text{Fo} = [5.208]^2 [0.8623] = 23.39$. The temperature at the brick's surface can be obtained by using Fig. 9–19.

$$T_s = 0.07 = \frac{T_s - T_\infty}{T_0 - T_\infty}$$

$$T_s = T_\infty + 0.07(T_0 - T_\infty)$$

$$= 15 + 0.07(0 - 15)$$

$$= 13.95°\ C$$

The temperature at the centerline of the brick is obtained from Fig. 9–18.

$$T_\ell = 0.26 = \frac{T_\ell - T_\infty}{T_0 - T_\infty}$$

$$T_\ell = T_\infty + 0.26(T_0 - T_\infty)$$

$$= 15 + 0.26(0 - 15)$$

$$= 11.10°\ C$$

The dimensionless heat loss from both sides of the plate is obtained from Fig. 9–20.

$$\frac{Q_{loss}}{Q_0} = \frac{Q_{loss}}{\rho c V(T_0 - T_\infty)} = 0.80$$

The heat loss per square meter of surface area is

$$Q_{loss} = 0.80\left[\frac{k}{\alpha}V(T_0 - T_\infty)\right]$$

$$= 0.80\left[\frac{0.72}{449.1 \times 10^{-9}}(0.075)(0 - 15)\right]$$

$$= -1.443 \times 10^6\ J/m^2$$

The negative sign indicates that the wall gains 1.443×10^6 J/m² of energy. The average wall temperature, T_m, is obtained by equating the heat gained by the wall to the increase in the internal energy of the wall.

$$Q = \rho c V(T_m - T_0)$$

$$1.443 \times 10^6 = \frac{0.72}{449.1 \times 10^{-9}}(0.075)(T_m - 0)$$

$$T_m = 12.00°\ C$$

COMMENT

The lumped parameter analysis could not be used for the determination of the average temperature of the slab without introducing an appreciable error. The Biot number used in the lumped parameter analysis is

$$\text{Bi} = \frac{h(V/A)}{k} = \frac{50(0.075)}{0.72}$$
$$= 5.208$$

The Fourier number is

$$\text{Fo} = \frac{\alpha t}{(V/A)^2} = \frac{449.1 \times 10^{-9}(10800)}{(0.075)^2}$$
$$= 0.8623$$

The dimensionless temperature for the lumped parameter analysis is

$$T = \frac{T - T_\infty}{T_0 - T_\infty}$$
$$= \exp(-\text{BiFo})$$
$$= \exp(-5.208(0.8623))$$
$$= 0.0112$$

The temperature is

$$T = T_\infty + (T_0 - T_\infty)(0.0112)$$
$$= 15 + (0 - 15)(0.0112)$$
$$= 14.83° \text{ C}$$

An approximate error of 2.83°C has been introduced when using the lumped parameter analysis.

9.4.2.C   Infinite Cylinder

The temperature distribution and the heat transfer from an *infinite cylinder* can be obtained in a manner similar to that described in Section 9.4.2.B for an infinite slab. The dimensionless radius is defined as the dimensional radial location divided by the outer radius of the cylinder, $R = r/r_o$. The outer radius of the cylinder is used as the characteristic dimension in the evaluation of the Biot and Fourier numbers. The dimensionless temperature at the center of the infinite cylinder, $R = 0$, may be obtained from Fig. 9–21, whereas the dimensionless temperature at the surface, $R = 1$, may be obtained from Fig. 9–22. The dimensionless heat loss per unit length is obtained by using Fig. 9–23 where

$$\frac{Q_{\text{loss}}}{Q_0} = \frac{Q_{\text{loss}}}{\rho c(\pi r_o^2)(T_0 - T_\infty)} \qquad (9\text{–}46)$$

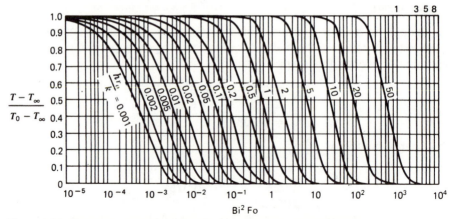

Figure 9-21   Temperature at centerline—infinite cylinder, $R = 0$. Used with permission.[12]

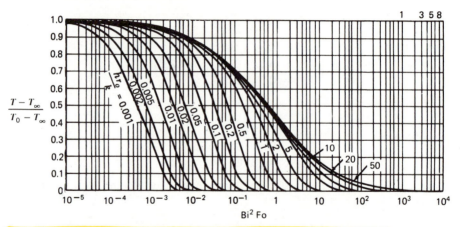

Figure 9-22   Temperature at surface—infinite cylinder, $R = 1$. Used with permission.[12]

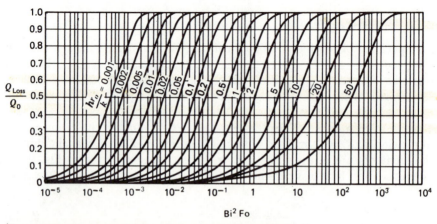

Figure 9-23   Heat loss—infinite cylinder. Used with permission.[12]

9.4.2.D Sphere

The heat transfer to or from a *sphere* is one dimensional if the boundary conditions surrounding the sphere are uniform. The characteristic length used is the outer radius of the sphere. The dimensionless temperatures at $R = r/r_o = 0$ and $R = 1$ can be determined from Figs. 9–24 and 9–25. The heat loss from the sphere is determined by using Fig. 9–26 where

$$\frac{Q_{loss}}{Q_0} = \frac{Q_{loss}}{\rho c(4\pi r_o^3/3)(T_0 - T_\infty)} \tag{9–47}$$

9.4.3 Multidimensional Configurations

When the geometrical configuration and thermal boundary conditions result in two- or three-dimensional heat transfer, it is necessary to use the complete transient heat conduction equation to determine the temperature distribution

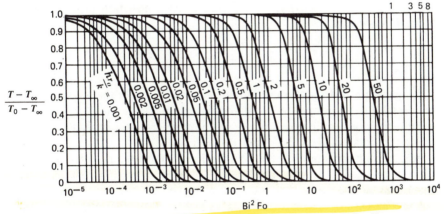

Figure 9-24 Temperature at centerline—sphere, $R = 0$. Used with permission.[12]

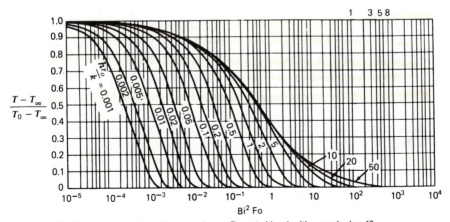

Figure 9-25 Temperature at surface—sphere, $R = 1$. Used with permission.[12]

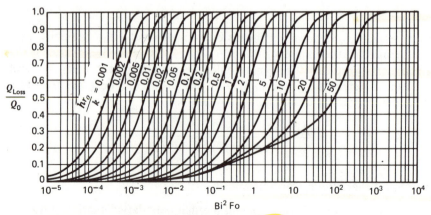

Figure 9-26 Heat loss—sphere. Used with permission.[12]

within the body. For a three-dimensional body with no heat generation and constant thermal conductivity, the heat conduction equation in the cartesian coordinate system is

$$\left[\frac{\partial^2 T}{\partial x^2} + \frac{\partial^2 T}{\partial y^2} + \frac{\partial^2 T}{\partial z^2}\right] = \frac{1}{\alpha}\frac{\partial T}{\partial t} \tag{9-48}$$

The temperature distribution in the body is obtained from the solution of this equation with the appropriate initial and boundary conditions. Analytic or numerical techniques can be used to obtain the temperature distribution and are described in detail in Refs. 3, 9, and 10.

Under certain conditions the one-dimensional solutions presented in Section 9.4.2 can be used. The multidimensional geometrical configurations are restricted to those that can be formed by using a semi-infinite solid, infinite slabs, or an infinite cylinder as illustrated in Fig. 9–27. Additional restrictions to the application of these techniques are:

1. All thermal boundary conditions must experience a simultaneous step change.
2. The initial temperature distribution in the body is uniform, T_0.
3. All fluid and surface temperatures defined in the boundary conditions after the step change must be equal, T_∞.
4. Uniform heat flux or nonlinear boundary conditions must not be present.

If these conditions are satisfied, the temperature distribution in the body can be obtained from the product of two or more one-dimensional solutions. The expression for the dimensionless temperature must be used. The solutions available are as follows.

Semi-infinite Solid.   The expressions for the two boundary conditions at $x = 0$ are:

Surface temperature:

$$x = 0 \quad \text{and} \quad T = T_\infty \tag{9-49}$$

$$T_s = \frac{T - T_\infty}{T_0 - T_\infty} = \text{erf}\left(\frac{x}{2\sqrt{\alpha t}}\right)$$

Convective boundary:

$$x = 0 \qquad h(T_\infty - T) = -k\frac{\partial T}{\partial x}$$

$$T_s = \frac{T - T_\infty}{T_0 - T_\infty} = \text{erf}\left(\frac{x}{2\sqrt{\alpha t}}\right) + \left[\exp\left(\frac{hx}{k} + \frac{h^2\alpha t}{k^2}\right)\right]$$

$$\cdot \left[1 - \text{erf}\left(\frac{x}{2\sqrt{\alpha t}} + \frac{h\sqrt{\alpha t}}{k}\right)\right] \tag{9-50}$$

Infinite Slab. Results shown in Figs. 9–18 and 9–19 are for convection boundary conditions. The dimensionless temperature for the infinite slab is defined as

$$T_1 = \left(\frac{T - T_\infty}{T_0 - T_\infty}\right)_{x,y, \text{ or } z}$$

At the plane of symmetry $X = 0$, an adiabatic surface T_1 is obtained from Fig. 9–18. At the surface T_1 is obtained from Fig. 9–19.

Infinite Cylinder. The dimensionless temperature for the infinite cylinder is defined as

$$T_c = \left(\frac{T - T_\infty}{T_0 - T_\infty}\right)_r$$

All results are for a convection boundary conditions. The value of T_c at the centerline is obtained from Fig. 9–21. The temperature at the surface is obtained from Fig. 9–22.

These solutions can be used to determine the dimensionless temperature in the parallelpiped shown in Fig. 9–27 by

$$T = \frac{T - T_\infty}{T_0 - T_\infty} = \left(\frac{T - T_\infty}{T_0 - T_\infty}\right)_x\left(\frac{T - T_\infty}{T_0 - T_\infty}\right)_y\left(\frac{T - T_\infty}{T_0 - T_\infty}\right)_z = T_1|_x T_1|_y T_1|_z \tag{9-51}$$

The dimensionless temperature in a finite cylinder is found using

$$T = \frac{T - T_\infty}{T_0 - T_\infty} = \left(\frac{T - T_\infty}{T_0 - T_\infty}\right)_x\left(\frac{T - T_\infty}{T_0 - T_\infty}\right)_c = T_1|_x T_c|_r \tag{9-52}$$

Two dimensional

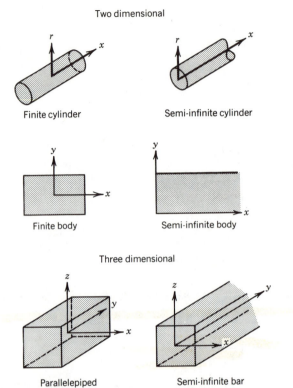

Finite cylinder   Semi-infinite cylinder

Finite body   Semi-infinite body

Three dimensional

Parallelepiped   Semi-infinite bar

Figure 9-27   Two- and three-dimensional geometric configurations.

EXAMPLE 9–7

A stainless-steel (AISI 304) cylinder 2 cm in diameter and 5 cm long is removed from a furnace and placed on the table as shown in Fig. E9–7. The cylinder is at a uniform temperature of 300° C when it is removed from the furnace. The surface in contact with the table during the cooling process is considered to be

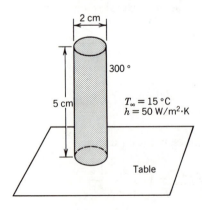

2 cm

300 °

5 cm

$T_\infty = 15\,°C$
$h = 50\ W/m^2·K$

Table

Figure E9-7   Stainless-steel cylinder.

perfectly insulated but the remaining surfaces are in contact with the surrounding air which is at a temperature of 15° C. Determine the maximum temperature of the cylinder 15 min after it has been removed from the furnace. The average heat transfer coefficient during the cooling process is 50 W/m²·K.

SOLUTION

The thermophysical properties of the stainless steel are obtained from Table A–15 at 430 K.

$$k = 17.08 \text{ W/m·K} \quad \text{and} \quad \alpha = 4.15 \times 10^{-6} \text{m}^2/\text{s}$$

The maximum temperature in the cylinder will occur at the center of the surface resting on the table. The product solution can be used to obtain the desired results. The one-dimensional solutions needed are:

Infinite Slab	**Infinite Cylinder**
$Bi = \dfrac{hL}{k} = \dfrac{50(0.05)}{17.08}$	$Bi = \dfrac{hr_o}{k} = \dfrac{50(0.01)}{17.08}$
$= 0.1464$	$= 0.029$
$Fo = \dfrac{\alpha t}{L^2}$	$Fo = \dfrac{\alpha t}{r_o^2}$
$= \dfrac{4.15 \times 10^{-6}(15)(60)}{(0.05)^2}$	$= \dfrac{4.15 \times 10^{-6}(15)(60)}{(0.01)^2}$
$= 1.494$	$= 37.35$
$Bi^2Fo = 0.032$	$Bi^2F_o = 0.0314$
From Fig. 9–18, $X = 0$	From Fig. 9–21, $R = 0$
$T_1 = 0.82$	$T_c = 0.13$

The maximum dimensionless temperature in the finite cylinder will be

$$T = \frac{T - T_\infty}{T_0 - T_\infty}$$

$$= T_1 T_c$$

$$= 0.82(0.13)$$

$$= 0.1066$$

This corresponds to a temperature of

$$T = T_\infty + (T_0 - T_\infty)(0.1066)$$
$$= 15 + (300 - 15)(0.1066)$$
$$= 45.38° \text{ C}$$

REFERENCES

1. Jakob, M., *Heat Transfer*, Vol. I, pp. 68–117, Wiley, New York, 1949.

2. Gebhart, B., *Heat Transfer*, 2nd ed., p. 6, McGraw-Hill, New York, 1971.

3. Myers, G. E., *Analytical Methods in Conductions Heat Transfer*, McGraw-Hill, New York, 1971.

4. Patankar, S. V., *Numerical Heat Transfer and Fluid Flow*, McGraw-Hill, New York, 1980.

5. Ozisik, M. N., *Basic Heat Transfer*, McGraw-Hill, New York, 1977.

6. Holman, J. P., *Heat Transfer*, 5th ed., McGraw-Hill, New York, 1981.

7. Kreith, F., and Black, W. Z., *Basic Heat Transfer*, 4th ed., Harper & Row, New York, 1980.

8. Karlekar, B. V., and Desmond, R. M., *Engineering Heat Transfer*, West, St. Paul, 1977.

9. Arpaci, V. S., *Conduction Heat Transfer*, Addison-Wesley, Reading, Mass., 1966.

10. Carslaw, H. S., and Jaeger, J. C., *Conduction of Heat in Solids*, 2nd. ed., Oxford University Press, London, 1959.

11. Incropera, F. P., and DeWitt, D. P., *Fundamentals of Heat Transfer*, Wiley, New York, 1981.

12. Grober, H., Erk, S., and Grigull, U., *Fundamentals of Heat Transfer*, McGraw-Hill, New York, 1961

PROBLEMS

9–1   A plain carbon steel pipe (5.25 cm inside diameter and 6.03 cm outside diameter) is covered with 2 cm thick six-ply asbestos corrugated paper. The temperature of the steam inside the pipe is 150° C and that of the surrounding air is 25° C. Estimate the outside surface temperature of the insulation and the rate of heat transfer per meter of pipe length.

$$h_{\text{steam}} = 1500 \text{ W/m}^2 \cdot \text{K} \qquad h_{\text{air}} = 5 \text{ W/m}^2 \cdot \text{K}$$

9–2   The outside wall of a building consists of an inner layer of gypsum plaster, 1.5 cm thick, placed on concrete blocks, three oval cores sand/gravel, which are 20 cm thick. The outside of the wall is face brick 10 cm thick. The heat transfer coefficients on the inside and outside surfaces of the wall are 8.35 and 34.10 W/m²·K, respectively. The outside air temperature is −10° C while the interior air temperature is 20° C. Determine:

(a) The rate of heat transfer per unit area.

(b) The surface temperature of the inner surface of the wall.

9–3 A plain carbon steel pipe with an inside diameter of 9 cm and a thickness of 0.5 cm carries hot water which is at a temperature of 150° C. The pipe is insulated with 5 cm thick 85% magnesia. The inside heat transfer coefficient is 7100 W/ m²·K. The convection heat transfer coefficient on the outside of the insulation is 57 W/m²·K. Scale builds up on the inside of the pipe yielding a fouling factor of 0.0005 m²·K/W. Determine:

(a) The overall heat transfer coefficient based on the outside surface area of the insulation, in watts per meter squared-degrees Kelvin.

(b) The rate of heat lost from the pipe per meter length if the outside air temperature is 20° C.

9–4 A Freon refrigerant flows through a stainless-steel (AISI 302) tube having an outside diameter of 0.953 cm and a thickness of 0.124 cm. The tube is covered with 1 cm thick R-12 polystyrene insulation (ρ = 35 kg/m³). The inside surface temperature of the tube is considered to be equal to that of the Freon, $-40°$ C. The temperature of the air surrounding the insulated tube is 15° C. Determine the rate of heat loss per meter of tubing length and the temperature of the outer surface of the insulation if the outside convection heat transfer coefficient is 10 W/m²·K.

9–5 An electrical heating element is shrunk in a hollow cylinder of amorphous carbon, Fig. P9–5. The outside of the carbon is in contact with 20° C air. The convection heat transfer coefficient is 40 W/m²·K. Determine the maximum electrical heating (watts per meter of cylinder length) which can be applied if the maximum allowable temperature of the carbon is 200° C. The resistance to the transfer of heat within the heating element may be neglected.

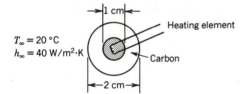

$T_\infty = 20 °C$
$h_\infty = 40 W/m^2\cdot K$

Figure P9-5 Heating element shrunk in hollow cylinder.

9–6 The ceiling of a house is composed of 1.5 cm plaster board. For a typical winter day, the room temperature is 21° C and the air temperature in the attic is $-10°$ C. Determine the amount of heat saved if the attic floor is covered with 1.5 cm plywood and the 9 cm space between this floor and plaster board is filled with a glass fiber blanket with a density of 16 kg/m³, Fig. P9–6. The convection film coefficients are both 40 W/m²·K.

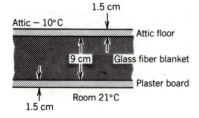

Figure P9-6 Ceiling of house.

9-7   A wall consists of two thin (2 mm) aluminum (2024-T6) plates separated by aluminum (2024-T6) structural support members as shown in Fig. P9–7. The space between the supports is filled with vermiculite ($\rho = 80$ kg/m³). The air on the right side of the wall is at 15° C and the convection film coefficient is 500 W/m²·K. The left side of the wall is exposed to a liquid at 170 K with a film convection coefficient of 3000 W/m²·K. Assume the heat flow to be one dimensional. Calculate the rate of heat transfer for a section 1 m high and 1 m wide in two ways:

(a)   Neglect the presence of the structural supports (entire space between plates is filled with vermiculite).

(b)   Consider the resistance to the transfer of heat between the two plates to be composed of that associated with the vermiculite and the structural support members which are connected in parallel.

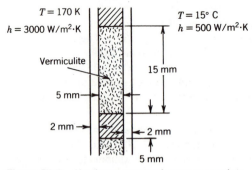

Figure P9-7   Aluminum structural support member.

9-8   Two pipes shown in the sketch Fig. P9–8 are buried deeply in the ground. If the surface temperature of the smaller pipe is 10° C while that of the larger pipe is 50° C, estimate the rate of heat transferred between the pipes per meter at length. The thermal conductivity of the soil is 0.52 W/m·K.

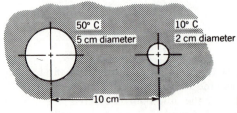

Figure P9-8   Buried pipes.

9-9   Radiative waste material is to be placed in a sphere which is then buried in soil ($k = 0.52$ W/m·K). The sphere has a diameter of 3 m and its center is buried 10 m below the surface of the ground. The rate of heat released at the start of the storage process is 1250 W. Estimate the surface temperature of the sphere if the surface temperature of the ground on a hot summer day is 33° C.

9-10   A small camp-type cooler has inside dimensions of 20 × 20 × 30 cm, see Fig. P9–10. The cooler is constructed of 3 cm polystyrene with a density of 56 kg/m³. The interior surface of the cooler is at 2° C. Estimate the rate of heat transfer if the outer exposed surfaces of the cooler are at 20° C. No heat is lost through the bottom of the cooler.

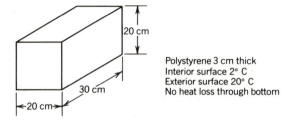

Figure P9-10 Polystyrene camp cooler.

Polystyrene 3 cm thick
Interior surface 2° C
Exterior surface 20° C
No heat loss through bottom

9–11   The rate of heat transfer and the temperature distribution in the straight rectangular fin shown in Fig. P9–11 is desired. Since two-dimensional effects may be present, the quantities are to be obtained using numerical techniques and the grid pattern shown in the figure. The root of the fin is held at a uniform temperature of 80° C while the surrounding air is at a temperature of 15° C. The fin is made of AISI 316 stainless steel with the dimensions indicated on the sketch. If the convection heat transfer coefficient is 30 W/m²·K, determine the rate of heat transfer per meter of fin length and the temperature distribution in the fin. Symmetry indicate that the top and bottom halves of the fin will have the same temperature distribution. Use the grid pattern shown in Fig. P9–11.

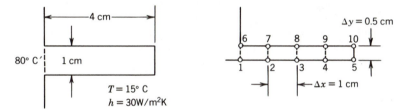

Figure P9-11 Straight rectangular fin.

9–12   A pure nichrome block of material is pressed into a block of hard vulcanized rubber, Fig. P9–12. The top and bottom of the rubber block are insulated while the sides of the rubber are exposed to the air, $h = 25$ W/m²·K and $T_\infty = 15°$ C. The nichrome block is an electrical conductor with an resistance internal heat generation of 100 kW/m³. Use numerical method to determine the temperatures at the nodes indicated in the diagram.

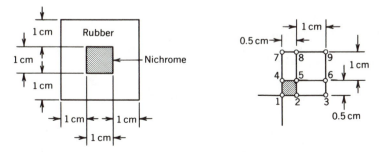

Figure P9-12 Nichrome bar pressed in hard rubber block.

9–13 The thermocouple shown in the sketch, Fig. P9–13, is to be used as the sensing element in a temperature control unit. The control unit is set to take corrective action if the fluid temperature is equal to or greater than 150° C. The normal operating fluid temperature is 100° C.

A malfunction in the system results in an instantaneous increase in the fluid temperature to 200° C. How long will it take the thermocouple to sense that the control unit must take corrective action if the convection heat transfer coefficient is 500 W/m²·K? The average thermocouple properties are:

$$k = 23 \text{ W/m·K}$$
$$\rho = 8920 \text{ kg/m}^3$$
$$c = .384 \text{ J/kg·K}$$

and the thermocouple diameter is 0.5 mm.

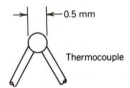

←—0.5 mm

Thermocouple

Figure P9-13 Thermocouple sensing element.

9–14 The fine thermocouple shown in Fig. P9–14 is initially at a temperature 15° C. The convection film coefficient is estimated to be 50 W/m²·K. It is submerged in a liquid that is at a temperature of 85° C. Determine the transient response of the thermocouple assuming its thermophysical properties are those given in Prob. 9–13.

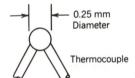

←— 0.25 mm
Diameter

Thermocouple

Figure P9-14 Small thermocouple.

9–15 A very long pyrex cylinder 0.5 cm in diameter is initially at a uniform temperature of 20° C. The cylinder is suddenly placed in an oil bath which is at 150° C. The average convection heat transfer coefficient is 150 W/m²·K. Estimate the temperature of the geometrical center of the cylinder 10 s after it was placed in the hot oil.

9–16 Soil is considered to be at a uniform temperature of 0° C. Over a relatively short period of time the temperature of the surrounding air is raised to 20° C. Estimate the temperature 3 cm below the ground surface after 1 hr. The convection heat transfer coefficient is 20 W/m²·K.

9–17 A fire test is to be conducted on a large mass of concrete initially at a uniform temperature of 15° C. The temperature of the surface is instantaneously raised to 500° C. Estimate the time required for the temperature at a depth of 30 cm to reach 100° C. The concrete may be considered to be a semi-infinite solid.

9–18 A flat plate, commercial bronze, 2 cm thick is initially at a temperature of 20° C. Hot gases at a temperature of 350° C are passed over the plate, Fig. P9–18. The

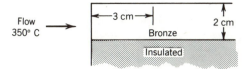

Figure P9-18 Flat bronze plate in hot gas stream.

local convection heat transfer coefficient at a location 3 cm from the leading edge is estimated to be 1000 W/m²·K. Conduction of heat within the plate in the direction of flow can be neglected.

(a) Estimate the surface temperature of the plate 3 cm from the leading edge 60 s after the gas flow starts.

(b) Estimate the time required for the bronze, 3 cm from the leading edge, to have a temperature of 346° C or greater (nearly steady-state conditions reached).

9–19 A hot water pipe for a home heating system is initially at a temperature of 15° C. The pump is started and water at 80° C is circulated through a plain carbon steel pipe which has an outside diameter of 3.340 cm and a thickness of 0.338 cm. The convection heat transfer coefficient on the water side is assumed to be 5000 W/m²·K.

If we touch the outside of the tube, assuming no heat is transferred between our fingers and the tube, how long must we wait after the hot water reaches our location before our fingers can sense heat (outside tube surface at 50° C)? Please state all assumptions made in your analysis.

9–20 Most foods are blanched in a water bath before they can be processed for canning. In the blanching process the minimum temperature of the food must be high enough to destroy the undesirable enzymes. The specific produce of concern is mushrooms which have an average spherical shape of 2 cm in diameter. If the minimum allowable temperature is 75° C, determine the minimum amount of time that the mushrooms must remain in the water bath. The initial temperature of the mushroom entering the water bath is 10° C and the bath temperature is 95° C. The thermodynamic properties of the mushrooms are approximately the same as those for water and the convection heat transfer coefficient is 1400 W/m²·K.

9–21 A plain carbon steel bar 15 × 15 × 15 cm is initially at a uniform temperature of 150° C. It is dropped into an oil bath which is maintained at a temperature of 35° C. The convection heat transfer coefficient is 600 W/m²·K. Determine the temperature of the center of the bar 10 min after it was placed in the oil using both the lumped parameter analysis and charts. Compare the answers.

9–22 A stainless-steel cylinder, AISI 316, 5 cm in diameter and 20 cm long is suspended in an agitated hot oil bath. The cylinder is initially at a temperature of 15° C. The average convection heat transfer coefficient during the heating process is 100 W/m²·K. Estimate the minimum bath temperature required to raise the temperature at the center of the cylinder to 100° C in 15 min.

10

THERMAL RADIATION HEAT TRANSFER

10.1 INTRODUCTION

In our discussion of the modes by which energy or heat can be transferred we have primarily been concerned with transfer process involving molecular transport of energy or energy transport associated with the movement of a fluid. These modes of heat transfer have been classified as conduction and convection. We shall now turn our attention to phenomena by which energy is transferred by *electromagnetic waves*. This is called thermal radiation heat transfer and was briefly mentioned in Chapter 1.

Since thermal radiation heat transfer is a wave phenomena, many factors must be considered when evaluating the rate at which energy transfer occurs. The distribution of the energy leaving a surface as thermal radiation is wavelength dependent. The spectral distribution of the radiation will depend upon the absolute temperature of the surface and the surface finish. When thermal radiation strikes a surface the amount of energy absorbed depends upon the spectral distribution of the incident radiation as well as the surface finish.

The wave characteristic of the energy transfer requires that consideration be given to the geometrical orientation of the surfaces involved in the radiation heat transfer process Direct radiant energy transfer is only possible between surfaces that "see" each other.

10.2 THERMAL RADIATION

All matter at a finite absolute temperature will emit radiation because of its molecular and atomic activity. The radiation is emitted in the form of electromagnetic waves and, for matter in an equilibrium state, is related to the internal energy of the matter. In a solid or liquid the radiation is basically a surface phenomena since it originates within a distance of 1×10^{-6} m of the surface. In a gas the radiation is emitted throughout the complete gaseous volume and a portion of it will be transmitted through the gas to the containing surface. Gaseous radiation is thus considered to be a volumetric phenomenon.

The basic theory of electromagnetic radiation can be studied from either the wave or quanta viewpoint. In most engineering treatments of the subject[1,2,3] the wave theory is used extensively. For the radiation emitted by a body, reference will be made to the *frequency*, ν, and *wavelength*, λ, of the radiation. These are related by

$$\lambda = \frac{c}{\nu} \tag{10-1}$$

where c is the velocity of light in the material. The velocity of light in a vacuum is $c_0 = 2.998 \times 10^8$ m/s. The index of refraction of a material is the ratio of the velocity of light in a vacuum to the velocity of light in the material.

$$\eta = \frac{c}{c_0} \tag{10-2}$$

The spectral distribution of electromagnetic radiation is presented in Fig. 10-1. The types of radiation have been classified according to their wavelengths. The units of wavelength commonly used are microns or micrometers, μm, 10^{-6} m, and angstroms, Å, 10^{-10} m. The visible light region of the spectrum extends between 0.4 and 0.7 μm and the region between 0.1 and 100 μm is called the thermal radiation region.

The thermal radiation emitted by matter can be broken down into its mon-

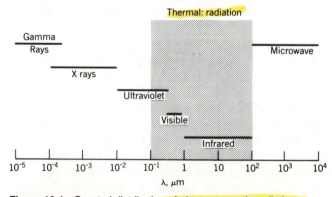

Figure 10-1   Spectral distribution of electromagnetic radiation.

ochromatic, single-wavelength, components. The spectral distribution of radiation for the surface of an ideal radiating body, called a *blackbody*, has been obtained by Planck[4]. The monochromatic rate of energy emitted to a hemispheric envelope over an ideal radiator is a function of wavelength and the temperature of the emitting surface. It is given by the following expression,

$$E_{\lambda,b} = \frac{C_1}{\lambda^5[\exp(C_2/\lambda T) - 1]} \quad \text{W/m}^2 \cdot \mu\text{m} \tag{10-3}$$

where $C_1 = 3.742 \times 10^8\,\text{W}\cdot\mu\text{m}^4/\text{m}^2$ and $C_2 = 1.439 \times 10^4\,\mu\text{m}\cdot\text{K}$. This expression is for a surface in a vacuum and must be modified if the index of refraction differs appreciably from 1. The monochromatic energy spectral distribution for the surface of an ideal radiating body at several different temperatures is shown in Fig. 10-2. The wavelength where the maximum monochromatic emission occurs, λ_{max}, decreases as the temperature of the ideal radiating body increases. The relationship between λ_{max} and temperature is given by Wien's displacement law

$$\lambda_{max} T = 2.90 \times 10^3\,\mu\text{m}\cdot\text{K} \tag{10-4}$$

The total rate of energy emitted to the hemispherical envelope over an ideal radiating body, blackbody, is obtained by integrating the monochromatic emission over the entire wavelength range.

$$E_b = \int_0^\infty E_{\lambda,b}\,d\lambda \tag{10-5}$$

The value of the integral is

$$E_b = \sigma T^4 \tag{10-6}$$

where σ is the Stefan–Boltzmann constant, $\sigma = 5.670 \times 10^{-8}\,\text{W/m}^2\cdot\text{K}^4$

It is often desirable to know the rate of energy radiating from a surface in a

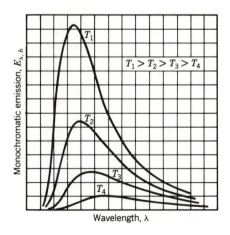

T_1

$T_1 > T_2 > T_3 > T_4$

T_2

T_3

T_4

Monochromatic emission, $E_{\lambda,b}$

Wavelength, λ

Figure 10-2 Spectral distribution of radiation for a blackbody (eq. 10-3).

certain wavelength interval. The rate of radiation emitted by the surface of a blackbody at an absolute temperature of T in the wavelength interval $0-\lambda_1$ can be determined using eq. 10–3.

$$E_{0-\lambda_1,b} = \int_0^{\lambda_1} \frac{C_1}{\lambda^5[\exp(C_2/\lambda T) - 1]} d\lambda \qquad (10\text{–}7)$$

A more convenient expression is obtained by expressing the emitted radiation within a wavelength interval as a fraction of the total rate of energy radiated by the surface of an ideal radiating body at the same temperature. The fraction of the radiation in the wavelength interval from 0 to λ_1 is obtained by dividing eq. 10–7 by eq. 10–6.

Table 10–1 Blackbody Radiation Functions

$\lambda T(\mu m \cdot K)$	$F_{[0-\lambda]}$	$\lambda T(\mu m \cdot K)$	$F_{[0-\lambda]}$
200	0.000000	6,200	0.754140
400	0.000000	6,400	0.769234
600	0.000000	6,600	0.783199
800	0.000016	6,800	0.796129
1,000	0.000321	7,000	0.808109
1,200	0.002134	7,200	0.819217
1,400	0.007790	7,400	0.829527
1,600	0.019718	7,600	0.839102
1,800	0.039341	7,800	0.848005
2,000	0.066728	8,000	0.856288
2,200	0.100888	8,500	0.874608
2,400	0.140256	9,000	0.890029
2,600	0.183120	9,500	0.903085
2,800	0.227897	10,000	0.914199
2,898	0.250108	10,500	0.923710
3,000	0.273232		
3,200	0.318102	11,000	0.931890
3,400	0.361735	11,500	0.939959
3,600	0.403607	12,000	0.945098
3,800	0.443382	13,000	0.955139
4,000	0.480877	14,000	0.962898
4,200	0.516014	15,000	0.969981
4,400	0.548796	16,000	0.973814
4,600	0.579280	18,000	0.980860
4,800	0.607559	20,000	0.985602
5,000	0.633747	25,000	0.992215
5,200	0.658970	30,000	0.995340
5,400	0.680360	40,000	0.997967
5,600	0.701046	50,000	0.998953
5,800	0.720158	75,000	0.999713
6,000	0.737818	100,000	0.999905

$$F_{[0-\lambda_1]} = \frac{E_{0-\lambda_1,b}}{E_b} = \int_0^{\lambda_1 T} \frac{C_1}{\sigma \lambda^5 T^5 [\exp(C_2/\lambda T) - 1]} d(\lambda T) \qquad (10\text{–}8)$$

The values of $E_{[0-\lambda_1]}$, as a function of λT, are presented in Table 10–1. The fraction of the radiation emitted by the surfaces of an ideal radiating body within the wavelength interval $\lambda_1 - \lambda_2$ may be obtained through use of Table 10–1.

$$F_{[\lambda_1-\lambda_2]} = F_{[0-\lambda_2]} - F_{[0-\lambda_1]} \qquad (10\text{–}9)$$

EXAMPLE 10–1

Solar radiation has approximately the same spectral distribution as an ideal radiating body at a temperature of 5800 K. Determine the amount of solar radiation which is in the visible range 0.40–0.70 μm.

SOLUTION

The total rate of energy radiated by an ideal radiating body at a temperature of 5800 K is obtained from eq. 10–6.

$$E_b = \sigma T^4 = 5.67 \times 10^{-8}(5800)^4$$

$$= 64.16 \times 10^6 \text{ W/m}^2$$

The fraction of this radiation in the visible range may be obtained using Table 10–1. The radiation contained in the wavelength intervals of 0–0.4 and 0–0.7 μm is

$0 \leqslant \lambda \leqslant 0.4$ $\lambda_1 T = 0.4(5800) = 2320$ $F_{[0-0.4]} = 0.1245$

$0 \leqslant \lambda \leqslant 0.7$ $\lambda_2 T = 0.7(5800) = 4060$ $F_{[0-0.7]} = 0.4914$

The fraction of the solar radiation contained in the visible range is

$$F_{[0.4-0.7]} = F_{[0-0.7]} - F_{[0-0.4]}$$

$$= 0.4914 - 0.1245$$

$$= 0.3669$$

The amount of radiation in the visible range, $0.4 < \lambda < 0.7$ μm is

$$E_{\lambda_1-\lambda_2,b} = F_{[\lambda_1-\lambda_2]} E_b$$

$$= 0.3669(64.16 \times 10^6)$$

$$= 23.54 \times 10^6 \text{ W/m}^2$$

This is shown schematically in Fig. E10–1.

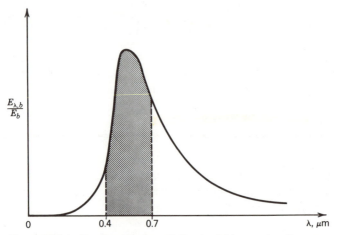

Figure E10-1   Fraction of solar radiation in visible wavelength range.

COMMENT

The wavelength at which the maximum rate of energy is emitted from the sun can be found by using Wien's displacement law, eq. 10–4,

$$\lambda_{max} = \frac{2.9 \times 10^3}{5800} = 0.5 \ \mu m$$

10.3   BASIC RADIATION PROPERTIES

10.3.1   Blackbody

A *blackbody* is an ideal body whose surface is a perfect absorber of incident radiation regardless of the wavelength or the direction of the radiation. Since there are no known surfaces that possess these characteristics, the concept of a blackbody is idealistic. Nevertheless, it is of value because it is a standard for the comparison of the radiation properties of real surfaces.

   It can be shown that a blackbody is also a perfect emitter of radiation in every direction and at every wavelength. For a given temperature, no surface can emit more energy, either total or monochromatic, than a blackbody. All the radiation characteristics presented in the previous section were associated with an ideal radiating surface or blackbody. The radiation characteristics of a blackbody are identified by the use of a subscript b.

10.3.2   Irradiation

The rate at which radiation strikes a surface is called *irradiation*. Directional characteristics of the radiation are important. The irradiation per unit area is identified by G, in watts per meter squared. The subscript λ will be used to denote the monochromatic rate of radiant energy striking the surface. The total radiation incident on a surface is obtained by integrating over the complete range of wavelengths.

$$G = \int_0^\infty G_\lambda \, d\lambda \tag{10--10}$$

10.3.3 Absorptivity, Reflectivity, and Transmissivity

When radiation is incident on a real surface some of the radiation is absorbed, some of it is reflected, and the remainder is transmitted through the body as shown in Fig. 10–3. The sum of these quantities must be equal to the total radiation incident on the surface, G.

It is convenient to express the amount of the incident radiation which is absorbed, reflected, or transmitted as a fraction of the total radiation incident on the surface. The following quantities are defined.

Absorptivity. The fraction of the total incident radiation that is absorbed by the surface. For a real body, the absorptivity usually varies with wavelength, so the monochromatic absorptivity is denoted by α_λ. The absorptivity is expressed in terms of the monochromatic absorptivity by

$$\alpha = \frac{\text{absorbed radiation}}{\text{incident radiation}} = \frac{1}{G} \int_0^\infty \alpha_\lambda G_\lambda \, d\lambda \tag{10--11}$$

Reflectivity. The fraction of the total incident radiation that is reflected by the surface. Again this property is a function of wavelength so ρ_λ is used to represent the monochromatic reflectivity of a surface and

$$\rho = \frac{\text{reflected radiation}}{\text{incident radiation}} = \frac{1}{G} \int_0^\infty \rho_\lambda G_\lambda \, d\lambda \tag{10--12}$$

There are two types of reflection of the electromagnetic waves, specular and diffuse. Specular reflections are present when the angle of incidence is equal to the angle of reflection. Diffuse radiation is present when the reflection is

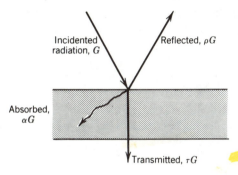

Incidented radiation, G

Reflected, ρG

Absorbed, αG

Transmitted, τG

Figure 10-3 Incident radiation on a surface.

uniformly distributed in all directions. These two types of reflection are shown in Fig. 10–4. A real body exhibits neither pure specular nor diffuse reflections. A highly polished surface will yield specular reflections while the reflections from a rough surface have a distinct diffuse characteristic.

Transmissivity. The fraction of the total incident radiation that is transmitted through the body. It also has a wavelength dependency. The monochromatic transmissivity is designated by τ_λ and the total transmissivity is

$$\tau = \frac{\text{transmitted radiation}}{\text{incident radiation}} = \frac{1}{G} \int_0^\infty \tau_\lambda \, G_\lambda \, d\lambda \qquad (10\text{–}13)$$

For most solid surfaces the transmissivity is equal to the zero since the bodies are usually opaque to the incident radiation.

The sum of the absorptivity, reflectivity, and transmissivity is equal to 1.

$$\alpha + \rho + \tau = 1 \qquad (10\text{–}14)$$

For an opaque body

$$\tau = 0$$

so

$$\alpha + \rho = 1 \qquad (10\text{–}15)$$

10.3.4 Emissivity

The total amount of energy radiated by the surface of a blackbody is given by eq. 10–6 and the monochromatic radiation emitted by the surface is given by eq. 10–3. A real body emits less radiation than a blackbody. The ratio of the actual energy emitted by a real body to that emitted by a blackbody at the same temperature is called the *emissivity*. The monochromatic emissivity is ε_λ and the total emissivity is obtained by integrating over the complete wavelength spectrum

$$\varepsilon = \frac{1}{E_b} \int_0^\infty \varepsilon_\lambda \, E_{\lambda,b} \, d\lambda \qquad (10\text{–}16)$$

The spectral distribution of radiation, as previously noted, is associated with the temperature of the radiating body. The radiation characteristics of a surface,

Figure 10-4 Specular and diffuse reflection.

absorptivity and emissivity, are also strongly dependent on the spectral distribution of the radiation. If the radiation incident on a surface at T_1 comes from a surface which is also at a temperature of T_1, the spectral distribution of the energy will be identical and the emissivity and absorptivity of the surface will be equal,

$$\varepsilon = \alpha \quad \text{and} \quad \varepsilon_\lambda = \alpha\lambda$$

This situation is shown schematically in Fig. 10–5.

10.3.5 Gray Body

The surface of a body whose monochromatic emissivity and absorptivity are independent of wavelength and direction is called a gray body

$$\varepsilon = \varepsilon_\lambda = \text{const}$$

and

$$\alpha = \alpha_\lambda = \text{const}$$

The radiation emitted by a gray body and the reflected radiation from the body are considered to be diffuse. The emissivity and the absorptivity of a gray body are equal,

$$\varepsilon = \alpha$$

10.3.6 Real Body

The radiation properties of the surface of a real body are different, in the strict sense of the word, from those of both a blackbody and gray body. The monochromatic emissivity of several real surfaces is shown in Fig. 10–6. The radiation emitted by a real body is not entirely diffuse, thus the emissivity of the body is dependent on the viewing angle. The directional variation in the emissivity for several different materials is shown in Fig. 10–7.

Since engineering calculations are of primary interest, it is important to rec-

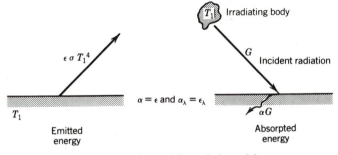

Figure 10-5 Equivalence of emissivity and absorptivity.

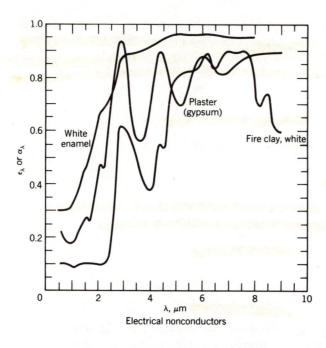

Electrical nonconductors

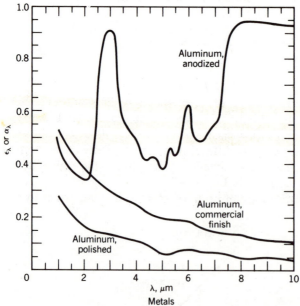

Metals

Figure 10-6 Spectral dependency of emissivity and absorptivity[2]. Used with permission. (*a*) Electrical nonconductors. (*b*) Metals.

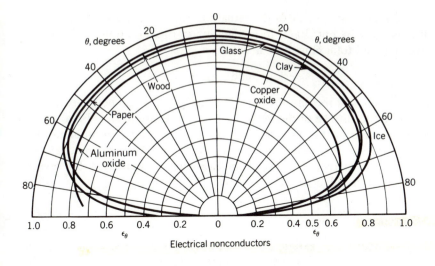

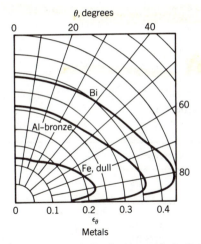

Figure 10-7 Directional total emissivity[2]. Used with permission[2]. (*a*) Electrical nonconductors. (*b*) Metals.

ognize when the radiation characteristics of the surfaces of a real body may be approximated by those of a gray body. In deciding if such approximations are possible, the spectral distribution of the radiation emitted by the body and the radiation irradiated on the body must be considered. For further illustration, refer to Fig. 10–6. If the major portion of the incident radiation striking an aluminum anodized surface falls in the wavelength range of 8 to 10 μm, the surface could be considered to behave as a gray body with an absorptivity of 0.93. No appreciable error would be introduced since the absorptivity is nearly constant in this wavelength range. If, however, the radiation incident on the surface was spread over the wavelength interval of 2 to 10 μm, the gray body approximation can still be used but with decreased accuracy. The average absorptivity of the surface is obtained using eq. 10–11.

The directional characteristic of the radiation from the surface of real bodies has been illustrated by Fig. 10–7. In recognition of this variation, a directional monochromatic and total emissivity is used.

$$\varepsilon_{\lambda,\theta} \quad \text{and} \quad \varepsilon_\theta$$

The tabulated values of the emissivity for the surface of a real body are usually those normal to the surface of the body, $\theta = 0°$. These are distinguished by the subscripts n, $\varepsilon_{\lambda,n}$ and ε_n. For electrical nonconductors, the variation in $\varepsilon_{\lambda,\theta}$ and ε_θ is less than $\pm 3\%$. For conductors, the variation may be somewhat greater, perhaps as large as $\pm 15\%$. Values of the normal total emissivity for several surfaces are given in Table 10–2.

10.3.7 Radiosity

The amount of thermal radiation leaving a body is called the *radiosity*. It is the sum of the incident radiation which is reflected and that which is emitted by the body. The radiosity for a gray body is shown schematically in Fig. 10–8. With the radiosity denoted by J, it may be expressed in terms of emissivity and reflectivity of the surface as

$$J = \varepsilon E_b + \rho G \tag{10–17}$$

The radiosity represents the rate of energy transferred per unit area and has the units of watts per meter squared.

Table 10–2 Normal Total Emissivity

Substance	Metals Surface temperature, K	ε_η
Aluminum		
Highly polished	480–870	0.038–0.06
Heavily oxidized	370–810	0.20–0.33
Brass		
Highly polished	530–640	0.028–0.031
Oxidized	480–810	0.60
Chromium, polished	310–1370	0.08–0.40
Copper		
Highly polished	310	0.02
Black oxidized	310	0.78
Gold, polished	400	0.018
Iron		
Highly polished	310–530	0.05–0.07
Wrought iron, polished	310–530	0.28

Table 10–2—Continued.

Substance	Metals Surface temperature, K	ε_η
Cast iron, freshly turned	310	0.44
Iron plate, rusted	293	0.61
Cast, iron, rough and strongly oxidized	310–530	0.95
Platinum, polished	500–900	0.054–0.104
Silver, polished	310–810	0.01–0.03
Stainless steel		
Type 310, smooth	1090	0.39
Type 316, polished	480–1310	0.24–0.31
Tins, polished	310	0.05
Tungsten, filament	3590	0.39
Nonmetals		
Asbestos		
Paper	310	0.93
Board	310	0.96
Brick		
White refractory	1370	0.29
Red, rough	310	0.93
Carbon, lampsoot	310	0.95
Concrete, rough	310	0.94
Ice, smooth	273	0.966
Marble, white	310	0.95
Paint		
Oil, all colors	373	0.92–0.96
Lead, red	370	0.93
Plaster	310	0.91
Rubber, hard	293	0.92
Snow	270	0.82
Water, deep	273–373	0.96
Wood		
Oak	295	0.90
Beech	340	0.94

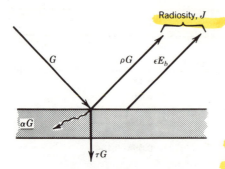

Figure 10-8 Energy balance on the surface of a gray body.

10.4 RADIATION HEAT TRANSFER BETWEEN TWO INFINITE PARALLEL SURFACES

The characteristics of the radiation emitted, reflected, absorbed, or transmitted by a surface have been discussed. These are to be used to determine the net rate of heat transferred by radiation between two surfaces that are at different temperatures. To simplify the calculations, the two bodies are assumed parallel and infinite so that all the radiation leaving one body will strike the second body.

Consider the two surfaces shown in Fig. 10–9 which are at T_1 and T_2. Since both of these surfaces are at a temperature greater than absolute zero, each surface will emit radiation. The total energy leaving surface 1 is its radiosity times its surface area, $J_1 A_1$, and that leaving surface 2 is $J_2 A_2$. The net rate of heat transfer between the two surfaces is

$$\dot{Q} = J_1 A_1 - J_2 A_2 = \frac{J_1 - J_2}{(1/A)} \tag{10–18}$$

since A_1 is equal to A_2.

If the surfaces are blackbodies, $\varepsilon_1 = \varepsilon_2 = 1$ and $\alpha_1 = \alpha_2 = 1$. The reflectivity and transmissivity are 0 and the radiosities for bodies 1 and 2 are

$$J_1 = \sigma T_1^4$$

and

$$J_2 = \sigma T_2^4$$

The net rate of heat transfer per unit area is

$$\frac{\dot{Q}}{A} = \sigma(T_1^4 - T_2^4) \tag{10–19}$$

When a surface is an opaque gray body, transmissivity equal to 0, the radiosity is

$$J = \varepsilon E_b + \rho G \tag{10–20}$$

$$= \varepsilon E_b + (1 - \varepsilon)G$$

This equation can be rearranged to obtain an expression for the irradiation

$$G = \frac{J - \varepsilon E_b}{(1 - \varepsilon)} \qquad (10\text{-}21)$$

The net rate of heat transfer from the surface of an opaque gray body can be expressed as the difference between the radiosity and the irradiation

$$\dot{Q} = A(J - G) = A\left[J - \left(\frac{J - \varepsilon E_b}{1 - \varepsilon}\right)\right]$$

$$= \frac{\varepsilon A}{1 - \varepsilon}(E_b - J)$$

$$= \frac{E_b - J}{[(1 - \varepsilon)/\varepsilon A]} \qquad (10\text{-}22)$$

If J is greater than E_b, $\dot{Q}$ will have a *negative* sign which indicates that the net rate of heat transfer is *to* the surface in question.

From Fig. 10-9, it is obvious that if both surfaces are opaque gray bodies, the rate of heat loss by body 1 is equal to that gained by body 2. This is expressed mathematically as

$$\dot{Q} = \frac{E_{b1} - J_1}{[(1 - \varepsilon_1)/\varepsilon_1 A]} = -\frac{E_{b2} - J_2}{[(1 - \varepsilon_2)/\varepsilon_2 A]} \qquad (10\text{-}23)$$

Equation 10-23 contains two unknowns, J_1 and J_2. These radiosities can be determined by solving eq. 10-18 and 10-23 simultaneously.

It is appropriate at this time to recall the analogy between heat transfer and the flow of electrical current which was described in Chapter 9. Equations 10-18 and 10-23 can be represented by equivalent electrical resistances and potential

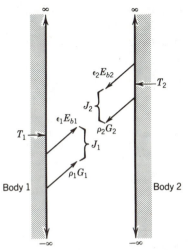

Body 1

Body 2

Figure 10-9   Two infinite parallel surfaces.

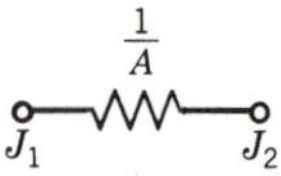

$$_1\dot{Q}_2 = \frac{J_1 - J_2}{(1/A)}$$ eq. 10-18

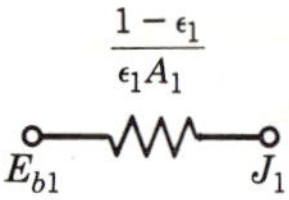

$$_1\dot{Q}_2 = \frac{E_{b1} - J_1}{(1 - \epsilon_1)/\epsilon_1 A_1}$$ eq. 10-23

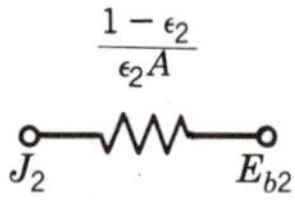

$$_1\dot{Q}_2 = \frac{J_2 - E_{b2}}{(1 - \epsilon_2)/\epsilon_2 A}$$ eq. 10-23

Figure 10-10   Resistances for radiation network.

differences as shown in Fig. 10–10. Radiation heat transfer between body 1 and 2 in Fig. 10–9 can be obtained by solving the radiation network formed by combining the resistances shown in Fig. 10–10. The equivalent radiation network for the heat transfer between two infinite parallel gray surfaces is shown in Fig. 10–11.

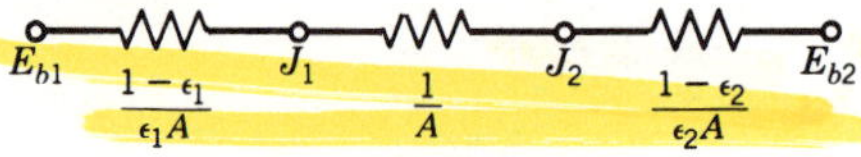

$$\dot{Q} = \frac{E_{b1} - E_{b2}}{[(1 - \epsilon_1)/\epsilon_1 A] + (1/A) + [1 - \epsilon_2)/\epsilon_1 A]}$$

Figure 10-11   Radiation network for infinite parallel surfaces (Figure 10-9).

EXAMPLE 10–2

Two very large parallel surfaces are held at uniform temperatures of 300° C (573.2 K) and 20° C, (293.2 K), respectively. The high-temperature surface has an emissivity of 0.8 and the emissivity of the low temperature surface is 0.1. Determine the rate of heat transfer by radiation between the two surfaces.

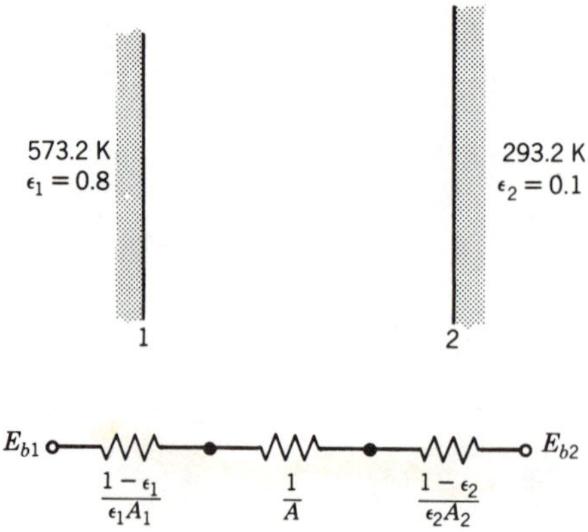

573.2 K $\epsilon_1 = 0.8$ 293.2 K $\epsilon_2 = 0.1$

1   2

E_{b1} $\frac{1 - \epsilon_1}{\epsilon_1 A_1}$ $\frac{1}{A}$ $\frac{1 - \epsilon_2}{\epsilon_2 A_2}$ E_{b2}

Figure E10-2   Radiation network.

SOLUTION

Assume that both surfaces are large enough for them to be considered to be infinite parallel plates. The radiation network for the system is shown in Fig. E10–2. The rate of radiation heat transfer per unit area, $A = 1 \text{ m}^2$, is obtained from

$$\dot{Q} = \frac{E_{b1} - E_{b2}}{\Sigma R}$$

$$= \frac{\sigma T_1^4 - \sigma T_2^4}{[(1 - \varepsilon_1)/\varepsilon_1 A] + (1/A) + [(1 - \varepsilon_2)/\varepsilon_2 A]}$$

$$= \frac{5.67 \times 10^{-8}[(573.2)^4 - (293.2)^4]}{[(1 - 0.8)/0.8(1)] + (1/1) + [(1 - 0.1)/0.1(1)]}$$

$$= 556.3 \text{ W/m}^2$$

EXAMPLE 10–3

A third plate with an emissivity of 0.3 is inserted between the two plates described in Example 10–2. Recalculate the rate of radiative heat transfer between the original two plates.

SOLUTION

The equivalent radiation network for this system is shown in Fig. E10–3. The radiative heat transfer per unit area is

$$\dot{Q} = \frac{E_{b1} - E_{b2}}{\Sigma R}$$

where

$$\Sigma R = \frac{1 - \varepsilon_1}{\varepsilon_1 A} + \frac{1}{A} + \frac{1 - \varepsilon_3}{\varepsilon_3 A} + \frac{1 - \varepsilon_3}{\varepsilon_3 A} + \frac{1}{A} + \frac{1 - \varepsilon_2}{\varepsilon_2 A}$$

$$= \frac{1 - 0.8}{0.8(1)} + \frac{1}{1} + \frac{1 - 0.3}{0.3(1)} + \frac{1 - 0.3}{0.3(1)} + \frac{1}{1} + \frac{1 - 0.1}{0.1(1)}$$

$$= 15.92 \text{ 1/m}^2$$

Thus

$$\dot{Q} = \frac{5.67 \times 10^{-8}[(573.2)^4 - (292.2)^4]}{15.92}$$

$$= 358.2 \text{ W/m}^2$$

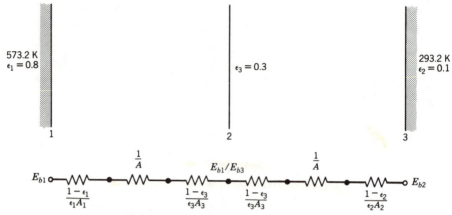

Figure E10-3   Radiation network.

COMMENT

The rate of heat transfer between the two plates was reduced by approximately one-third by inserting the plate between the two surfaces. If the emissivity of the inserted plate was 0.1, the rate of heat transfer would be reduced to 194.9 W/m².

The temperature of the third plate may be obtained using the radiation network which indicates

$$\dot{Q} = \frac{E_{b1} - E_{b3}}{[(1 - \varepsilon_1)/\varepsilon_1 A] + (1/A) + [(1 - \varepsilon_3)/\varepsilon_3 A]}$$

$$358 = \frac{5.67 \times 10^{-8}[(573.2)^4 - (T_3)^4]}{[(1 - 0.8)/0.8(1)] + (1/1) + [(1 - 0.3)/0.3(1)]}$$

$$T_3 = 540.4\text{K}(267.2° \text{ C})$$

10.5 SHAPE FACTORS

Thermal radiation is transmitted via electromagnetic waves that travel in straight lines. The geometrical orientation of surfaces will greatly influence the magnitude of the rate of radiation heat transfer between the surfaces. In order to take these factors into consideration, a shape factor is introduced. The shape factor, $F_{i,j}$, is defined as the fraction of the radiation which leaves surface i and is incident on surface j. For the two surfaces shown in Fig. 10–12, the radiation leaving surface 1, $J_1 A_1$, which strikes surface 2 is denoted by $F_{1,2} J_1 A_1$ while that leaving surface 2, $J_2 A_2$, and striking surface 1 is $F_{2,1} J_2 A_2$.

If both the surfaces shown in Fig. 10–12 are blackbodies, the radiosities are

$$J_1 = E_{b1} = \sigma T_1^4$$

$$J_2 = E_{b2} = \sigma T_2^4$$

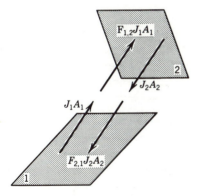

Figure 10-12 Radiation heat transfer between two opaque surfaces.

The net rate of radiation heat transfer between the two bodies is

$$\dot{Q} = F_{1,2}E_{b1}A_1 - F_{2,1}E_{b2}A_2 \tag{10-24}$$

If both surfaces are at the same temperature no heat will be transferred so

$$0 = F_{1,2}E_{b1}A_1 - F_{2,1}E_{b2}A_2 \tag{10-25}$$

Since $E_{b1} = E_{b2}$, an important relationship between the shape factors is obtained, namely

$$A_1F_{1,2} = A_2F_{2,1}$$

or in general form

$$A_iF_{i,j} = A_jF_{j,i} \tag{10-26}$$

This expression contains only geometrical factors. The relationship, often referred to as the *reciprocity relationship*, is valid as long as the radiation is diffuse regardless of the other radiation characteristics of the surface.

The shape factors for several configurations are shown in Figs. 10–13 through 10–15. A more complete set of shape factors relationships is available in Refs. 1, 2, and 3. There are two important relationships that are useful in the determination of the shape factor if curves are not available for the configuration desired. One of these, the reciprocity relationship, has already been presented, eq. 10–26. The second is extremely useful in multisurface systems and is based upon the definition of the shape factor. All the surfaces seen from the radiating surface will receive radiation from the surface. The sum of all the shape factors for a given radiating surface must be equal to 1. For example, if surface 1 is enclosed and sees n surfaces then

$$\sum_{j=1}^{n} F_{1,j} = F_{1,1} + F_{1,2} + F_{1,3} + \cdots F_{1,n} = 1 \tag{10-27}$$

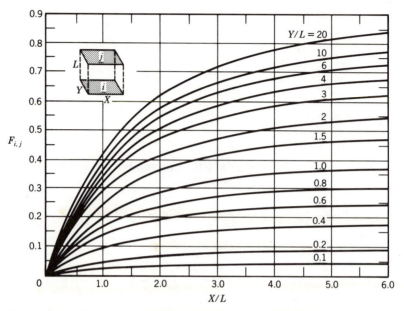

The shape factor $F_{1,1}$ will usually be zero. The exception is when surface 1 sees itself, it is a concave surface. Two examples will be presented to illustrate the use of these relationships.

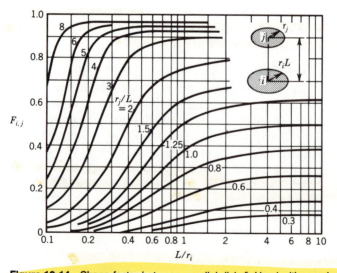

Figure 10-14   Shape factor between parallel disks[5]. Used with permission.

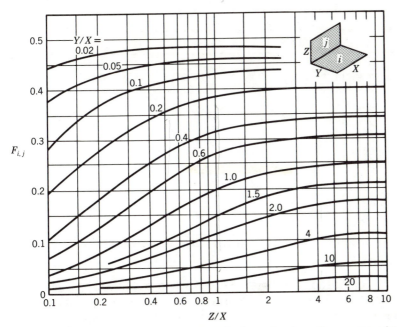

Figure 10-15 Shape factor between perpendicular rectangles with common edge[5]. Used with permission.

EXAMPLE 10–4

Determine the shape factor $F_{1,2}$ for the configuration shown in Fig. E10–4.

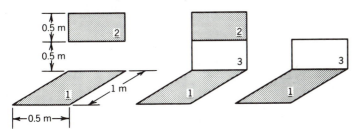

Figure E10-4 Determination of shape factor between surfaces 1 and 2.

SOLUTION

The shape factor cannot be determined directly from the charts that have been presented. It can, however, be determined if a third surface connecting 1 and 2 is introduced. The fraction of the radiant energy leaving surface 1 which strikes the combined surfaces 2 and 3 is

$$F_{1,23} = F_{1,2} + F_{1,3}$$

Shape factors $F_{1,23}$ and $F_{1,3}$ can be obtained from Fig. 10–15.

Shape Factor

		Z/X	Y/X	Fig. 10–15
$F_{1,23}$	$X = 0.5m$ $Y = 1m$ $Z = 1m$	$\dfrac{1.0}{0.5} = 2$	$\dfrac{1.0}{0.5} = 2$	$F_{1,23} = 0.14$
$F_{1,3}$	$X = 0.5m$ $Y = 1m$ $Z = 0.5m$	$\dfrac{0.5}{0.5} = 1$	$\dfrac{1.0}{0.5} = 2$	$F_{1,3} = 0.12$

The shape factor $F_{1,2}$ is

$$F_{1,2} = F_{1,23} - F_{1,3}$$
$$= 0.14 - 0.12$$
$$= 0.02$$

EXAMPLE 10–5

A container is composed of two parallel disks and a connecting cylindrical surface as shown in Fig. E10–5. Determine the fraction of the radiant energy leaving surface 3 which strikes itself because of its concave shape.

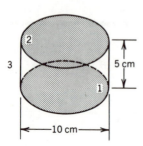

Figure E10-5 Cylindrical enclosure.

SOLUTION

The areas of the three surfaces are:

$$A_1 = A_2 = \frac{\pi d^2}{4} = \frac{\pi(0.1)^2}{4} = 7.854 \times 10^{-3} m^2$$

$$A_3 = \pi dL = \pi(0.1)(0.05) = 15.71 \times 10^{-3} m^2$$

The following relationships exist between the shape factors associated with each surface.

$$F_{1,2} + F_{1,3} = 1, \quad F_{2,1} + F_{2,3} = 1, \quad \text{and} \quad F_{3,1} + F_{3,2} + F_{3,3} = 1$$

By symmetry

$$F_{1,2} = F_{2,1}, \quad F_{1,3} = F_{2,3}, \quad \text{and} \quad F_{3,1} = F_{3,2}$$

Figure 10–14 is used to find $F_{1,2}$ and $F_{2,1}$.

$$\frac{r_2}{L} = \frac{0.05}{0.05} = 1 \qquad \frac{L}{r_1} = \frac{0.05}{0.05} = 1$$

$$F_{1,2} = 0.38$$

Therefore,

$$F_{1,3} = 1 - F_{1,2} = 1 - 0.38$$

$$= 0.62$$

and $F_{2,3} = 0.62$.

The reciprocity relationship eq. 10–26 is used now to find $F_{3,1}$ and $F_{3,2}$.

$$A_1 F_{1,3} = A_3 F_{3,1}$$

so

$$F_{3,1} = \frac{A_1}{A_3} F_{1,3}$$

$$= \frac{7.854 \times 10^{-3}}{15.71 \times 10^{-3}} (0.62)$$

$$= 0.310$$

and

$$F_{3,2} = 0.310$$

The fraction of the radiant energy leaving surface 3 which strikes itself is

$$F_{3,3} = 1 - F_{3,1} - F_{3,2}$$

$$= 1 - 0.31 - 0.31$$

$$= 0.38$$

10.6   RADIATION HEAT TRANSFER BETWEEN ANY TWO GRAY SURFACES

In Section 10.4, the heat transfer between two infinite plates was discussed. The radiation network and the corresponding equation shown in Fig. 10–11 was used to calculate the total heat transfer between the two gray surfaces. Since both surfaces were infinite, the shape factor $F_{1,2} = F_{2,1} = 1$. Now a general approach will be developed, based upon the electrical analogy, for the calculation of the rate of radiant heat transfer between opaque gray surfaces.

The net rate of heat transfer between any two opaque gray surfaces can be expressed as

$$\dot{Q} = F_{1,2} J_1 A_1 - F_{2,1} J_2 A_2 \tag{10-28}$$

This can be rearranged by the reciprocity relationship, $A_1 F_{1,2} = A_2 F_{2,1}$ to obtain

$$\dot{Q} = \frac{J_1 - J_2}{(1/A_1 F_{1,2})} \tag{10-29}$$

or

$$\dot{Q} = \frac{J_1 - J_2}{(1/A_2 F_{2,1})}$$

The net rate of heat transfer between two gray surfaces can be obtained by using a radiation network composed of two classes of resistances. One class of resistance will be associated with the *geometrical* relationship between the surfaces. These will contain the shape factors and be expressed as

$$R_s = \frac{1}{A_i F_{i,j}} \tag{10-30}$$

The other class of resistance, which will be called a *surface characteristic* resistance, will account for the radiation characteristics of the surface and will be expressed as

$$R_\varepsilon = \frac{1 - \varepsilon_i}{\varepsilon_i A_i} \tag{10-31}$$

The equivalent radiation network for the configuration shown in Fig. 10–12 is presented in Fig. 10–16. The rate of radiation heat transfer between opaque gray surfaces 1 and 2 is

$$\dot{Q} = \frac{E_{b1} - E_{b2}}{\Sigma R} \tag{10-32}$$

$$= \frac{\sigma(T_1^4 - T_2^4)}{[(1 - \varepsilon_1)/\varepsilon_1 A_1] + (1/A_1 F_{1,2}) + [(1 - \varepsilon_2)/\varepsilon_2 A_2]}$$

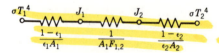

Figure 10-16 Radiation network for heat transfer between two opaque gray surfaces.

EXAMPLE 10–6

Determine the rate of heat transfer from a small electrically heated sphere placed in a closed evacuated cylinder, Fig. E10–6. The sphere is 10 cm in diameter with an emissivity of 0.8 and is maintained at a uniform temperature of 300° C (572.2 K). The inside surface of the cylinder, surface area 0.5 m², has an emissivity of 0.2 and is maintained at a uniform temperature of 20° C (293.2 K).

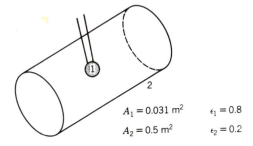

$A_1 = 0.031 \ m^2$ $\epsilon_1 = 0.8$

$A_2 = 0.5 \ m^2$ $\epsilon_2 = 0.2$

$$E_{b1} \ \text{—} \underset{\dfrac{1-\epsilon_1}{\epsilon_1 A_1}}{\text{\Large\char`\~}} \text{—} \underset{\dfrac{1}{A_1 F_{1,2}}}{\text{\Large\char`\~}} \text{—} \underset{\dfrac{1-\epsilon_2}{\epsilon_2 A_2}}{\text{\Large\char`\~}} \text{—} \ E_{b2}$$

Figure E10-6   Sphere within cylindrical enclosure.

SOLUTION

The equivalent radiation network is shown in Fig. E10–6. The surface area of the sphere is calculated and found to be 0.031 m². The shape factor $F_{1,2} = 1$ since all the radiation emitted by the sphere strikes the inner surface of the cylinder. The rate of heat transfer is obtained by

$$\dot{Q} = \frac{E_{b1} - E_{b2}}{\Sigma R}$$

$$= \frac{\sigma(T_1^4 - T_2^4)}{[(1 - \epsilon_1)/\epsilon_1 A_1] + (1/A_1 F_{1,2}) + [(1 - \epsilon_2)/\epsilon_2 A_2]}$$

$$= \frac{5.67 \times 10^{-8}[(573.2)^4 - (293.2)^4]}{[(1 - 0.8)/0.8(0.031)] + [1/(0.031)(1)] + [(1 - 0.2)/0.2(0.5)]}$$

$$= \frac{5.67 \times 10^{-8}[(573.2)^4 - (293.2)^4]}{48.32}$$

$$= 118.0 \ W$$

10.7   RADIATION HEAT TRANSFER IN AN ENCLOSURE

In most enclosures more than two surfaces are present. The rate of radiant heat transfer between the surfaces can be obtained by extending the general procedures developed in the previous section for a two surface system. A radiation network approach will be used.

The geometrical resistances, R_s, will be connected to junctions whose potentials are proportional to the radiosity of the surfaces. The criterion used is that if a surface can "see" another surface, a resistance must be inserted between the two junctions which are associated with the radiosities of the surfaces.

Each *internal surface* of the enclosure will be at a uniform temperature. Some of the surfaces will be at specified temperatures while the temperatures of the other surfaces are not specified. The temperature of these surfaces will be determined by the radiation interchange between the surface and the rest of the enclosure and by the rate of heat transferred by the *exterior surface* of the body exposed to the surroundings. If the exterior surface is insulated and heat conduction within the body is neglected, the surface is called a nonconducting reradiating surface. The radiation incident on this type of surface is equal to the radiation leaving the surface.

$$G = J$$

The net rate of heat transfer from the surface is zero and indicates that

$$J = E_b = \sigma T^4$$

There is no need to place a surface resistance between the radiosity junction and the blackbody junction for a nonconducting reradiating surface because both junctions are at the same potential.

A general procedure to be followed in the formation of a radiation network for an enclosure is to:

1. Designate a radiosity junction for each of the isothermal surfaces of the enclosure.
2. Connect the radiosity junctions with geometrical resistances if the surfaces interchange radiation with each other.
3. Locate blackbody junctions for all surfaces that are at specified temperatures.
4. Connect all blackbody junctions to the corresponding radiosity junctions using the appropriate surface characteristic resistance.

The general procedure outlined above will be applied to determine the rate of heat transfer between the internal surfaces forming the conic enclosure shown in Fig. 10–17. The temperatures of surface 1 and 2 are specified whereas surface 3 is a nonconducting reradiating wall. The steps in the formation of the network are illustrated in Fig. 10–18.

The unknown quantities in the radiation network are the radiosities. They are determined by using conventional circuit theory based upon an energy bal-

Figure 10-17 Conic enclosure.

ance written for each radiation junction in the system. It can be seen that three equations are formed and that the set of equations contains three unknowns, J_1, J_2, and J_3. These equations are solved simultaneously to obtain the values of the radiosities. The net rate of radiative heat transfer from a surface, or the rate of heat transferred between two surfaces, can be obtained. For the network shown in Fig. 10–18, the net amount of heat transfer from surface 1 is

$$\dot{Q}_1 = \frac{\sigma T_1^4 - J_1}{[(1 - \varepsilon_1)/\varepsilon_1 A_1]}$$

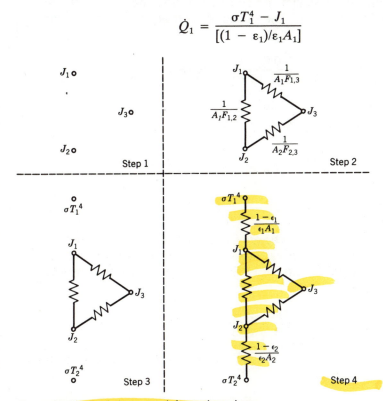

Figure 10-18 Radiation network for conic enclosure.

and the amount of heat transferred directly between surfaces 1 and 2 is

$$\dot{Q}_D = \frac{J_1 - J_2}{(1/A_1 F_{1,2})}$$

EXAMPLE 10–7

The top surface of the container described in Example 10–5 is held at a uniform temperature of 250° C (523.2 K), while the bottom surface is maintained at a temperature of 60° C (332.2 K). The surface that connects the disks is nonconducting reradiating. The emissivity of all three surfaces is 0.6. Determine the rate of radiative heat transfer between the top and bottom disks and estimate the temperature of the nonconducting reradiating surface.

SOLUTION

The radiation network for the determination of the radiation heat transfer between the surfaces of the container is shown in Fig. E10–7a. The values of the shape factors can be obtained from the calculation performed for Example 10–5. The values of the resistance in the network are

$$\frac{1 - \varepsilon_1}{\varepsilon_1 A_1} = \frac{1 - 0.6}{0.6(7.854 \times 10^{-3})} = 84.88 \ 1/m^2$$

$$\frac{1 - \varepsilon_2}{\varepsilon_2 A_2} = \frac{1 - 0.6}{0.6(7.854 \times 10^{-3})} = 84.88 \ 1/m^2$$

$$\frac{1}{A_1 F_{1,3}} = \frac{1}{A_2 F_{2,3}} = \frac{1}{(7.854 \times 10^{-3})(0.62)} = 205.4 \ 1/m^2$$

$$\frac{1}{A_1 F_{1,2}} = \frac{1}{(7.854 \times 10^{-3})(0.38)} = 335.1 \ 1/m^2$$

The value of the resistances are shown in Fig. E10–7b. The network obtained by using an equivalent resistance for the resistances connected in parallel is shown in Fig. E10–7c. The equivalent resistance is

$$R_e = \frac{410.8(335.1)}{410.8 + 335.1} = 184.6 \ 1/m^2$$

The rate of heat transfer between the top and bottom surfaces of the container is determined using

$$\dot{Q} = \frac{E_{b1} - E_{b2}}{\Sigma R}$$

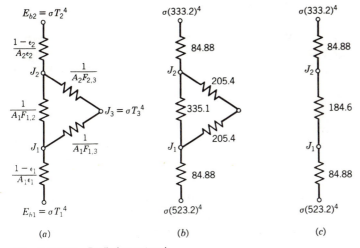

Figure E10-7 Radiation network.

The sum of the resistance between the two surfaces is

$$\Sigma R = 84.88 + 184.6 + 84.88 = 354.4 \ 1/m^2$$

The rate of heat transfer is

$$\dot{Q} = \frac{5.67 \times 10^{-8}[(523.2)^4 - (333.2)^4]}{354.4} = 10.02 \ W$$

The radiosities, J_1 and J_2, can be obtained using

$$\dot{Q} = \frac{E_{b1} - J_1}{[(1 - \varepsilon_1)/\varepsilon_1 A_1]}$$

or

$$10.02 = \frac{5.67 \times 10^{-8}(523.2)^4 - J_1}{84.4}$$

$$J_1 = 3398 \ W/m^2$$

and

$$\dot{Q} = \frac{J_2 - E_{b2}}{[(1 - \varepsilon_2)/\varepsilon_2 A_2]}$$

yielding $J_2 = 1549 \ W/m^2$. The value of J_3, which is equal to σT_3^4, is obtained using

$$\frac{J_1 - J_2}{(1/A_1 F_{1,3}) + (1/A_2 F_{2,3})} = \frac{J_1 - \sigma T_3^4}{(1/A_1 F_{1,3})}$$

yielding $T_3 = 457.0 \ K \ (183.8° \ C)$.

COMMENT

A portion of the total rate of heat transfer between the top and bottom passes directly between the two surfaces while the remainder of the heat is exchanged with the nonconducting reradiating surface before reaching the top or bottom surface.

The direct rate of heat transfer is

$$\dot{Q}_D = \frac{J_1 - J_2}{(1/A_1F_{1,2})} = \frac{3398 - 1549}{335.1} = 5.518 \text{ W}$$

The indirect is

$$\dot{Q}_{ID} = \frac{J_1 - J_2}{(1/A_1F_{1,3}) + (1/A_2F_{2,3})} = \frac{3398 - 1549}{205.4 + 205.4} = 4.501 \text{ W}$$

REFERENCES

1. Siegel, R., and Howell, J. R., *Thermal Radiation Heat Transfer*, 2nd ed., Hemisphere Publishing Corporation, Washington, 1981.
2. Sparrow, E. M., and Cess, R. D., *Radiation Heat Transfer*, Brooks/Cole Publishing Company, Belmont, Calif., 1966.
3. Hottel, H. C., and Sarofim, A. F., *Radiative Transfer*, McGraw-Hill, New York, 1967.
4. Plank, M., *The Theory of Heat Radiation*. Dover, New York, 1959.
5. Incropera, F. P., and DeWitt, D. P., *Fundamentals of Heat Transfer*, Wiley, New York, 1981.

PROBLEMS

10–1 An ideal radiating body, blackbody, is maintained at a temperature of 400° C. Determine:
    (a) The total rate of energy emitted by the body.
    (b) The fraction of the energy that occurs between the wavelengths of 0.5 and 3.5 μm.

10–2 The surface of an ideal radiating body, blackbody, is maintained at a uniform temperature of 15° C. Determine:
    (a) The total rate of energy emitted by the body.
    (b) The wavelength at which the maximum monochromatic emission occurs.
    (c) The maximum monochromatic rate of energy emitted by the body.

10–3 The sun is considered to radiate as a blackbody at 5800 K. What portion of the energy is in the ultraviolet, visible, and infrared regions?

10–4 Solar radiation has approximately the same spectral distribution as that of a blackbody at a temperature of 5800 K. The windows of a house can be fabricated using either plain glass or tinted glass. The spectral transmissivities of the two glasses are shown in Fig. P10–4. Estimate the amount of energy, in watts per meter squared, which is blocked out by tinting the glass.

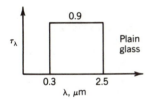
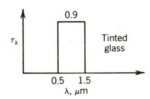

Figure P10-4 Radiation characteristics of glass.

10–5   A gray body has an emissivity of 0.6. Plot the spectral distribution of emitted radiation, monochromatic rate of energy emitted, by the gray body if it is at a temperature of:
(a)   300 K
(b)   500 K

10–6   Two very large parallel surfaces are held at uniform temperatures of 300 and 20° C, respectively. The high-temperature surface is a blackbody while the emissivity of the low temperature surface is 0.1. Determine the amount of radiation, radiosity, leaving the low-temperature body and net rate of radiation heat transfer between the two surfaces.

10–7   Determine the shape factor F_{1-2} for the following surfaces.

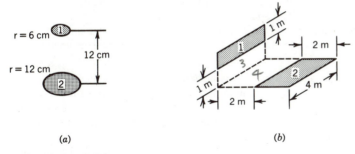

(a)   (b)

Figure P10-7 Surfaces.

10–8   Determine the shape factor F_{1-3} for the enclosure which is a 1-m cube. All vertical sides collectively represent surface 3.

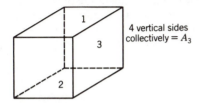

Figure P10-8 Enclosure.

10–9   Two very large parallel plates, 2 × 2 m, are held at uniform temperatures of 200 and 15° C, respectively. The air surrounding the plate is at 15° C. Assume that all the radiant energy leaving each plate strikes the other plate (shape factor equals one). The emissivity of each plate is 0.8. Determine:

(a) The net heat transfer by radiation.

(b) The amount of heat lost by the hot plate due to natural convection.

(c) The total amount of heat lost by the hot plate.

Air

$T_\infty = 15^\circ$ C

200° C   15° C   **Figure P10-9** Infinite parallel plates.

10–10   Two equal parallel disks, 50 cm in diameter, are located opposite each other and 25 cm apart. Calculate the rate of heat transfer by radiation between the two disks if the surroundings is at a temperature of 15° C. The two disks are at temperatures of 200 and 60° C and the emissivity of the disks is 0.82.

10–11   Two parallel white refratory rectangular surfaces, 1 × 4 m, are located directly opposite each other and 1 m apart. They are connected by a rough red refractory surface which may be considered to be a nonconducting reradiating surface. If the two surfaces are at 1000 and 500° C. Determine:

(a) The direct radiation exchange between the two rectangular surfaces.

(b) The total rate of radiation heat transfer between the two rectangular surfaces.

10–12   A room 6 × 4 m and 2.5 m high is heated by embedding electrical heating elements in the ceiling which maintain the surface temperature of the ceiling at 32° C. The walls and the floor are considered to be at a temperature of 15° C. The walls and ceiling are plaster while the floor is made of oak. Determine:

(a) The appropriate radiation network.

(b) The rate of heat transfer by radiation between the ceiling and the walls and floor.

10–13   A "dead" air space is used in the construction of the walls of a house as shown in Fig. P10–13. Since the space is only 9 cm thick, the two containing surfaces may be considered to be infinite plates. The brick outside wall has an inner surface temperature of 35° C in the summer while the temperature of the inner wall, polished aluminum foil, is at a temperature of 18° C. Determine the rate of heat transfer by radiation across the air space.

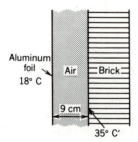

Aluminum foil
18° C

Air   Brick

9 cm

35° C′   **Figure P10-13** "Dead" air space in house wall.

10–14   A rectangular enclosure is shown in Fig. P10–14. Surface 1 is at a uniform temperature of 200° C while surface 2 is at a uniform temperature of 50° C. All other surfaces are nonconducting reradiating. $\varepsilon_1 = 0.80$ and $\varepsilon_2 = 0.4$.

    (a)   Draw the appropriate radiation network.

    (b)   Estimate the temperature of the nonconducting reradiating surfaces.

    (c)   Determine the "direct" rate of heat transfer between surfaces 1 and 2.

    (c)   Determine the total or net rate of heat transfer between surfaces 1 and 2.

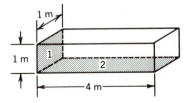

Figure P10-14   Rectangular enclosure.

10–15   An oxidized brass plate, 20 cm square, at a uniform temperature of 300° C is separated from a cooled oxidized copper plate, 20 cm square and a uniform temperature of 20° C, by a thin highly polished aluminum radiation shield. The two plates and the shield are encased by nonconducting reradiating walls. The spaces between the plates and shield are evacuated. Determine the total rate of heat transfer and estimate the temperature of the shield.

APPENDIX

Thermodynamic and Thermophysical Properties of Matter

REFERENCES

1. Van Wylen, G. J., and Sonntag, R. E., *Fundamentals of Classical Thermodynamics,* 2nd ed., S. I. Version, Wiley, New York, 1978.

2. Keenan, J. H., Keyes, F. G., Hill, P. G., and Moore, J. G., *Steam Tables,* Wiley, New York, 1969.

3. Incropera, F. P., and DeWitt, D. P., *Fundamentals of Heat Transfer,* Wiley, New York, 1981.

4. Fox, R. W., and McDonald, A. T., *Introduction to Fluid Dynamics,* 2nd ed., Wiley, New York, 1978.

Table A–1 Properties of Water[1]

Table A–1.1 Saturated Liquid–Saturated Vapor: Temperature Table

Temp. T,°C	Press. P, kPa	Specific volume, m³/kg		Internal energy, kJ/kg			Enthalpy, kJ/kg			Entropy, kJ/kg·K		
		Sat. liquid v_f	Sat. vapor v_g	Sat. liquid u_f	Evap. u_{fg}	Sat. vapor u_g	Sat. liquid h_f	Evap. h_{fg}	Sat. vapor h_g	Sat. liquid s_f	Evap. s_{fg}	Sat. vapor s_g
0.01	0.6113	0.001 000	206.14	.00	2375.3	2375.3	.01	2501.3	2501.4	.0000	9.1562	9.1562
5	0.8721	0.001 000	147.12	20.97	2361.3	2382.3	20.98	2489.6	2510.6	.0761	8.9496	9.0257
10	1.2276	0.001 000	106.38	42.00	2347.2	2389.2	42.01	2477.7	2519.8	.1510	8.7498	8.9008
15	1.7051	0.001 001	77.93	62.99	2333.1	2396.1	62.99	2465.9	2528.9	.2245	8.5569	8.7814
20	2.339	0.001 002	57.79	83.95	2319.0	2402.9	83.96	2454.1	2538.1	.2966	8.3706	8.6672
25	3.169	0.001 003	43.36	104.88	2304.9	2409.8	104.89	2442.3	2547.2	.3674	8.1905	8.5580
30	4.246	0.001 004	32.89	125.78	2290.8	2416.6	125.79	2430.5	2556.3	.4369	8.0164	8.4533
35	5.628	0.001 006	25.22	146.67	2276.7	2423.4	146.68	2418.6	2565.3	.5053	7.8478	8.3531
40	7.384	0.001 008	19.52	167.56	2262.6	2430.1	167.57	2406.7	2574.3	.5725	7.6845	8.2570
45	9.593	0.001 010	15.26	188.44	2248.4	2436.8	188.45	2394.8	2583.2	.6387	7.5261	8.1648
50	12.349	0.001 012	12.03	209.32	2234.4	2443.5	209.33	2382.7	2592.1	.7038	7.3725	8.0763
55	15.758	0.001 015	9.568	230.21	2219.9	2450.1	230.23	2370.7	2600.9	.7679	7.2234	7.9913
60	19.940	0.001 017	7.671	251.11	2205.5	2456.6	251.13	2358.5	2609.6	.8312	7.0784	7.9096
65	25.03	0.001 020	6.197	272.02	2191.1	2463.1	272.06	2346.2	2618.3	.8935	6.9375	7.8310
70	31.19	0.001 023	5.042	292.95	2176.6	2469.6	292.98	2333.8	2626.8	.9549	6.8004	7.7553
75	38.58	0.001 026	4.131	313.90	2162.0	2475.9	313.93	2321.4	2635.3	1.0155	6.6669	7.6824
80	47.39	0.001 029	3.407	334.86	2147.4	2482.2	334.91	2308.8	2643.7	1.0753	6.5369	7.6122
85	57.83	0.001 033	2.828	355.84	2132.6	2488.4	355.90	2296.0	2651.9	1.1343	6.4102	7.5445
90	70.14	0.001 036	2.361	376.85	2117.7	2494.5	376.92	2283.2	2660.1	1.1925	6.2866	7.4791
95	84.55	0.001 040	1.982	397.88	2102.7	2500.6	397.96	2270.2	2668.1	1.2500	6.1659	7.4159

[1] Adapted from Joseph H. Keenan, Frederick G. Keyes, Philip G. Hill, and Joan G. Moore, *Steam Tables*, New York: Wiley, 1969.

Table A–1.1 (continued)

Temp. T, °C	Press. P, MPa	Specific volume, m³/kg		Internal energy, kJ/kg			Enthalpy, kJ/kg			Entropy, kJ/kg·K		
		Sat. liquid v_f	Sat. vapor v_g	Sat. liquid u_f	Evap. u_{fg}	Sat. vapor u_g	Sat. liquid h_f	Evap. h_{fg}	Sat. vapor h_g	Sat. liquid s_f	Evap. s_{fg}	Sat. vapor s_g
100	0.101 35	0.001 044	1.6729	418.94	2087.6	2506.5	419.04	2257.0	2676.1	1.3069	6.0480	7.3549
105	0.120 82	0.001 048	1.4194	440.02	2072.3	2512.4	440.15	2243.7	2683.8	1.3630	5.9328	7.2958
110	0.143 27	0.001 052	1.2102	461.14	2057.0	2518.1	461.30	2230.2	2691.5	1.4185	5.8202	7.2387
115	0.169 06	0.001 056	1.0366	482.30	2041.4	2523.7	482.48	2216.5	2699.0	1.4734	5.7100	7.1833
120	0.198 53	0.001 060	0.8919	503.50	2025.8	2529.3	503.71	2202.6	2706.3	1.5276	5.6020	7.1296
125	0.2321	0.001 065	0.7706	524.74	2009.9	2534.6	524.99	2188.5	2713.5	1.5813	5.4962	7.0775
130	0.2701	0.001 070	0.6685	546.02	1993.9	2539.9	546.31	2174.2	2720.5	1.6344	5.3925	7.0269
135	0.3130	0.001 075	0.5822	567.35	1977.7	2545.0	567.69	2159.6	2727.3	1.6870	5.2907	6.9777
140	0.3613	0.001 080	0.5089	588.74	1961.3	2550.0	589.13	2144.7	2733.9	1.7391	5.1908	6.9299
145	0.4154	0.001 085	0.4463	610.18	1944.7	2554.9	610.63	2129.6	2740.3	1.7907	5.0926	6.8833
150	0.4758	0.001 091	0.3928	631.68	1927.9	2559.5	632.20	2114.3	2746.5	1.8418	4.9960	6.8379
155	0.5431	0.001 096	0.3468	653.24	1910.8	2564.1	653.84	2098.6	2752.4	1.8925	4.9010	6.7935
160	0.6178	0.001 102	0.3071	674.87	1893.5	2568.4	675.55	2082.6	2758.1	1.9427	4.8075	6.7502
165	0.7005	0.001 108	0.2727	696.56	1876.0	2572.5	697.34	2066.2	2763.5	1.9925	4.7153	6.7078
170	0.7917	0.001 114	0.2428	718.33	1858.1	2576.5	719.21	2049.5	2768.7	2.0419	4.6244	6.6663
175	0.8920	0.001 121	0.2168	740.17	1840.0	2580.2	741.17	2032.4	2773.6	2.0909	4.5347	6.6256
180	1.0021	0.001 127	0.194 05	762.09	1821.6	2583.7	763.22	2015.0	2778.2	2.1396	4.4461	6.5857
185	1.1227	0.001 134	0.174 09	784.10	1802.9	2587.0	785.37	1997.1	2782.4	2.1879	4.3586	6.5465
190	1.2544	0.001 141	0.156 54	806.19	1783.8	2590.0	807.62	1978.8	2786.4	2.2359	4.2720	6.5079
195	1.3978	0.001 149	0.141 05	828.37	1764.4	2592.8	829.98	1960.0	2790.0	2.2835	4.1863	6.4698
200	1.5538	0.001 157	0.127 36	850.65	1744.7	2595.3	852.45	1940.7	2793.2	2.3309	4.1014	6.4323
205	1.7230	0.001 164	0.115 21	873.04	1724.5	2597.5	875.04	1921.0	2796.0	2.3780	4.0172	6.3952
210	1.9062	0.001 173	0.104 41	895.53	1703.9	2599.5	897.76	1900.7	2798.5	2.4248	3.9337	6.3585

215	2.104	0.001181	0.09479	918.14	1682.9	2601.1	920.62	1879.9	2800.5	2.4714	3.8507	6.3221
220	2.318	0.001190	0.08619	940.87	1661.5	2602.4	943.62	1858.5	2802.1	2.5178	3.7683	6.2861
225	2.548	0.001199	0.07849	963.73	1639.6	2603.3	966.78	1836.5	2803.3	2.5639	3.6863	6.2503
230	2.795	0.001209	0.07158	986.74	1617.2	2603.9	990.12	1813.8	2804.0	2.6099	3.6047	6.2146
235	3.060	0.001219	0.06537	1009.89	1594.2	2604.1	1013.62	1790.5	2804.2	2.6558	3.5233	6.1791
240	3.344	0.001229	0.05976	1033.21	1570.8	2604.0	1037.32	1766.5	2803.8	2.7015	3.4422	6.1437
245	3.648	0.001240	0.05471	1056.71	1546.7	2603.4	1061.23	1741.7	2803.0	2.7472	3.3612	6.1083
250	3.973	0.001251	0.05013	1080.39	1522.0	2602.4	1085.36	1716.2	2801.5	2.7927	3.2802	6.0730
255	4.319	0.001263	0.04598	1104.28	1496.7	2600.9	1109.73	1689.8	2799.5	2.8383	3.1992	6.0375
260	4.688	0.001276	0.04221	1128.39	1470.6	2599.0	1134.37	1662.5	2796.9	2.8838	3.1181	6.0019
265	5.081	0.001289	0.03877	1152.74	1443.9	2596.6	1159.28	1634.4	2793.6	2.9294	3.0368	5.9662
270	5.499	0.001302	0.03564	1177.36	1416.3	2593.7	1184.51	1605.2	2789.7	2.9751	2.9551	5.9301
275	5.942	0.001317	0.03279	1202.25	1387.9	2590.2	1210.07	1574.9	2785.0	3.0208	2.8730	5.8938
280	6.412	0.001332	0.03017	1227.46	1358.7	2586.1	1235.99	1543.6	2779.6	3.0668	2.7903	5.8571
285	6.909	0.001348	0.02777	1253.00	1328.4	2581.4	1262.31	1511.0	2773.3	3.1130	2.7070	5.8199
290	7.436	0.001366	0.02557	1278.92	1297.1	2576.0	1289.07	1477.1	2766.2	3.1594	2.6227	5.7821
295	7.993	0.001384	0.02354	1305.2	1264.7	2569.9	1316.3	1441.8	2758.1	3.2062	2.5375	5.7437
300	8.581	0.001404	0.02167	1332.0	1231.0	2563.0	1344.0	1404.9	2749.0	3.2534	2.4511	5.7045
305	9.202	0.001425	0.019948	1359.3	1195.9	2555.2	1372.4	1366.4	2738.7	3.3010	2.3633	5.6643
310	9.856	0.001447	0.018350	1387.1	1159.4	2546.4	1401.3	1326.0	2727.3	3.3493	2.2737	5.6230
315	10.547	0.001472	0.016867	1415.5	1121.1	2536.6	1431.0	1283.5	2714.5	3.3982	2.1821	5.5804
320	11.274	0.001499	0.015488	1444.6	1080.9	2525.5	1461.5	1238.6	2700.1	3.4480	2.0882	5.5362
330	12.845	0.001561	0.012996	1505.3	993.7	2498.9	1525.3	1140.6	2665.9	3.5507	1.8909	5.4417
340	14.586	0.001638	0.010797	1570.3	894.3	2464.6	1594.2	1027.9	2622.0	3.6594	1.6763	5.3357
350	16.513	0.001740	0.008813	1641.9	776.6	2418.4	1670.6	893.4	2563.9	3.7777	1.4335	5.2112
360	18.651	0.001893	0.006945	1725.2	626.3	2351.5	1760.5	720.5	2481.0	3.9147	1.1379	5.0526
370	21.03	0.002213	0.004925	1844.0	384.5	2228.5	1890.5	441.6	2332.1	4.1106	0.6865	4.7971
374.14	22.09	0.003155	0.003155	2029.6	0	2029.6	2099.3	0	2099.3	4.4298	0	4.4298

Table A–1.2   Saturated Liquid–Saturated Vapor: Pressure Table

Press. P, kPa	Temp. T, °C	Specific volume, m³/kg Sat. liquid v_f	Sat. vapor v_g	Internal energy, kJ/kg Sat. liquid u_f	Evap. u_{fg}	Sat. vapor u_g	Enthalpy, kJ/kg Sat. liquid h_f	Evap. h_{fg}	Sat. vapor h_g	Entropy, kJ/kg·K Sat. liquid s_f	Evap. s_{fg}	Sat. vapor s_g
0.6113	0.01	0.001 000	206.14	.00	2375.3	2375.3	.01	2501.3	2501.4	.0000	9.1562	9.1562
1.0	6.98	0.001 000	129.21	29.30	2355.7	2385.0	29.30	2484.9	2514.2	.1059	8.8697	8.9756
1.5	13.03	0.001 001	87.98	54.71	2338.6	2393.3	54.71	2470.6	2525.3	.1957	8.6322	8.8279
2.0	17.50	0.001 001	67.00	73.48	2326.0	2399.5	73.48	2460.0	2533.5	.2607	8.4629	8.7237
2.5	21.08	0.001 002	54.25	88.48	2315.9	2404.4	88.49	2451.6	2540.0	.3120	8.3311	8.6432
3.0	24.08	0.001 003	45.67	101.04	2307.5	2408.5	101.05	2444.5	2545.5	.3545	8.2231	8.5776
4.0	28.96	0.001 004	34.80	121.45	2293.7	2415.2	121.46	2432.9	2554.4	.4226	8.0520	8.4746
5.0	32.88	0.001 005	28.19	137.81	2282.7	2420.5	137.82	2423.7	2561.5	.4764	7.9187	8.3951
7.5	40.29	0.001 008	19.24	168.78	2261.7	2430.5	168.79	2406.0	2574.8	.5764	7.6750	8.2515
10	45.81	0.001 010	14.67	191.82	2246.1	2437.9	191.83	2392.8	2584.7	.6493	7.5009	8.1502
15	53.97	0.001 014	10.02	225.92	2222.8	2448.7	225.94	2373.1	2599.1	.7549	7.2536	8.0085
20	60.06	0.001 017	7.649	251.38	2205.4	2456.7	251.40	2358.3	2609.7	.8320	7.0766	7.9085
25	64.97	0.001 020	6.204	271.90	2191.2	2463.1	271.93	2346.3	2618.2	.8931	6.9383	7.8314
30	69.10	0.001 022	5.229	289.20	2179.2	2468.4	289.23	2336.1	2625.3	.9439	6.8247	7.7686
40	75.87	0.001 027	3.993	317.53	2159.5	2477.0	317.58	2319.2	2636.8	1.0259	6.6441	7.6700
50	81.33	0.001 030	3.240	340.44	2143.4	2483.9	340.49	2305.4	2645.9	1.0910	6.5029	7.5939
75	91.78	0.001 037	2.217	384.31	2112.4	2496.7	384.39	2278.6	2663.0	1.2130	6.2434	7.4564
MPa												
0.100	99.63	0.001 043	1.6940	417.36	2088.7	2506.1	417.46	2258.0	2675.5	1.3026	6.0568	7.3594
0.125	105.99	0.001 048	1.3749	444.19	2069.3	2513.5	444.32	2241.0	2685.4	1.3740	5.9104	7.2844
0.150	111.37	0.001 053	1.1593	466.94	2052.7	2519.7	467.11	2226.5	2693.6	1.4336	5.7897	7.2233
0.175	116.06	0.001 057	1.0036	486.80	2038.1	2524.9	486.99	2213.6	2700.6	1.4849	5.6868	7.1717
0.200	120.23	0.001 061	0.8857	504.49	2025.0	2529.5	504.70	2201.9	2706.7	1.5301	5.5970	7.1271
0.225	124.00	0.001 064	0.7933	520.47	2013.1	2533.6	520.72	2191.3	2712.1	1.5706	5.5173	7.0878

0.250	127.44	0.001 067	0.7187	535.10	2002.1	2537.2	535.37	2181.5	2716.9	1.6072	5.4455	7.0527
0.275	130.60	0.001 070	0.6573	548.59	1991.9	2540.5	548.89	2172.4	2721.3	1.6408	5.3801	7.0209
0.300	133.55	0.001 073	0.6058	561.15	1982.4	2543.6	561.47	2163.8	2725.3	1.6718	5.3201	6.9919
0.325	136.30	0.001 076	0.5620	572.90	1973.5	2546.4	573.25	2155.8	2729.0	1.7006	5.2646	6.9652
0.350	138.88	0.001 079	0.5243	583.95	1965.0	2548.9	584.33	2148.1	2732.4	1.7275	5.2130	6.9405
0.375	141.32	0.001 081	0.4914	594.40	1956.9	2551.3	594.81	2140.8	2735.6	1.7528	5.1647	6.9175
0.40	143.63	0.001 084	0.4625	604.31	1949.3	2553.6	604.74	2133.8	2738.6	1.7766	5.1193	6.8959
0.45	147.93	0.001 088	0.4140	622.77	1934.9	2557.6	623.25	2120.7	2743.9	1.8207	5.0359	6.8565
0.50	151.86	0.001 093	0.3749	639.68	1921.6	2561.2	640.23	2108.5	2748.7	1.8607	4.9606	6.8213
0.55	155.48	0.001 097	0.3427	655.32	1909.2	2564.5	655.93	2097.0	2753.0	1.8973	4.8920	6.7893
0.60	158.85	0.001 101	0.3157	669.90	1897.5	2567.4	670.56	2086.3	2756.8	1.9312	4.8288	6.7600
0.65	162.01	0.001 104	0.2927	683.56	1886.5	2570.1	684.28	2076.0	2760.3	1.9627	4.7703	6.7331
0.70	164.97	0.001 108	0.2729	696.44	1876.1	2572.5	697.22	2066.3	2763.5	1.9922	4.7158	6.7080
0.75	167.78	0.001 112	0.2556	708.64	1866.1	2574.7	709.47	2057.0	2766.4	2.0200	4.6647	6.6847
0.80	170.43	0.001 115	0.2404	720.22	1856.6	2576.8	721.11	2048.0	2769.1	2.0462	4.6166	6.6628
0.85	172.96	0.001 118	0.2270	731.83	1847.4	2578.7	732.22	2039.4	2771.6	2.0710	4.5711	6.6421
0.90	175.38	0.001 121	0.2150	741.95	1838.6	2580.5	742.83	2031.1	2773.9	2.0946	4.5280	6.6226
0.95	177.69	0.001 124	0.2042	751.95	1830.2	2582.1	753.02	2023.1	2776.1	2.1172	4.4869	6.6041
1.00	179.91	0.001 127	0.194 44	761.68	1822.0	2583.6	762.81	2015.3	2778.1	2.1387	4.4478	6.5865
1.10	184.09	0.001 133	0.177 53	780.09	1806.3	2586.4	781.34	2000.4	2781.7	2.1792	4.3744	6.5536
1.20	187.99	0.001 139	0.163 33	797.29	1791.5	2588.8	798.65	1986.2	2784.8	2.2166	4.3067	6.5233
1.30	191.64	0.001 144	0.151 25	813.44	1775.5	2591.0	814.93	1972.7	2787.6	2.2515	4.2438	6.4953
1.40	195.07	0.001 149	0.140 84	828.70	1764.1	2592.8	830.30	1959.7	2790.0	2.2842	4.1850	6.4693
1.50	198.32	0.001 154	0.131 77	843.16	1751.3	2594.5	844.89	1947.3	2792.2	2.3150	4.1298	6.4448
1.75	205.76	0.001 166	0.113 49	876.46	1721.4	2597.8	878.50	1917.9	2796.4	2.3851	4.0044	6.3896
2.00	212.42	0.001 177	0.099 63	906.44	1693.8	2600.3	908.79	1890.7	2799.5	2.4474	3.8935	6.3409
2.25	218.45	0.001 187	0.088 75	933.83	1668.2	2602.0	936.49	1865.2	2801.7	2.5035	3.7937	6.2972
2.5	223.99	0.001 197	0.079 98	959.11	1644.0	2603.1	962.11	1841.0	2803.1	2.5547	3.7028	6.2575
3.0	233.90	0.001 217	0.066 68	1004.78	1599.3	2604.1	1008.42	1795.7	2804.2	2.6457	3.5412	6.1869

Table A–1.2 (continued)

Press. P, MPa	Temp. T, °C	Specific volume, m³/kg		Internal energy, kJ/kg			Enthalpy, kJ/kg			Entropy, kJ/kg·K		
		Sat. liquid v_f	Sat. vapor v_g	Sat. liquid u_f	Evap. u_{fg}	Sat. vapor u_g	Sat. liquid h_f	Evap. h_{fg}	Sat. vapor h_g	Sat. liquid s_f	Evap. s_{fg}	Sat. vapor s_g
3.5	242.60	0.001 235	0.057 07	1045.43	1558.3	2603.7	1049.75	1753.7	2803.4	2.7253	3.4000	6.1253
4	250.40	0.001 252	0.049 78	1082.31	1520.0	2602.3	1087.31	1714.1	2801.4	2.7964	3.2737	6.0701
5	263.99	0.001 286	0.039 44	1147.81	1449.3	2597.1	1154.23	1640.1	2794.3	2.9202	3.0532	5.9734
6	275.64	0.001 319	0.032 44	1205.44	1384.3	2589.7	1213.35	1571.0	2784.3	3.0267	2.8625	5.8892
7	285.88	0.001 351	0.027 37	1257.55	1323.0	2580.5	1267.00	1505.1	2772.1	3.1211	2.6922	5.8133
8	295.06	0.001 384	0.023 52	1305.57	1264.2	2569.8	1316.64	1441.3	2758.0	3.2068	2.5364	5.7432
9	303.40	0.001 418	0.020 48	1350.51	1207.3	2557.8	1363.26	1378.9	2742.1	3.2858	2.3915	5.6772
10	311.06	0.001 452	0.018 026	1393.04	1151.4	2544.4	1407.56	1317.1	2724.7	3.3596	2.2544	5.6141
11	318.15	0.001 489	0.015 987	1433.7	1096.0	2529.8	1450.1	1255.5	2705.6	3.4295	2.1233	5.5527
12	324.75	0.001 527	0.014 263	1473.0	1040.7	2513.7	1491.3	1193.6	2684.9	3.4962	1.9962	5.4924
13	330.93	0.001 567	0.012 780	1511.1	985.0	2496.1	1531.5	1130.7	2662.2	3.5606	1.8718	5.4323
14	336.75	0.001 611	0.011 485	1548.6	928.2	2476.8	1571.1	1066.5	2637.6	3.6232	1.7485	5.3717
15	342.24	0.001 658	0.010 337	1585.6	869.8	2455.5	1610.5	1000.0	2610.5	3.6848	1.6249	5.3098
16	347.44	0.001 711	0.009 306	1622.7	809.0	2431.7	1650.1	930.6	2580.6	3.7461	1.4994	5.2455
17	352.37	0.001 770	0.008 364	1660.2	744.8	2405.0	1690.3	856.9	2547.2	3.8079	1.3698	5.1777
18	357.06	0.001 840	0.007 489	1698.9	675.4	2374.3	1732.0	777.1	2509.1	3.8715	1.2329	5.1044
19	361.54	0.001 924	0.006 657	1739.9	598.1	2338.1	1776.5	688.0	2464.5	3.9388	1.0839	5.0228
20	365.81	0.002 036	0.005 834	1785.6	507.5	2293.0	1826.3	583.4	2409.7	4.0139	0.9130	4.9269
21	369.89	0.002 207	0.004 952	1842.1	388.5	2230.6	1884.4	446.2	2334.6	4.1075	0.6938	4.8013
22	373.80	0.002 742	0.003 568	1961.9	125.2	2087.1	2022.2	143.4	2165.6	4.3110	0.2216	4.5327
22.09	374.14	0.003 155	0.003 155	2029.6	0	2029.6	2099.3	0	2099.3	4.4298	0	4.4298

Table A–1.3 Superheated Vapor

T	P = 0.010 MPa (45.81)				P = 0.050 MPa (81.33)				P = 0.10 MPa (99.63)			
	v	u	h	s	v	u	h	s	v	u	h	s
Sat.	14.674	2437.9	2584.7	8.1502	3.240	2483.9	2645.9	7.5939	1.6940	2506.1	2675.5	7.3594
50	14.869	2443.9	2592.6	8.1749								
100	17.196	2515.5	2687.5	8.4479	3.418	2511.6	2682.5	7.6947	1.6958	2506.7	2676.2	7.3614
150	19.512	2587.9	2783.0	8.6882	3.889	2585.6	2780.1	7.9401	1.9364	2582.8	2776.4	7.6134
200	21.825	2661.3	2879.5	8.9038	4.356	2659.9	2877.7	8.1580	2.172	2658.1	2875.3	7.8343
250	24.136	2736.0	2977.3	9.1002	4.820	2735.0	2976.0	8.3556	2.406	2733.7	2974.3	8.0333
300	26.445	2812.1	3076.5	9.2813	5.284	2811.3	3075.5	8.5373	2.639	2810.4	3074.3	8.2158
400	31.063	2968.9	3279.6	9.6077	6.209	2968.5	3278.9	8.8642	3.103	2967.9	3278.2	8.5435
500	35.679	3132.3	3489.1	9.8978	7.134	3132.0	3488.7	9.1546	3.565	3131.6	3488.1	8.8342
600	40.295	3302.5	3705.4	10.1608	8.057	3302.2	3705.1	9.4178	4.028	3301.9	3704.7	9.0976
700	44.911	3479.6	3928.7	10.4028	8.981	3479.4	3928.5	9.6599	4.490	3479.2	3928.2	9.3398
800	49.526	3663.8	4159.0	10.6281	9.904	3663.6	4158.9	9.8852	4.952	3663.5	4158.6	9.5652
900	54.141	3855.0	4396.4	10.8396	10.828	3854.9	4396.3	10.0967	5.414	3854.8	4396.1	9.7767
1000	58.757	4053.0	4640.6	11.0393	11.751	4052.8	4640.5	10.2964	5.875	4052.8	4640.3	9.9764
1100	63.372	4257.5	4891.2	11.2287	12.674	4257.4	4891.1	10.4859	6.337	4257.3	4891.0	10.1659
1200	67.987	4467.9	5147.8	11.4091	13.597	4467.8	5147.7	10.6662	6.799	4467.7	5147.6	10.3463
1300	72.602	4683.7	5409.7	11.5811	14.521	4683.6	5409.6	10.8382	7.260	4683.5	5409.5	10.5183

T	P = 0.20 MPa (120.23)				P = 0.30 MPa (133.55)				P = 0.40 MPa (143.63)			
	v	u	h	s	v	u	h	s	v	u	h	s
Sat.	0.8857	2529.5	2706.7	7.1272	0.6058	2543.6	2725.3	6.9919	0.4625	2553.6	2738.6	6.8959
150	0.9596	2576.9	2768.8	7.2795	0.6339	2570.8	2761.0	7.0778	0.4708	2564.5	2752.8	6.9299
200	1.0803	2654.4	2870.5	7.5066	0.7163	2650.7	2865.6	7.3115	0.5342	2646.8	2860.5	7.1706
250	1.1988	2731.2	2971.0	7.7086	0.7964	2728.7	2967.6	7.5166	0.5951	2726.1	2964.2	7.3789
300	1.3162	2808.6	3071.8	7.8926	0.8753	2806.7	3069.3	7.7022	0.6548	2804.8	3066.8	7.5662
400	1.5493	2966.7	3276.6	8.2218	1.0315	2965.6	3275.0	8.0330	0.7726	2964.4	3273.4	7.8985
500	1.7814	3130.8	3487.1	8.5133	1.1867	3130.0	3486.0	8.3251	0.8893	3129.2	3484.9	8.1913
600	2.013	3301.4	3704.0	8.7770	1.3414	3300.8	3703.2	8.5892	1.0055	3300.2	3702.4	8.4558
700	2.244	3478.8	3927.6	9.0194	1.4957	3478.4	3927.1	8.8319	1.1215	3477.9	3926.5	8.6987
800	2.475	3663.1	4158.2	9.2449	1.6499	3662.9	4157.8	9.0576	1.2372	3662.4	4157.3	8.9244
900	2.706	3854.5	4395.8	9.4566	1.8041	3854.2	4395.4	9.2692	1.3529	3853.9	4395.1	9.1362

Table A–1.3 (continued)

	P = 0.20 MPa (120.23)				P = 0.30 MPa (133.55)				P = 0.40 MPa (143.63)			
T	v	u	h	s	v	u	h	s	v	u	h	s
1000	2.937	4052.5	4640.0	9.6563	1.9581	4052.3	4639.7	9.4690	1.4685	4052.0	4639.4	9.3360
1100	3.168	4257.0	4890.7	9.8458	2.1121	4256.8	4890.4	9.6585	1.5840	4256.5	4890.2	9.5256
1200	3.399	4467.5	5147.3	10.0262	2.2661	4467.2	5147.1	9.8389	1.6996	4467.0	5146.8	9.7060
1300	3.630	4683.2	5409.3	10.1982	2.4201	4683.0	5409.0	10.0110	1.8151	4682.8	5408.8	9.8780

	P = 0.50 MPa (151.86)				P = 0.60 MPa (158.85)				P = 0.80 MPa (170.43)			
T	v	u	h	s	v	u	h	s	v	u	h	s
Sat.	0.3749	2561.2	2748.7	6.8213	0.3157	2567.4	2756.8	6.7600	0.2404	2576.8	2769.1	6.6628
200	0.4249	2642.9	2855.4	7.0592	0.3520	2638.9	2850.1	6.9665	0.2608	2630.6	2839.3	6.8158
250	0.4744	2723.5	2960.7	7.2709	0.3938	2720.9	2957.2	7.1816	0.2931	2715.5	2950.0	7.0384
300	0.5226	2802.9	3064.2	7.4599	0.4344	2801.0	3061.6	7.3724	0.3241	2797.2	3056.5	7.2328
350	0.5701	2882.6	3167.7	7.6329	0.4742	2881.2	3165.7	7.5464	0.3544	2878.2	3161.7	7.4089
400	0.6173	2963.2	3271.9	7.7938	0.5137	2962.1	3270.3	7.7079	0.3843	2959.7	3267.1	7.5716
500	0.7109	3128.4	3483.9	8.0873	0.5920	3127.6	3482.8	8.0021	0.4433	3126.0	3480.6	7.8673
600	0.8041	3299.6	3701.7	8.3522	0.6697	3299.1	3700.9	8.2674	0.5018	3297.9	3699.4	8.1333
700	0.8969	3477.5	3925.9	8.5952	0.7472	3477.0	3925.3	8.5107	0.5601	3476.2	3924.2	8.3770
800	0.9896	3662.1	4156.9	8.8211	0.8245	3661.8	4156.5	8.7367	0.6181	3661.1	4155.6	8.6033
900	1.0822	3853.6	4394.7	9.0329	0.9017	3853.4	4394.4	8.9486	0.6761	3852.8	4393.7	8.8153
1000	1.1747	4051.8	4639.1	9.2328	0.9788	4051.5	4638.8	9.1485	0.7340	4051.0	4638.2	9.0153
1100	1.2672	4256.3	4889.9	9.4224	1.0559	4256.1	4889.6	9.3381	0.7919	4255.6	4889.1	9.2050
1200	1.3596	4466.8	5146.6	9.6029	1.1330	4466.5	5146.3	9.5185	0.8497	4466.1	5145.9	9.3855
1300	1.4521	4682.5	5408.6	9.7749	1.2101	4682.3	5408.3	9.6906	0.9076	4681.8	5407.9	9.5575

	P = 1.00 MPa (179.91)				P = 1.20 MPa (187.99)				P = 1.40 MPa (195.07)			
T	v	u	h	s	v	u	h	s	v	u	h	s
Sat.	0.194 44	2583.6	2778.1	6.5865	0.163 33	2588.8	2784.8	6.5233	0.140 84	2592.8	2790.0	6.4693
200	0.2060	2621.9	2827.9	6.6940	0.169 30	2612.8	2815.9	6.5898	0.143 02	2603.1	2803.3	6.4975
250	0.2327	2709.9	2942.6	6.9247	0.192 34	2704.2	2935.0	6.8294	0.163 50	2698.3	2927.2	6.7467
300	0.2579	2793.2	3051.2	7.1229	0.2138	2789.2	3045.8	7.0317	0.182 28	2785.2	3040.4	6.9534
350	0.2825	2875.2	3157.7	7.3011	0.2345	2872.2	3153.6	7.2121	0.2003	2869.2	3149.5	7.1360
400	0.3066	2957.3	3263.9	7.4651	0.2548	2954.9	3260.7	7.3774	0.2178	2952.5	3257.5	7.3026
500	0.3541	3124.4	3478.5	7.7622	0.2946	3122.8	3476.3	7.6759	0.2521	3121.1	3474.1	7.6027

600	0.4011	3296.8	3697.9	8.0290	0.3339	3295.6	3696.3	7.9435	0.2860	3294.4	3694.8	7.8710
700	0.4478	3475.3	3923.1	8.2731	0.3729	3474.4	3922.0	8.1881	0.3195	3473.6	3920.8	8.1160
800	0.4943	3660.4	4154.7	8.4996	0.4118	3659.7	4153.8	8.4148	0.3528	3659.0	4153.0	8.3431
900	0.5407	3852.2	4392.9	8.7118	0.4505	3851.6	4392.2	8.6272	0.3861	3851.1	4391.5	8.5556
1000	0.5871	4050.5	4637.6	8.9119	0.4892	4050.0	4637.0	8.8274	0.4192	4049.5	4636.4	8.7559
1100	0.6335	4255.1	4888.6	9.1017	0.5278	4254.6	4888.0	9.0172	0.4524	4254.1	4887.5	8.9457
1200	0.6798	4465.6	5145.4	9.2822	0.5665	4465.1	5144.9	9.1977	0.4855	4464.7	5144.4	9.1262
1300	0.7261	4681.3	5407.4	9.4543	0.6051	4680.9	5407.0	9.3698	0.5186	4680.4	5406.5	9.2984
	P = 1.60 MPa (201.41)				P = 1.80 MPa (207.15)				P = 2.00 MPa (212.42)			
Sat.	0.123 80	2596.0	2794.0	6.4218	0.110 42	2598.4	2797.1	6.3794	0.099 63	2600.3	2799.5	6.3409
225	0.132 87	2644.7	2857.3	6.5518	0.116 73	2636.6	2846.7	6.4808	0.103 77	2628.3	2835.8	6.4147
250	0.141 84	2692.3	2919.2	6.6732	0.124 97	2686.0	2911.0	6.6066	0.111 44	2679.6	2902.5	6.5453
300	0.158 62	2781.1	3034.8	6.8844	0.140 21	2776.9	3029.2	6.8226	0.125 47	2772.6	3023.5	6.7664
350	0.174 56	2866.1	3145.4	7.0694	0.154 57	2863.0	3141.2	7.0100	0.138 57	2859.8	3137.0	6.9563
400	0.190 05	2950.1	3254.2	7.2374	0.168 47	2947.7	3250.9	7.1794	0.151 20	2945.2	3247.6	7.1271
500	0.2203	3119.5	3472.0	7.5390	0.195 50	3117.9	3469.8	7.4825	0.175 68	3116.2	3467.6	7.4317
600	0.2500	3293.3	3693.2	7.8080	0.2220	3292.1	3691.7	7.7523	0.199 60	3290.9	3690.1	7.7024
700	0.2794	3472.7	3919.7	8.0535	0.2482	3471.8	3918.5	7.9983	0.2232	3470.9	3917.4	7.9487
800	0.3086	3658.3	4152.1	8.2808	0.2742	3657.6	4151.2	8.2258	0.2467	3657.0	4150.3	8.1765
900	0.3377	3850.5	4390.8	8.4935	0.3001	3849.9	4390.1	8.4386	0.2700	3849.3	4389.4	8.3895
1000	0.3668	4049.0	4635.8	8.6938	0.3260	4048.5	4635.2	8.6391	0.2933	4048.0	4634.6	8.5901
1100	0.3958	4253.7	4887.0	8.8837	0.3518	4253.2	4886.4	8.8290	0.3166	4252.7	4885.9	8.7800
1200	0.4248	4464.2	5143.9	9.0643	0.3776	4463.7	5143.4	9.0096	0.3398	4463.3	5142.9	8.9607
1300	0.4538	4679.9	5406.0	9.2364	0.4034	4679.5	5405.6	9.1818	0.3631	4679.0	5405.1	9.1329
	P = 2.50 MPa (223.99)				P = 3.00 MPa (233.90)				P = 3.50 MPa (242.60)			
Sat.	0.079 98	2603.1	2803.1	6.2575	0.066 68	2604.1	2804.2	6.1869	0.057 07	2603.7	2803.4	6.1253
225	0.080 27	2605.6	2806.3	6.2639								
250	0.087 00	2662.6	2880.1	6.4085	0.070 58	2644.0	2855.8	6.2872	0.058 72	2623.7	2829.2	6.1749
300	0.098 90	2761.6	3008.8	6.6438	0.081 14	2750.1	2993.5	6.5390	0.068 42	2738.0	2977.5	6.4461
350	0.109 76	2851.9	3126.3	6.8403	0.090 53	2843.7	3115.3	6.7428	0.076 78	2835.3	3104.0	6.6579
400	0.120 10	2939.1	3239.3	7.0148	0.099 36	2932.8	3230.9	6.9212	0.084 53	2926.4	3222.3	6.8405
450	0.130 14	3025.5	3350.8	7.1746	0.107 87	3020.4	3344.0	7.0834	0.091 96	3015.3	3337.2	7.0052
500	0.139 98	3112.1	3462.1	7.3234	0.116 19	3108.0	3456.5	7.2338	0.099 18	3103.0	3450.9	7.1572
600	0.159 30	3288.0	3686.3	7.5960	0.132 43	3285.0	3682.3	7.5085	0.113 24	3282.1	3678.4	7.4339

Table A–1.3 (continued)

T	P = 2.50 MPa (223.99)				P = 3.0 MPa (233.90)				P = 3.50 MPa (242.60)			
	v	u	h	s	v	u	h	s	v	u	h	s
700	0.178 32	3468.7	3914.5	7.8435	0.148 38	3466.5	3911.7	7.7571	0.126 99	3464.3	3908.8	7.6837
800	0.197 16	3655.3	4148.2	8.0720	0.164 14	3653.5	4145.9	7.9862	0.140 56	3651.8	4143.7	7.9134
900	0.215 90	3847.9	4387.6	8.2853	0.179 80	3846.5	4385.9	8.1999	0.154 02	3845.0	4384.1	8.1276
1000	0.2346	4046.7	4633.1	8.4861	0.195 41	4045.4	4631.6	8.4009	0.167 43	4044.1	4630.1	8.3288
1100	0.2532	4251.5	4884.6	8.6762	0.210 98	4250.3	4883.3	8.5912	0.180 80	4249.2	4881.9	8.5192
1200	0.2718	4462.1	5141.7	8.8569	0.226 52	4460.9	5140.5	8.7720	0.194 15	4459.8	5139.3	8.7000
1300	0.2905	4677.8	5404.0	9.0291	0.242 06	4676.6	5402.8	8.9442	0.207 49	4675.5	5401.7	8.8723

T	P = 4.0 MPa (250.40)				P = 4.5 MPa (257.49)				P = 5.0 MPa (263.99)			
	v	u	h	s	v	u	h	s	v	u	h	s
Sat.	0.049 78	2602.3	2801.4	6.0701	0.044 06	2600.1	2798.3	6.0198	0.039 44	2597.1	2794.3	5.9734
275	0.054 57	2667.9	2886.2	6.2285	0.047 30	2650.3	2863.2	6.1401	0.041 41	2631.3	2838.2	6.0544
300	0.058 84	2725.3	2960.7	6.3615	0.051 35	2712.0	2943.1	6.2828	0.045 32	2698.0	2924.5	6.2084
350	0.066 45	2826.7	3092.5	6.5821	0.058 40	2817.8	3080.6	6.5131	0.051 94	2808.7	3068.4	6.4493
400	0.073 41	2919.9	3213.6	6.7690	0.064 75	2913.3	3204.7	6.7047	0.057 81	2906.6	3195.7	6.6459
450	0.080 02	3010.2	3330.3	6.9363	0.070 74	3005.0	3323.3	6.8746	0.063 30	2999.7	3316.2	6.8186
500	0.086 43	3099.5	3445.3	7.0901	0.076 51	3095.3	3439.6	7.0301	0.068 57	3091.0	3433.8	6.9759
600	0.098 85	3279.1	3674.4	7.3688	0.087 65	3276.0	3670.5	7.3110	0.078 69	3273.0	3666.5	7.2589
700	0.110 95	3462.1	3905.9	7.6198	0.098 47	3459.9	3903.0	7.5631	0.088 49	3457.6	3900.1	7.5122
800	0.122 87	3650.0	4141.5	7.8502	0.109 11	3648.3	4139.3	7.7942	0.098 11	3646.6	4137.1	7.7440
900	0.134 69	3843.6	4382.3	8.0647	0.119 65	3842.2	4380.6	8.0091	0.107 62	3840.7	4378.8	7.9593
1000	0.146 45	4042.9	4628.7	8.2662	0.130 13	4041.6	4627.2	8.2108	0.117 07	4040.4	4625.7	8.1612
1100	0.158 17	4248.0	4880.6	8.4567	0.140 56	4246.8	4879.3	8.4015	0.126 48	4245.6	4878.0	8.3520
1200	0.169 87	4458.6	5138.1	8.6376	0.150 98	4457.5	5136.9	8.5825	0.135 87	4456.3	5135.7	8.5331
1300	0.181 56	4674.3	5400.5	8.8100	0.161 39	4673.1	5399.4	8.7549	0.145 26	4672.0	5398.2	8.7055

	P = 6.0 MPa (275.64)				P = 7.0 MPa (285.88)				P = 8.0 MPa (295.06)			
	v	u	h	s	v	u	h	s	v	u	h	s
Sat.	0.032 44	2589.7	2784.3	5.8892	0.027 37	2580.5	2772.1	5.8133	0.023 52	2569.8	2758.0	5.7432
300	0.036 16	2667.2	2884.2	6.0674	0.029 47	2632.2	2838.4	5.9305	0.024 26	2590.9	2785.0	5.7906
350	0.042 23	2789.6	3043.0	6.3335	0.035 24	2769.4	3016.0	6.2283	0.029 95	2747.7	2987.3	6.1301
400	0.047 39	2892.9	3177.2	6.5408	0.039 93	2878.6	3158.1	6.4478	0.034 32	2863.8	3138.3	6.3634
450	0.052 14	2988.9	3301.8	6.7193	0.044 16	2978.0	3287.1	6.6327	0.038 17	2966.7	3272.0	6.5551
500	0.056 65	3082.2	3422.2	6.8803	0.048 14	3073.4	3410.3	6.7975	0.041 75	3064.3	3398.3	6.7240
550	0.061 01	3174.6	3540.6	7.0288	0.051 95	3167.2	3530.9	6.9486	0.045 16	3159.8	3521.0	6.8778
600	0.065 25	3266.9	3658.4	7.1677	0.055 65	3260.7	3650.3	7.0894	0.048 45	3254.4	3642.0	7.0206
700	0.073 52	3453.1	3894.2	7.4234	0.062 83	3448.5	3888.3	7.3476	0.054 81	3443.9	3882.4	7.2812
800	0.081 60	3643.1	4132.7	7.6566	0.069 81	3639.5	4128.2	7.5822	0.060 97	3636.0	4123.8	7.5173
900	0.089 58	3837.8	4375.3	7.8727	0.076 69	3835.0	4371.8	7.7991	0.067 02	3832.1	4368.3	7.7351
1000	0.097 49	4037.8	4622.7	8.0751	0.083 50	4035.3	4619.8	8.0020	0.073 01	4032.8	4616.9	7.9384
1100	0.105 36	4243.3	4875.4	8.2661	0.090 27	4240.9	4872.8	8.1933	0.078 96	4238.6	4870.3	8.1300
1200	0.113 21	4454.0	5133.3	8.4474	0.097 03	4451.7	5130.9	8.3747	0.084 89	4449.5	5128.5	8.3115
1300	0.121 06	4669.6	5396.0	8.6199	0.103 77	4667.3	5393.7	8.5473	0.090 80	4665.0	5391.5	8.4842

	P = 9.0 MPa (303.40)				P = 10.0 MPa (311.06)				P = 12.5 MPa (327.89)			
	v	u	h	s	v	u	h	s	v	u	h	s
Sat.	0.020 48	2557.8	2742.1	5.6772	0.018 026	2544.4	2724.7	5.6141	0.013 495	2505.1	2673.8	5.4624
325	0.023 27	2646.6	2856.0	5.8712	0.019 861	2610.4	2809.1	5.7568				
350	0.025 80	2724.4	2956.6	6.0361	0.022 42	2699.2	2923.4	5.9443	0.016 126	2624.5	2826.2	5.7118
400	0.029 93	2848.4	3117.8	6.2854	0.026 41	2832.4	3096.5	6.2120	0.020 00	2789.3	3039.3	6.0417
450	0.033 50	2955.2	3256.6	6.4844	0.029 75	2943.4	3240.9	6.4190	0.022 99	2912.5	3199.8	6.2719
500	0.036 77	3055.2	3386.1	6.6576	0.032 79	3045.8	3373.7	6.5966	0.025 60	3021.7	3341.8	6.4618
550	0.039 87	3152.2	3511.0	6.8142	0.035 64	3144.6	3500.9	6.7561	0.028 01	3125.0	3475.2	6.6290
600	0.042 85	3248.1	3633.7	6.9589	0.038 37	3241.7	3625.3	6.9029	0.030 29	3225.4	3604.0	6.7810
650	0.045 74	3343.6	3755.3	7.0943	0.041 01	3338.2	3748.2	7.0398	0.032 48	3324.4	3730.4	6.9218
700	0.048 57	3439.3	3876.5	7.2221	0.043 58	3434.7	3870.5	7.1687	0.034 60	3422.9	3855.3	7.0536
800	0.054 09	3632.5	4119.3	7.4596	0.048 59	3628.9	4114.8	7.4077	0.038 69	3620.0	4103.6	7.2965
900	0.059 50	3829.2	4364.8	7.6783	0.053 49	3826.3	4361.2	7.6272	0.042 67	3819.1	4352.5	7.5182
1000	0.064 85	4030.3	4614.0	7.8821	0.058 32	4027.8	4611.0	7.8315	0.046 58	4021.6	4603.8	7.7237
1100	0.070 16	4236.3	4867.7	8.0740	0.063 12	4234.0	4865.1	8.0237	0.050 45	4228.2	4858.8	7.9165
1200	0.075 44	4447.2	5126.2	8.2556	0.067 89	4444.9	5123.8	8.2055	0.054 30	4439.3	5118.0	8.0987
1300	0.080 72	4662.7	5389.2	8.4284	0.072 65	4660.5	5387.0	8.3783	0.058 13	4654.8	5381.4	8.2717

Table A–1.3 (continued)

T	P = 15.0 MPa (342.24)				P = 17.5 MPa (354.75)				P = 20.0 MPa (365.81)			
	v	u	h	s	v	u	h	s	v	u	h	s
Sat.	0.010 337	2455.5	2610.5	5.3098	0.007 920	2390.2	2528.8	5.1419	0.005 834	2293.0	2409.7	4.9269
350	0.011 470	2520.4	2692.4	5.4421								
400	0.015 649	2740.7	2975.5	5.8811	0.012 447	2685.0	2902.9	5.7213	0.009 942	2619.3	2818.1	5.5540
450	0.018 445	2879.5	3156.2	6.1404	0.015 174	2844.2	3109.7	6.0184	0.012 695	2806.2	3060.1	5.9017
500	0.020 80	2996.6	3308.6	6.3443	0.017 358	2970.3	3274.1	6.2383	0.014 768	2942.9	3238.2	6.1401
550	0.022 93	3104.7	3448.6	6.5199	0.019 288	3083.9	3421.4	6.4230	0.016 555	3062.4	3393.5	6.3348
600	0.024 91	3208.6	3582.3	6.6776	0.021 06	3191.5	3560.1	6.5866	0.018 178	3174.0	3537.6	6.5048
650	0.026 80	3310.3	3712.3	6.8224	0.022 74	3296.0	3693.9	6.7357	0.019 693	3281.4	3675.3	6.6582
700	0.028 61	3410.9	3840.1	6.9572	0.024 34	3398.7	3824.6	6.8736	0.021 13	3386.4	3809.0	6.7993
800	0.032 10	3610.9	4092.4	7.2040	0.027 38	3601.8	4081.1	7.1244	0.023 85	3592.7	4069.7	7.0544
900	0.035 46	3811.9	4343.8	7.4279	0.030 31	3804.7	4335.1	7.3507	0.026 45	3797.5	4326.4	7.2830
1000	0.038 75	4015.4	4596.6	7.6348	0.033 16	4009.3	4589.5	7.5589	0.028 97	4003.1	4582.5	7.4925
1100	0.042 00	4222.6	4852.6	7.8283	0.035 97	4216.9	4846.4	7.7531	0.031 45	4211.3	4840.2	7.6874
1200	0.045 23	4433.8	5112.3	8.0108	0.038 76	4428.3	5106.6	7.9360	0.033 91	4422.8	5101.0	7.8707
1300	0.048 45	4649.1	5376.0	8.1840	0.041 54	4643.5	5370.5	8.1093	0.036 36	4638.0	5365.1	8.0442

T	P = 25.0 MPa				P = 30.0 MPa				P = 35.0 MPa			
	v	u	h	s	v	u	h	s	v	u	h	s
375	0.001 973	1798.7	1848.0	4.0320	0.001 789	1737.8	1791.5	3.9305	0.001 700	1702.9	1762.4	3.8722
400	0.006 004	2430.1	2580.2	5.1418	0.002 790	2067.4	2151.1	4.4728	0.002 100	1914.1	1987.6	4.2126
425	0.007 881	2609.2	2806.3	5.4723	0.005 303	2455.1	2614.2	5.1504	0.003 428	2253.4	2373.4	4.7747
450	0.009 162	2720.7	2949.7	5.6744	0.006 735	2619.3	2821.4	5.4424	0.004 961	2498.7	2672.4	5.1962
500	0.011 123	2884.3	3162.4	5.9592	0.008 678	2820.7	3081.1	5.7905	0.006 927	2751.9	2994.4	5.6282

(continued)

T	v	u	h	s	v	u	h	s	v	u	h	s
550	0.012 724	3017.5	3335.6	6.1765	0.010 168	2970.3	3275.4	6.0342	0.008 345	2921.0	3213.0	5.9026
600	0.014 137	3137.9	3491.4	6.3602	0.011 446	3100.5	3443.9	6.2331	0.009 527	3062.0	3395.5	6.1179
650	0.015 433	3251.6	3637.4	6.5229	0.012 596	3221.0	3598.9	6.4058	0.010 575	3189.8	3559.9	6.3010
700	0.016 646	3361.3	3777.5	6.6707	0.013 661	3335.8	3745.6	6.5606	0.011 533	3309.8	3713.5	6.4631
800	0.018 912	3574.3	4047.1	6.9345	0.015 623	3555.5	4024.2	6.8332	0.013 278	3536.7	4001.5	6.7450
900	0.021 045	3783.0	4309.1	7.1680	0.017 448	3768.5	4291.9	7.0718	0.014 883	3754.0	4274.9	6.9886
1000	0.023 10	3990.9	4568.5	7.3802	0.019 196	3978.8	4554.7	7.2867	0.016 410	3966.7	4541.1	7.2064
1100	0.025 12	4200.2	4828.2	7.5765	0.020 903	4189.2	4816.3	7.4845	0.017 895	4178.3	4804.6	7.4057
1200	0.027 11	4412.0	5089.9	7.7605	0.022 589	4401.3	5079.0	7.6692	0.019 360	4390.7	5068.3	7.5910
1300	0.029 10	4626.9	5354.4	7.9342	0.024 266	4616.0	5344.0	7.8432	0.020 815	4605.1	5333.6	7.7653

P = 40.0 MPa

T	v	u	h	s
375	0.001 641	1677.1	1742.8	3.8290
400	0.001 908	1854.6	1930.9	4.1135
425	0.002 532	2096.9	2198.1	4.5029
450	0.003 693	2365.1	2512.8	4.9459
500	0.005 622	2678.4	2903.3	5.4700
550	0.006 984	2869.7	3149.1	5.7785
600	0.008 094	3022.6	3346.4	6.0114
650	0.009 063	3158.0	3520.6	6.2054
700	0.009 941	3283.6	3681.2	6.3750
800	0.011 523	3517.8	3978.7	6.6662
900	0.012 962	3739.4	4257.9	6.9150
1000	0.014 324	3954.6	4527.6	7.1356
1100	0.015 642	4167.4	4793.1	7.3364
1200	0.016 940	4380.1	5057.7	7.5224
1300	0.018 229	4594.3	5323.5	7.6969

P = 50.0 MPa

T	v	u	h	s
375	0.001 559	1638.6	1716.6	3.7639
400	0.001 730	1788.1	1874.6	4.0031
425	0.002 007	1959.7	2060.0	4.2734
450	0.002 486	2159.6	2284.0	4.5884
500	0.003 892	2525.5	2720.1	5.1726
550	0.005 118	2763.6	3019.5	5.5485
600	0.006 112	2942.0	3247.6	5.8178
650	0.006 966	3093.5	3441.8	6.0342
700	0.007 727	3230.5	3616.8	6.2189
800	0.009 076	3479.8	3933.6	6.5290
900	0.010 283	3710.3	4224.4	6.7882
1000	0.011 411	3930.5	4501.1	7.0146
1100	0.012 496	4145.7	4770.5	7.2184
1200	0.013 561	4359.1	5037.2	7.4058
1300	0.014 616	4572.8	5303.6	7.5808

P = 60.0 MPa

T	v	u	h	s
375	0.001 503	1609.4	1699.5	3.7141
400	0.001 634	1745.4	1843.4	3.9318
425	0.001 817	1892.7	2001.7	4.1626
450	0.002 085	2053.9	2179.0	4.4121
500	0.002 956	2390.6	2567.9	4.9321
550	0.003 956	2658.8	2896.2	5.3441
600	0.004 834	2861.1	3151.2	5.6452
650	0.005 595	3028.8	3364.5	5.8829
700	0.006 272	3177.2	3553.5	6.0824
800	0.007 459	3441.5	3889.1	6.4109
900	0.008 508	3681.0	4191.5	6.6805
1000	0.009 480	3906.4	4475.2	6.9127
1100	0.010 409	4124.1	4748.6	7.1195
1200	0.011 317	4338.2	5017.2	7.3083
1300	0.012 215	4551.4	5284.3	7.4837

Table A–2 Properties of Freon 12 (Dichlorodifluoromethane)

Table A–2.1 Saturated Liquid–Saturated Vapor: Temperature Table

Temp., °C	Abs. press. P, MPa	Specific volume, m³/kg			Enthalpy, kJ/kg			Entropy, kJ/kg·K		
		Sat. liquid v_f	Evap. v_{fg}	Sat. vapor v_g	Sat. liquid h_f	Evap. h_{fg}	Sat. vapor h_g	Sat. liquid s_f	Evap. s_{fg}	Sat. vapor s_g
−90	0.0028	0.000 608	4.414 937	4.415 545	−43.243	189.618	146.375	−0.2084	1.0352	0.8268
−85	0.0042	0.000 612	3.036 704	3.037 316	−38.968	187.608	148.640	−0.1854	0.9970	0.8116
−80	0.0062	0.000 617	2.137 728	2.138 345	−34.688	185.612	150.924	−0.1630	0.9609	0.7979
−75	0.0088	0.000 622	1.537 030	1.537 651	−30.401	183.625	153.224	−0.1411	0.9266	0.7855
−70	0.0123	0.000 627	1.126 654	1.127 280	−26.103	181.640	155.536	−0.1197	0.8940	0.7744
−65	0.0168	0.000 632	0.840 534	0.841 166	−21.793	179.651	157.857	−0.0987	0.8630	0.7643
−60	0.0226	0.000 637	0.637 274	0.637 910	−17.469	177.653	160.184	−0.0782	0.8334	0.7552
−55	0.0300	0.000 642	0.490 358	0.491 000	−13.129	175.641	162.512	−0.0581	0.8051	0.7470
−50	0.0391	0.000 648	0.382 457	0.383 105	−8.772	173.611	164.840	−0.0384	0.7779	0.7396
−45	0.0504	0.000 654	0.302 029	0.302 682	−4.396	171.558	167.163	−0.0190	0.7519	0.7329
−40	0.0642	0.000 659	0.241 251	0.241 910	−0.000	169.479	169.479	−0.0000	0.7269	0.7269
−35	0.0807	0.000 666	0.194 732	0.195 398	4.416	167.368	171.784	0.0187	0.7027	0.7214
−30	0.1004	0.000 672	0.158 703	0.159 375	8.854	165.222	174.076	0.0371	0.6795	0.7165
−25	0.1237	0.000 679	0.130 487	0.131 166	13.315	163.037	176.352	0.0552	0.6570	0.7121
−20	0.1509	0.000 685	0.108 162	0.108 847	17.800	160.810	178.610	0.0730	0.6352	0.7082

−15	0.1826	0.000 693	0.090 326	0.091 018	22.312	158.534	180.846	0.0906	0.6141	0.7046
−10	0.2191	0.000 700	0.075 946	0.076 646	26.851	156.207	183.058	0.1079	0.5936	0.7014
−5	0.2610	0.000 708	0.064 255	0.064 963	31.420	153.823	185.243	0.1250	0.5736	0.6986
0	0.3086	0.000 716	0.054 673	0.055 389	36.022	151.376	187.397	0.1418	0.5542	0.6960
5	0.3626	0.000 724	0.046 761	0.047 485	40.659	148.859	189.518	0.1585	0.5351	0.6937
10	0.4233	0.000 733	0.040 180	0.040 914	45.337	146.265	191.602	0.1750	0.5165	0.6916
15	0.4914	0.000 743	0.034 671	0.035 413	50.058	143.586	193.644	0.1914	0.4983	0.6897
20	0.5673	0.000 752	0.030 028	0.030 780	54.828	140.812	195.641	0.2076	0.4803	0.6879
25	0.6516	0.000 763	0.026 091	0.026 854	59.653	137.933	197.586	0.2237	0.4626	0.6863
30	0.7449	0.000 774	0.022 734	0.023 508	64.539	134.936	199.475	0.2397	0.4451	0.6848
35	0.8477	0.000 786	0.019 855	0.020 641	69.494	131.805	201.299	0.2557	0.4277	0.6834
40	0.9607	0.000 798	0.017 373	0.018 171	74.527	128.525	203.051	0.2716	0.4104	0.6820
45	1.0843	0.000 811	0.015 220	0.016 032	79.647	125.074	204.722	0.2875	0.3931	0.6806
50	1.2193	0.000 826	0.013 344	0.014 170	84.868	121.430	206.298	0.3034	0.3758	0.6792
55	1.3663	0.000 841	0.011 701	0.012 542	90.201	117.565	207.766	0.3194	0.3582	0.6777
60	1.5259	0.000 858	0.010 253	0.011 111	95.665	113.443	209.109	0.3355	0.3405	0.6760
65	1.6988	0.000 877	0.008 971	0.009 847	101.279	109.024	210.303	0.3518	0.3224	0.6742
70	1.8858	0.000 897	0.007 828	0.008 725	107.067	104.255	211.321	0.3683	0.3038	0.6721
75	2.0874	0.000 920	0.006 802	0.007 723	113.058	99.068	212.126	0.3851	0.2845	0.6697
80	2.3046	0.000 946	0.005 875	0.006 821	119.291	93.373	212.665	0.4023	0.2644	0.6667
85	2.5380	0.000 976	0.005 029	0.006 005	125.818	87.047	212.865	0.4201	0.2430	0.6631
90	2.7885	0.001 012	0.004 246	0.005 258	132.708	79.907	212.614	0.4385	0.2200	0.6585
95	3.0569	0.001 056	0.003 508	0.004 563	140.068	71.658	211.726	0.4579	0.1946	0.6526
100	3.3440	0.001 113	0.002 790	0.003 903	148.076	61.768	209.843	0.4788	0.1655	0.6444

Table A–2.2 Superheated Vapor (Freon 12)

Temp., °C	0.05 MPa v, m³/kg	0.05 MPa h, kJ/kg	0.05 MPa s, kJ/kg·K	0.10 MPa v, m³/kg	0.10 MPa h, kJ/kg	0.10 MPa s, kJ/kg·K	0.15 MPa v, m³/kg	0.15 MPa h, kJ/kg	0.15 MPa s, kJ/kg·K
−20.0	0.341 857	181.042	0.7912	0.167 701	179.861	0.7401			
−10.0	0.356 227	186.757	0.8133	0.175 222	185.707	0.7628	0.114 716	184.619	0.7318
0.0	0.370 508	192.567	0.8350	0.182 647	191.628	0.7849	0.119 866	190.660	0.7543
10.0	0.384 716	198.471	0.8562	0.189 994	197.628	0.8064	0.124 932	196.762	0.7763
20.0	0.398 863	204.469	0.8770	0.197 277	203.707	0.8275	0.129 930	202.927	0.7977
30.0	0.412 959	210.557	0.8974	0.204 506	209.866	0.8482	0.134 873	209.160	0.8186
40.0	0.427 012	216.733	0.9175	0.211 691	216.104	0.8684	0.139 768	215.463	0.8390
50.0	0.441 030	222.997	0.9372	0.218 839	222.421	0.8883	0.144 625	221.835	0.8591
60.0	0.455 017	229.344	0.9565	0.225 955	228.815	0.9078	0.149 450	228.277	0.8787
70.0	0.468 978	235.774	0.9755	0.233 044	235.285	0.9269	0.154 247	234.789	0.8980
80.0	0.482 917	242.282	0.9942	0.240 111	241.829	0.9457	0.159 020	241.371	0.9169
90.0	0.496 838	248.868	1.0126	0.247 159	248.446	0.9642	0.163 774	248.020	0.9354

Temp., °C	0.20 MPa v, m³/kg	0.20 MPa h, kJ/kg	0.20 MPa s, kJ/kg·K	0.25 MPa v, m³/kg	0.25 MPa h, kJ/kg	0.25 MPa s, kJ/kg·K	0.30 MPa v, m³/kg	0.30 MPa h, kJ/kg	0.30 MPa s, kJ/kg·K
0.0	0.088 608	189.669	0.7320	0.069 752	188.644	0.7139	0.057 150	187.583	0.6984
10.0	0.092 550	195.878	0.7543	0.073 024	194.969	0.7366	0.059 984	194.034	0.7216
20.0	0.096 418	202.135	0.7760	0.076 218	201.322	0.7587	0.062 734	200.490	0.7440
30.0	0.100 228	208.446	0.7972	0.079 350	207.715	0.7801	0.065 418	206.969	0.7658
40.0	0.103 989	214.814	0.8178	0.082 431	214.153	0.8010	0.068 049	213.480	0.7869
50.0	0.107 710	221.243	0.8381	0.085 470	220.642	0.8214	0.070 635	220.030	0.8075
60.0	0.111 397	227.735	0.8578	0.088 474	227.185	0.8413	0.073 185	226.627	0.8276
70.0	0.115 055	234.291	0.8772	0.091 449	233.785	0.8608	0.075 705	233.273	0.8473
80.0	0.118 690	240.910	0.8962	0.094 398	240.443	0.8800	0.078 200	239.971	0.8665
90.0	0.122 304	247.593	0.9149	0.097 327	247.160	0.8987	0.080 673	246.723	0.8853
100.0	0.125 901	254.339	0.9332	0.100 238	253.936	0.9171	0.083 127	253.530	0.9038
110.0	0.129 483	261.147	0.9512	0.103 134	260.770	0.9352	0.085 566	260.391	0.9220

	0.40 MPa			0.50 MPa			0.60 MPa		
20.0	0.045 836	198.762	0.7199	0.035 646	196.935	0.6999			
30.0	0.047 971	205.428	0.7423	0.037 464	203.814	0.7230	0.030 422	202.116	0.7063
40.0	0.050 046	212.095	0.7639	0.039 214	210.656	0.7452	0.031 966	209.154	0.7291
50.0	0.052 072	218.779	0.7849	0.040 911	217.484	0.7667	0.033 450	216.141	0.7511
60.0	0.054 059	225.488	0.8054	0.042 565	224.315	0.7875	0.034 887	223.104	0.7723
70.0	0.056 014	232.230	0.8253	0.044 184	231.161	0.8077	0.036 285	230.062	0.7929
80.0	0.057 941	239.012	0.8448	0.045 774	238.031	0.8275	0.037 653	237.027	0.8129
90.0	0.059 846	245.837	0.8638	0.047 340	244.932	0.8467	0.038 995	244.009	0.8324
100.0	0.061 731	252.707	0.8825	0.048 886	251.869	0.8656	0.040 316	251.016	0.8514
110.0	0.063 600	259.624	0.9008	0.050 415	258.845	0.8840	0.041 619	258.053	0.8700
120.0	0.065 455	266.590	0.9187	0.051 929	265.862	0.9021	0.042 907	265.124	0.8882
130.0	0.067 298	273.605	0.9364	0.053 430	272.923	0.9198	0.044 181	272.231	0.9061

	0.70 MPa			0.80 MPa			0.90 MPa		
40.0	0.026 761	207.580	0.7148	0.022 830	205.924	0.7016	0.019 744	204.170	0.6982
50.0	0.028 100	214.745	0.7373	0.024 068	213.290	0.7248	0.020 912	211.765	0.7131
60.0	0.029 387	221.854	0.7590	0.025 247	220.558	0.7469	0.022 012	219.212	0.7358
70.0	0.030 632	228.931	0.7799	0.026 380	227.766	0.7682	0.023 062	226.564	0.7575
80.0	0.031 843	235.997	0.8002	0.027 477	234.941	0.7888	0.024 072	233.856	0.7785
90.0	0.033 027	243.066	0.8199	0.028 545	242.101	0.8088	0.025 051	241.113	0.7987
100.0	0.034 189	250.146	0.8392	0.029 588	249.260	0.8283	0.026 005	248.355	0.8184
110.0	0.035 332	257.247	0.8579	0.030 612	256.428	0.8472	0.026 937	255.593	0.8376
120.0	0.036 458	264.374	0.8763	0.031 619	263.613	0.8657	0.027 851	262.839	0.8562
130.0	0.037 572	271.531	0.8943	0.032 612	270.820	0.8838	0.028 751	270.100	0.8745
140.0	0.038 673	278.720	0.9119	0.033 592	278.055	0.9016	0.029 639	277.381	0.8923
150.0	0.039 764	285.946	0.9292	0.034 563	285.320	0.9189	0.030 515	284.687	0.9098

Table A–2.2 (continued)

Temp., °C	1.00 MPa v, m³/kg	1.00 MPa h, kJ/kg	1.00 MPa s, kJ/kg·K	1.20 MPa v, m³/kg	1.20 MPa h, kJ/kg	1.20 MPa s, kJ/kg·K	1.40 MPa v, m³/kg	1.40 MPa h, kJ/kg	1.40 MPa s, kJ/kg·K
50.0	0.018 366	210.162	0.7021	0.014 483	206.661	0.6812			
60.0	0.019 410	217.810	0.7254	0.015 463	214.805	0.7060	0.012 579	211.457	0.6876
70.0	0.020 397	225.319	0.7476	0.016 368	222.687	0.7293	0.013 448	219.822	0.7123
80.0	0.021 341	232.739	0.7689	0.017 221	230.398	0.7514	0.014 247	227.891	0.7355
90.0	0.022 251	240.101	0.7895	0.018 032	237.995	0.7727	0.014 997	235.766	0.7575
100.0	0.023 133	247.430	0.8094	0.018 812	245.518	0.7931	0.015 710	243.512	0.7785
110.0	0.023 993	254.743	0.8287	0.019 567	252.993	0.8129	0.016 393	251.170	0.7988
120.0	0.024 835	262.053	0.8475	0.020 301	260.441	0.8320	0.017 053	258.770	0.8183
130.0	0.025 661	269.369	0.8659	0.021 018	267.875	0.8507	0.017 695	266.334	0.8373
140.0	0.026 474	276.699	0.8839	0.021 721	275.307	0.8689	0.018 321	273.877	0.8558
150.0	0.027 275	284.047	0.9015	0.022 412	282.745	0.8867	0.018 934	281.411	0.8738
160.0	0.028 068	291.419	0.9187	0.023 093	290.195	0.9041	0.019 535	288.946	0.8914

Temp., °C	1.60 MPa v, m³/kg	1.60 MPa h, kJ/kg	1.60 MPa s, kJ/kg·K	1.80 MPa v, m³/kg	1.80 MPa h, kJ/kg	1.80 MPa s, kJ/kg·K	2.00 MPa v, m³/kg	2.00 MPa h, kJ/kg	2.00 MPa s, kJ/kg·K
70.0	0.011 208	216.650	0.6959	0.009 406	213.049	0.6794			
80.0	0.011 984	225.177	0.7204	0.010 187	222.198	0.7057	0.008 704	218.859	0.6909
90.0	0.012 698	233.390	0.7433	0.010 884	230.835	0.7298	0.009 406	228.056	0.7166
100.0	0.013 366	241.397	0.7651	0.011 526	239.155	0.7524	0.010 035	236.760	0.7402
110.0	0.014 000	249.264	0.7859	0.012 126	247.264	0.7739	0.010 615	245.154	0.7624
120.0	0.014 608	257.035	0.8059	0.012 697	255.228	0.7944	0.011 159	253.341	0.7835
130.0	0.015 195	264.742	0.8253	0.013 244	263.094	0.8141	0.011 676	261.384	0.8037
140.0	0.015 765	272.406	0.8440	0.013 772	270.891	0.8332	0.012 172	269.327	0.8232
150.0	0.016 320	280.044	0.8623	0.014 284	278.642	0.8518	0.012 651	277.201	0.8420
160.0	0.016 864	287.669	0.8801	0.014 784	286.364	0.8698	0.013 116	285.027	0.8603
170.0	0.017 398	295.290	0.8975	0.015 272	294.069	0.8874	0.013 570	292.822	0.8781
180.0	0.017 923	302.914	0.9145	0.015 752	301.767	0.9046	0.014 013	300.598	0.8955

T	2.50 MPa			3.00 MPa			3.50 MPa		
90.0	0.006 595	219.562	0.6823						
100.0	0.007 264	229.852	0.7103	0.005 231	220.529	0.6770			
110.0	0.007 837	239.271	0.7352	0.005 886	232.068	0.7075	0.004 324	222.121	0.6750
120.0	0.008 351	248.192	0.7582	0.006 419	242.208	0.7336	0.004 959	234.875	0.7078
130.0	0.008 827	256.794	0.7798	0.006 887	251.632	0.7573	0.005 456	245.661	0.7349
140.0	0.009 273	265.180	0.8003	0.007 313	260.620	0.7793	0.005 884	255.524	0.7591
150.0	0.009 697	273.414	0.8200	0.007 709	269.319	0.8001	0.006 270	264.846	0.7814
160.0	0.010 104	281.540	0.8390	0.008 083	277.817	0.8200	0.006 626	273.817	0.8023
170.0	0.010 497	289.589	0.8574	0.008 439	286.171	0.8391	0.006 961	282.545	0.8222
180.0	0.010 879	297.583	0.8752	0.008 782	294.422	0.8575	0.007 279	291.100	0.8413
190.0	0.011 250	305.540	0.8926	0.009 114	302.597	0.8753	0.007 584	299.528	0.8597
200.0	0.011 614	313.472	0.9095	0.009 436	310.718	0.8927	0.007 878	307.864	0.8775

T	4.00 MPa		
120.0	0.003 736	224.863	0.6771
130.0	0.004 325	238.443	0.7111
140.0	0.004 781	249.703	0.7386
150.0	0.005 172	259.904	0.7630
160.0	0.005 522	269.492	0.7854
170.0	0.005 845	278.684	0.8063
180.0	0.006 147	287.602	0.8262
190.0	0.006 434	296.326	0.8453
200.0	0.006 708	304.906	0.8636
210.0	0.006 972	313.380	0.8813
220.0	0.007 228	321.774	0.8985
230.0	0.007 477	330.108	0.9152

Table A–3 Generalized Compressibility Chart[1]

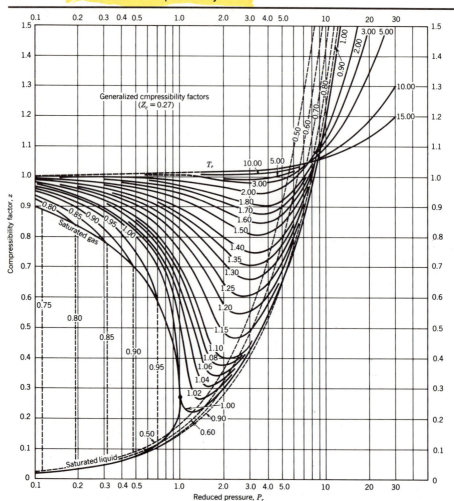

Table A–4 Equations of State

Name of equation	Equation	Constants*	Comments
van der Waals	$(P + \dfrac{a}{v^2})(v - b) = RT$	$a = \dfrac{27}{64}\dfrac{R^2 T_{CR}^2}{P_{CR}} \quad b = \dfrac{RT_{CR}}{8P_{CR}}$	Lacks accuracy—mainly of historical interest
Redlich–Kwong	$P = \dfrac{RT}{v - b} - \dfrac{a}{T^{1/2}v(v - b)}$	$a = 0.4278\,\dfrac{R^2 T_{CR}^{2.5}}{P_{CR}}$ $b = 0.0867\,\dfrac{RT_{CR}}{P_{CR}}$	Good at high pressures and temperatures near and above the critical
Beattie–Bridgeman	$P = \dfrac{RT(1 - e)(v + B)}{v^2} - \dfrac{A}{v^2}$	$e = \dfrac{c}{vT^3}$ $A = A_0\left(1 - \dfrac{a}{v}\right)$ $B = B_0\left(1 - \dfrac{b}{v}\right)$	High accuracy but requires that five constants (A_0, B_0, a, b, and c) be determined experimentally for each substance
Virial	$Pv = a + bP + cP^2 + dP^3 + \cdots$	$a, b, c, d, \cdots$ are functions of temperature and can be determined from statistical mechanics for each substance	High accuracy but coefficients, a, b, c, etc. are functions of temperature (not constants)

* Subscript $_{CR}$ refers to the critical state.

Table A–5   Constant-Pressure Specific Heats For Several Ideal Gases.[1] $\bar{c}_p = $ kJ/kmol·K, $\theta = T(K)/100$

Gas		Range, K	Max. error, %
N_2	$\bar{c}_p = 39.060 - 512.79\theta^{-1.5} + 1072.7\theta^{-2} - 820.40\theta^{-3}$	300–3500	0.43
O_2	$\bar{c}_p = 37.432 + 0.020102\theta^{1.5} - 178.57\theta^{-1.5} + 236.88\theta^{-2}$	300–3500	0.30
H_2	$\bar{c}_p = 56.505 - 702.74\theta^{-0.75} + 1165.0\theta^{-1} - 560.70\theta^{-1.5}$	300–3500	0.60
CO	$\bar{c}_p = 69.145 - 0.70463\theta^{0.75} - 200.77\theta^{-0.5} + 176.76\theta^{-0.75}$	300–3500	0.42
OH	$\bar{c}_p = 81.546 - 59.350\theta^{0.25} + 17.329\theta^{0.75} - 4.2660\theta$	300–3500	0.43
NO	$\bar{c}_p = 59.283 - 1.7096\theta^{0.5} - 70.613\theta^{-0.5} + 74.889\theta^{-1.5}$	300–3500	0.34
H_2O	$\bar{c}_p = 143.05 - 183.54\theta^{0.25} + 82.751\theta^{0.5} - 3.6989\theta$	300–3500	0.43
CO_2	$\bar{c}_p = -3.7357 + 30.529\theta^{0.5} - 4.1034\theta + 0.024198\theta^2$	300–3500	0.19
NO_2	$\bar{c}_p = 46.045 + 216.10\theta^{-0.5} - 363.66\theta^{-0.75} + 232.550\theta^{-2}$	300–3500	0.26
CH_4	$\bar{c}_p = -672.87 + 439.74\theta^{0.25} - 24.875\theta^{0.75} + 323.88\theta^{-0.5}$	300–2000	0.15
C_2H_4	$\bar{c}_p = -95.395 + 123.15\theta^{0.5} - 35.641\theta^{0.75} + 182.77\theta^{-3}$	300–2000	0.07
C_2H_6	$\bar{c}_p = 6.895 + 17.26\theta - 0.6402\theta^2 + 0.00728\theta^3$	300–1500	0.83
C_3H_8	$\bar{c}_p = -4.042 + 30.46\theta - 1.571\theta^2 + 0.03171\theta^3$	300–1500	0.40
C_4H_{10}	$\bar{c}_p = 3.954 + 37.12\theta - 1.833\theta^2 + 0.03498\theta^3$	300–1500	0.54

From T. C. Scott and R. E. Sonntag, Univ. of Michigan, unpublished (1971), except C_2H_6, C_3H_8, C_4H_{10} from K. A. Kobe, *Petroleum Refiner* **28**, No. 2, 113 (1949).

Table A–6 Temperature–entropy diagram for water[2]

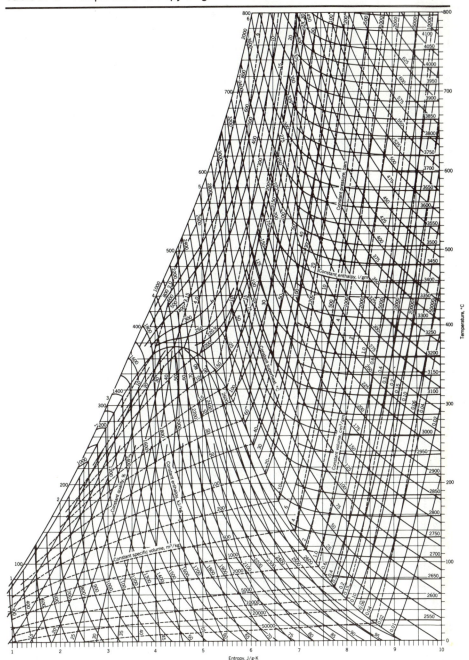

Entropy, J/g·K

Temperature, °C

Table A–7 Properties of Several Ideal Gases

Gas	Chemical formula	Molecular weight	R, kJ/kg·K	$c_p{}^*$ kJ/kg·K	c_v kJ/kg·K	γ	Critical temperature, T_c K	Critical pressure P_c MPa
Air	—	28.97	0.287 00	1.0035	0.7165	1.400	—	—
Argon	Ar	39.948	0.208 13	0.5203	0.3122	1.667	151	4.86
Butane	C_4H_{10}	58.124	0.143 04	1.7164	1.5734	1.091	425.2	3.80
Carbon dioxide	CO_2	44.01	0.188 92	0.8418	0.6529	1.289	304.2	7.39
Carbon monoxide	CO	28.01	0.296 83	1.0413	0.7445	1.400	133	3.50
Ethane	C_2H_6	30.07	0.276 50	1.7662	1.4897	1.186	305.5	4.88
Ethylene	C_2H_4	28.054	0.296 37	1.5482	1.2518	1.237	282.4	5.12
Helium	He	4.003	2.077 03	5.1926	3.1156	1.667	5.3	0.23
Hydrogen	H_2	2.016	4.124 18	14.2091	10.0849	1.409	33.3	1.30
Methane	CH_4	16.04	0.518 35	2.2537	1.7354	1.299	191.1	4.64
Neon	Ne	20.183	0.411 95	1.0299	0.6179	1.667	44.5	2.73
Nitrogen	N_2	28.013	0.296 80	1.0416	0.7448	1.400	126.2	3.39
Octane	C_8H_{18}	114.23	0.072 79	1.7113	1.6385	1.044	—	—
Oxygen	O_2	31.999	0.259 83	0.9216	0.6618	1.393	154.8	5.08
Propane	C_3H_8	44.097	0.188 55	1.6794	1.4909	1.126	370	4.26

* c_p, c_v, and γ are at 300 K.

Table A–8   Thermophysical Properties of Air[3]

$T,$ K	$\rho,$ kg/m³	$c_p,$ kJ/kg · K	$\mu \times 10^7,$ N · s/m²	$\nu \times 10^6,$ m²/s	$k \times 10^3,$ W/m · K	$\alpha \times 10^6,$ m²/s	Pr
200	1.7458	1.007	132.5	7.590	18.1	10.3	0.737
250	1.3947	1.006	159.6	11.44	22.3	15.9	0.720
300	1.1614	1.007	184.6	15.89	26.3	22.5	0.707
350	0.9950	1.009	208.2	20.92	30.0	29.9	0.700
400	0.8711	1.014	230.1	26.41	33.8	38.3	0.690
450	0.7740	1.021	250.7	32.39	37.3	47.2	0.686
500	0.6964	1.030	270.1	38.79	40.7	56.7	0.684
550	0.6329	1.040	288.4	45.57	43.9	66.7	0.683
600	0.5804	1.051	305.8	52.69	46.9	76.9	0.685
650	0.5356	1.063	322.5	60.21	49.7	87.3	0.690
700	0.4975	1.075	338.8	68.10	52.4	98.0	0.695
750	0.4643	1.087	354.6	76.37	54.9	109.	0.702
800	0.4354	1.099	369.8	84.93	57.3	120.	0.709
850	0.4097	1.110	384.3	93.80	59.6	131.	0.716
900	0.3868	1.121	398.1	102.9	62.0	143.	0.720
950	0.3666	1.131	411.3	112.2	64.3	155.	0.723
1000	0.3482	1.141	424.4	121.9	66.7	168.	0.726

Formulas for interpolation (T = absolute temperature)

$$\rho = \frac{348.59}{T} \qquad (\sigma = 9 \times 10^{-4})$$

$$f\{T\} = A + BT + CT^2 + DT^3$$

$f\{T\}$	A	B	C	D	Standard deviation, σ
c_p	1.0507	-3.645×10^{-4}	8.388×10^{-7}	-3.848×10^{-10}	4×10^{-4}
$\mu \times 10^7$	13.554	0.6738	-3.808×10^{-4}	1.183×10^{-7}	0.4192
$k \times 10^3$	-2.450	0.1130	-6.287×10^{-5}	1.891×10^{-8}	0.1198
$\alpha \times 10^6$	-11.064	7.04×10^{-2}	1.528×10^{-4}	-4.476×10^{-8}	0.4417
Pr		0.8650	-8.488×10^{-4}	1.234×10^{-6}	-5.232×10^{-10}	1.623×10^{-3}

Table A–9 Thermophysical Properties of Saturated Water (liquid)

T, K	ρ, kg/m^3	c_p, kJ/kg·K	$\mu \times 10^6$ N·s/m^2	k, W/m·K	Pr	$\beta \times 10^6$, K^{-1}
273.15	1000	4.217	1750	0.569	12.97	−68.05
275.0	1000	4.211	1652	0.574	12.12	−32.74
280	1000	4.198	1422	0.582	10.26	46.04
285	1000	4.189	1225	0.590	8.70	114.1
290	999	4.184	1080	0.598	7.56	174.0
295	998	4.181	959	0.606	6.62	227.5
300	997	4.179	855	0.613	5.83	276.1
305	995	4.178	769	0.620	5.18	320.6
310	993	4.178	695	0.628	4.62	361.9
315	991	4.179	631	0.634	4.16	400.4
320	989	4.180	577	0.640	3.77	436.7
325	987	4.182	528	0.645	3.42	471.2
330	984	4.184	489	0.650	3.15	504.0
335	982	4.186	453	0.656	2.89	535.5
340	979	4.188	420	0.660	2.66	566.0
345	977	4.191	389	0.664	2.46	595.4
350	974	4.195	365	0.668	2.29	624.2
355	971	4.199	343	0.671	2.15	652.3
360	967	4.203	324	0.674	2.02	679.9
365	963	4.209	306	0.677	1.90	707.1
370	961	4.214	289	0.679	1.79	728.7
373.15	958	4.217	279	0.680	1.73	750.1
400	937	4.256	217	0.688	1.34	896
450	890	4.40	152	0.678	0.99	
500	831	4.66	118	0.642	0.86	
550	756	5.24	97	0.580	0.88	
600	649	7.00	81	0.497	1.14	
647.3	315	00	45	0.238	00	

Formulas for interpolation (T = absolute temperature)
$$f\{T\} = A + BT + CT^2 + DT^3$$

$f\{T\}$	A	B	C	D	Standard deviation, σ
		273.15 < T < 373.15 K			
ρ	766.17	1.80396	−3.4589 × 10^{-3}		0.5868
c_p	5.6158	−9.0277 × 10^{-3}	14.177 × 10^{-6}		4.142 × 10^{-3}
k	−0.4806	5.84704 × 10^{-3}	−0.733188 × 10^{-5}		0.481 × 10^{-3}
		273.15 < T < 320 K			
$\mu \times 10^6$	0.239179 × 10^6	−2.23748 × 10^3	7.03318	−7.40993 × 10^{-3}	4.0534 × 10^{-6}
$\beta \times 10^6$	−57.2544 × 10^3	530.421	−1.64882	1.73329 × 10^{-3}	1.1498 × 10^{-6}
		320 < T < 373.15 K			
$\mu \times 10^6$	35.6602 × 10^3	−272.757	0.707777	−0.618833 × 10^{-3}	1.0194 × 10^{-6}
$\beta \times 10^6$	−11.1377 × 10^3	84.0903	−0.208544	0.183714 × 10^{-3}	1.2651 × 10^{-6}

Table A–10 Thermophysical Properties of Oil[3]

T, K	ρ, kg/m³	c_p, kJ/kg·K	$\mu \times 10^2$, N·s/m²	$\nu \times 10^6$, m²/s	$k \times 10^3$, W/m·K	$\alpha \times 10^7$, m²/s	Pr	$\beta \times 10^3$, K⁻¹
Engine oil (unused)								
273	899.1	1.796	385	4,280	147	0.910	47,000	0.70
280	895.3	1.827	217	2,430	144	0.880	27,500	0.70
290	890.0	1.868	99.9	1,120	145	0.872	12,900	0.70
300	884.1	1.909	48.6	550	145	0.859	6,400	0.70
310	877.9	1.951	25.3	288	145	0.847	3,400	0.70
320	871.8	1.993	14.1	161	143	0.823	1,965	0.70
330	865.8	2.035	8.36	96.6	141	0.800	1,205	0.70
340	859.9	2.076	5.31	61.7	139	0.779	793	0.70
350	853.9	2.118	3.56	41.7	138	0.763	546	0.70
360	847.8	2.161	2.52	29.7	138	0.753	395	0.70
370	841.8	2.206	1.86	22.0	137	0.738	300	0.70
380	836.0	2.250	1.41	16.9	136	0.723	233	0.70
390	830.6	2.294	1.10	13.3	135	0.709	187	0.70
400	825.1	2.337	0.874	10.6	134	0.695	152	0.70
410	818.9	2.381	0.698	8.52	133	0.682	125	0.70
420	812.1	2.427	0.564	6.94	133	0.675	103	0.70
430	806.5	2.471	0.470	5.83	132	0.662	88	0.70

Table A–11   Physical Properties of Common Liquids[4]

Liquid	Specific gravity (−)
Benzene	0.879
Carbon tetrachloride	1.595
Castor oil	0.969
Gasoline	0.72
Glycerine	1.26
Heptane	0.684
Kerosene	0.82
Lubricating oil	0.88
Mercury	13.55
Octane	0.702
Sea water	1.052
Fresh water	1.000

Table A–12   Vapor Pressure of Some Liquids

Temperature, °C	Water, kPa	Mercury, kpA × 10⁵	Kerosene, kPa	Methyl alcohol, kPa
0	0.6068	2.48		
5	0.8721		2.275	2.758
10	1.2276	6.55	2.551	7.102
15	1.7051			9.653
20	2.339	15.86	3.516	12.618
25	3.169		4.137	16.410
30	4.246	37.23	4.826	21.374
35	5.628		5.516	26.891
40	7.384	80.67	6.136	
45	9.593		6.619	
50	12.349		7.102	
55	15.758		7.516	
60	19.940			
65	25.03			
70	31.19			

Table A–13 Dynamic (absolute) viscosity of common fluids as a function of temperature[4]

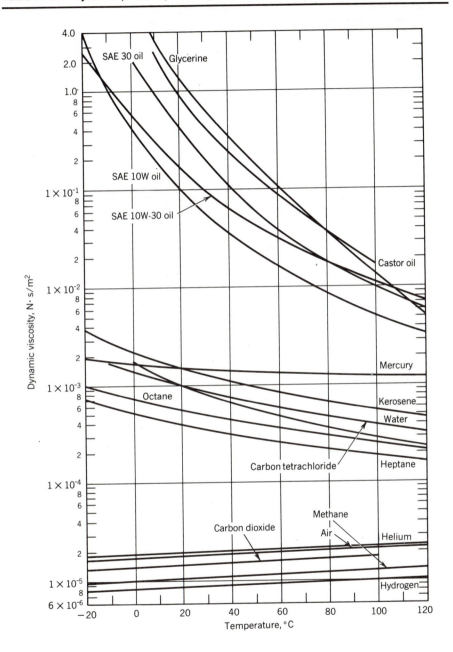

Table A–14   Kinematic viscosity of common fluids (at atmospheric pressure) as a function of temperature[4]

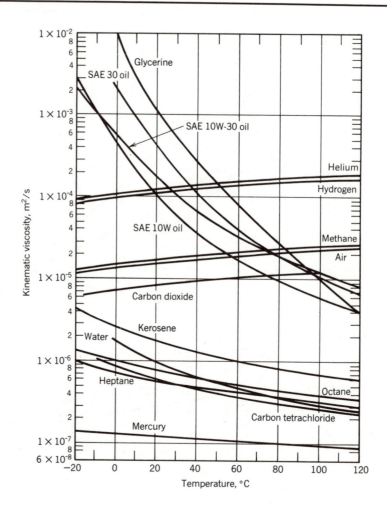

Table A–15 Thermophysical Properties of Selected Metallic Solids[3]

Composition	Melting point, K	Properties at 300 K				Properties at various temperatures, K							
		ρ, kg/m³	c_p, J/kg·K	k, W/m·K	$\alpha \times 10^6$, m²/s	k, W/m·K				c_p, J/kg·K			
						100	200	400	600	100	200	400	600
Aluminum													
Pure	933	2,702	903	237	97.1	302	237	240	231	482	796	949	1,033
Alloy 2024–T6 (4.5% Cu, 1.5% Mg, 0.6% Mn)	775	2,770	875	177	73.0	65	163	186	186	473	787	925	1,042
Alloy 195, cast (4.5% Cu)	—	2,790	883	168	68.2	—	—	174	185	—	—	—	—
Chromium	2,118	7,160	449	93.7	29.1	159	111	90.9	80.7	192	384	484	542
Copper													
Pure	1,358	8,933	385	401	117	482	413	393	379	252	356	397	417
Commercial bronze (90% Cu, 10% Al)	1,293	8,800	420	52	14	—	42	52	59	—	785	460	545
Phosphor gear bronze (89% Cu, 11% Sn)	1,104	8,780	355	54	17	—	41	65	74	—	—	—	—
Cartridge brass (70% Cu, 30% Zn)	1,188	8,530	380	110	33.9	75	95	137	149	—	360	395	425
Constantan (55% Cu, 45% Ni)	1,493	8,920	384	23	6.71	17	19	—	—	237	362	—	—
Iron													
Pure	1,810	7,870	447	80.2	23.1	134	94.0	69.5	54.7	216	384	490	574
Armco (99.75% pure)	—	7,870	447	72.7	20.7	95.6	80.6	65.7	53.1	215	384	490	574

Table A–15 (continued)

Composition	Melting point, K	Properties at 300 K				Properties at various temperatures, K								
						k, W/m·K				c_p, J/kg·K			
		ρ, kg/m³	c_p, J/kg·K	k, W/m·K	α × 10⁶, m²/s	100	200	400	600	100	200	400	600
Carbon steels													
Plain carbon (Mn ≤ 1%, Si ≤ 0.1%)	—	7,854	434	60.5	17.7	—	—	56.7	48.0	—	—	487	559
AISI 1010	—	7,832	434	63.9	18.8	—	—	58.7	48.8	—	—	487	559
Carbon–silicon (Mn ≤ 1%, 0.1% < Si ≤ 0.6%)	—	7,817	446	51.9	14.9	—	—	49.8	44.0	—	—	501	582
Carbon—man-ganese—silicon (1% < Mn ≤ 1.65%, 0.1% < Si ≤ 0.6%)	—	8,131	434	41.0	11.6	—	—	42.2	39.7	—	—	487	559
Chromium (low) steels													
½ Cr–¼ Mo–Si (0.18% C, 0.65% Cr), 0.23% Mo, 0.6% Si)	—	7,822	444	37.7	10.9	—	—	38.2	36.7	—	—	492	575

Material													
1 Cr–½ Mo (0.16% C, 1% Cr, 0.54% Mo, 0.39% Si)	—	7,858	442	42.3	12.2	—	—	42.0	39.1	—	—	492	575
1 Cr–V (0.2% C, 1.02% Cr, 0.15% V)	—	7,836	443	48.9	14.1	—	—	46.8	42.1	—	—	492	575
Stainless steels													
AISI 302	—	8,055	480	15.1	3.91	9.2	—	17.3	20.0	—	—	512	559
AISI 304	1,670	7,900	477	14.9	3.95	—	12.6	16.6	19.8	272	402	515	557
AISI 316	—	8,238	468	13.4	3.48	—	—	15.2	18.3	—	—	504	550
AISI 347	—	7,978	480	14.2	3.71	—	—	15.8	18.9	—	—	513	559
Lead	601	11,340	129	35.3	24.1	39.7	36.7	34.0	31.4	118	125	132	142
Magnesium	923	1,740	1,024	156	87.6	169	159	153	149	649	934	1,074	1,170
Molybdenum	2,894	10,240	251	138	53.7	179	143	134	126	141	224	261	275
Nickel													
Pure	1,728	8,900	444	90.7	23.0	164	107	80.2	65.6	232	383	485	592
Nichrome (80% Ni, 20% Cr)	1,672	8,400	420	12	3.4	—	—	14	16	—	—	480	525
Inconel X–750 (73% Ni, 15% Cr, 6.7% Fe)	1,665	8,510	439	11.7	3.1	8.7	10.3	13.5	17.0	—	372	473	510

Table A–16 Thermophysical Properties of Nonmetals

Table A–16.1 Thermophysical Properties of Common Materials[3]

Description/composition	Temperature, K	Density ρ, kg/m^3	Thermal conductivity k, W/m · K	Specific heat c_p, J/kg · K
Asphalt	300	2115	0.062	920
Bakelite	300	1300	1.4	1465
Brick, refractory				
Carborundum	872	—	18.5	—
	1672	—	11.0	—
Chrome brick	473	3010	2.3	835
	823	—	2.5	—
	1173	—	2.0	—
Diatomaceous silica, fired	478	—	0.25	—
	1145	—	0.30	—
Fire clay, burnt 1600 K	773	2050	1.0	960
	1073	—	1.1	—
	1373	—	1.1	—
Fire clay, burnt 1725 K	773	2325	1.3	960
	1073	—	1.4	—
	1373	—	1.4	—
Fire clay brick	478	2645	1.0	960
	922	—	1.5	—
	1478	—	1.8	—
Magnesite	478	—	3.8	1130
	922	—	2.8	—
	1478	—	1.9	—
Clay	300	1460	1.3	880
Coal, anthracite	300	1350	0.26	1260
Concrete (stone mix)	300	2300	1.4	880
Cotton	300	80	0.06	1300
Foodstuffs				
Banana (75.7% water content)	300	980	0.481	3350
Apple, red (75% water content)	300	840	0.513	3600
Cake, batter	300	720	0.223	—
Cake, fully baked	300	280	0.121	—
Chicken meat, white (74.4% water content)	233	—	1.49	—
	273	—	0.476	—
	293	—	0.489	—
Glass				
Plate (soda lime)	300	2500	1.4	750
Pyrex	300	2225	1.4	835
Ice	273	920	0.188	2040
	253	—	0.203	1945
Leather (sole)	300	998	0.013	—
Paper	300	930	0.011	1340
Paraffin	300	900	0.020	2890
Rock				
Granite, Barre	300	2630	2.79	775
Limestone, Salem	300	2320	2.15	810

Table A–16.1 (continued)

Description/composition	Temperature, K	Density ρ, kg/m³	Thermal conductivity k, W/m · K	Specific heat c_p, J/kg · K
Marble, Halston	300	2680	2.80	830
Quartzite, Sioux	300	2640	5.38	1105
Sandstone, Berea	300	2150	2.90	745
Rubber, vulcanized				
Soft	300	1100	0.012	2010
Hard	300	1190	0.013	—
Sand	300	1515	0.027	800
Soil	300	2050	0.52	1840
Snow	273	110	0.049	—
	273	500	0.190	—
Teflon	300	2200	0.35	—
	400	—	0.45	—
Tissue, human				
Skin	300	—	0.37	—
Fat layer (adipose)	300	—	0.2	—
Muscle	300	—	0.41	—
Wood, cross grain				
Balsa	300	140	0.055	—
Cypress	300	465	0.097	—
Fir	300	415	0.11	2720
Oak	300	545	0.17	2385
Yellow pine	300	640	0.15	2805
White pine	300	435	0.11	—
Wood, radial				
Oak	300	545	0.19	2385
Fir	300	420	0.14	2720

Table A–16.2   Thermophysical Properties of Structural Building Materials[3]

Description/composition	Typical properties at 300 K		
	Density ρ, kg/m^3	Thermal conductivity k, W/m·K	Specific heat c_p, J/kg·K
Building boards			
Asbestos–cement board	1920	0.58	—
Gypsum or plaster board	800	0.17	—
Plywood	545	0.12	1215
Sheathing, regular density	290	0.055	1300
Acoustic tile	290	0.058	1340
Hardboard, siding	640	0.094	1170
Hardboard, high-density	1010	0.15	1380
Particle board, low-density	590	0.078	1300
Particle board, high-density	1000	0.170	1300
Woods			
Hardwoods (oak, maple)	720	0.16	1255
Softwards (fir, pine)	510	0.12	1380
Masonry materials			
Cement mortar	1860	0.72	780
Brick, common	1920	0.72	835
Brick, face	2083	1.3	—
Clay tile, hollow			
one cell deep, 10 cm thick	—	0.52	—
three cell deep, 30 cm thick	—	0.69	—
Concrete block, three oval cores			
sand/gravel, 20 cm thick	—	1.0	—
cinder aggregate, 20 cm thick	—	0.67	—
Concrete block, rectangular core			
two-core, 20 cm thick, 16 kg	—	1.1	—
same with filled cores	—	0.60	—
Plastering materials			
Cement plaster, sand aggregate	1860	0.72	—
Gypsum plaster, sand aggregate	1860	0.22	1085
Gypsum plaster, vermiculite aggregate	720	0.25	—

Table A–16.3 Thermophysical Properties of Insulating Building Materials[3]

Description/composition	Typical properties at 300 K		
	Density ρ, kg/m³	Thermal conductivity k, W/m·K	Specific heat c_p, J/kg·K
Blanket and batt			
Glass fiber, paper faced	16	0.046	—
	28	0.038	—
	40	0.035	—
Glass fiber, coated, duct liner	32	0.038	835
Board and slab			
Cellular glass	145	0.058	1000
Glass fiber, organic bonded	105	0.036	795
Polystrene, expanded			
Extruded (R-12)	55	0.027	1210
Molded beads	16	0.040	1210
Mineral fiberboard, roofing material	265	0.049	—
Wood, shredded/cemented	350	0.087	1590
Cork	120	0.039	1800
Loose fill			
Cork, granulated	160	0.045	—
Diatomaceous silica, coarse powder	350	0.069	—
	400	0.091	—
Diatomaceous silica, fine powder	200	0.052	—
	275	0.061	—
Glass fiber, poured or blown	16	0.043	835
Vermiculite, flakes	80	0.068	835
	160	0.063	1000
Formed/foamed-in-place			
Mineral wood granules with asbestos/inorganic binders, sprayed	190	0.046	—
Polyvinyl acetate cork mastic, sprayed or troweled	—	0.100	—
Urethane, two-part mixture, rigid foam	70	0.026	1045
Reflective			
Aluminium foil separating fluffy glass mats, 10–12 layers, evacuated for cryogenic application (150 K)	40	0.00016	—
Aluminum foil and glass paper laminate, 75–150 layers, evacuated, for cryogenic application (150 K)	120	0.000017	—
Typical silica powder, evacuated	160	0.0017	—

Table A–16.4 Thermophysical Properties of Industrial Insulation[3]

Description/Composition	Max. service temp., K	Typical density, ρ kg/m³	Typical thermal conductivity, k (W/m·K), at various temperatures (K)														
			200	215	230	240	255	270	285	300	310	365	420	530	645	750
Blankets																
Blanket, mineral fiber, metal reinforced	920	96–192									0.038	0.046	0.056	0.078		
	815	40–96									0.035	0.045	0.058	0.088		
Blanket, mineral fiber, glass, fine fiber, organic bonded	450	10				0.036	0.038	0.040	0.043	0.048	0.052	0.076				
		12				0.035	0.036	0.039	0.042	0.046	0.049	0.069				
		16				0.033	0.035	0.036	0.039	0.042	0.046	0.062				
		24				0.030	0.032	0.033	0.036	0.039	0.040	0.053				
		32				0.029	0.030	0.032	0.033	0.036	0.038	0.048				
		48				0.027	0.029	0.030	0.032	0.033	0.035	0.045				
Blanket, alumina–silica fiber	1530	48												0.071	0.105	0.150
		64												0.059	0.087	0.125
		96												0.052	0.076	0.100
		128												0.049	0.068	0.091
Felt, semirigid, organic bonded	480	50–125	0.023	0.025	0.026	0.027	0.029	0.035	0.036	0.038	0.039	0.051	0.063			
	730	50						0.030	0.032	0.033	0.035	0.051	0.079			
Felt, laminated, no binder	920	120											0.051	0.065	0.087	

Blocks, boards, and pipe insulations

Asbestos paper, laminated and corrugated											
four-ply	420	190	0.078	0.082	0.098						
six ply	420	255	0.071	0.074	0.085						
eight-ply	420	300	0.068	0.071	0.082						
Magnesia, 85%	590	185	0.051	0.055	0.061						
Calcium silicate	920	190	0.055	0.059	0.063	0.075	0.089	0.104			
Cellular glass	700	145	0.046	0.048	0.051	0.052	0.055	0.058	0.062	0.069	0.079
Diatomaceous silica	1145	345	0.092	0.098	0.104						
	1310	385	0.101	0.100	0.115						
Polystyrene rigid											
Extruded (R-12)	350	56	0.023	0.023	0.023	0.023	0.025	0.026	0.027	0.029	
Extruded (R-12)	350	35	0.023	0.023	0.025	0.025	0.026	0.027	0.029		
Molded beads	350	16	0.026	0.029	0.030	0.033	0.035	0.036	0.038	0.040	
Rubber, rigid foamed	340	70	0.029	0.030	0.032	0.033					
Loose fill											
Cellulose, wood, or paper pulp		45	0.038	0.039	0.042						
Perlite, expanded		105	0.036	0.039	0.042	0.043	0.046	0.049	0.051	0.053	0.056
Vermiculite, expanded		122	0.056	0.058	0.061	0.063	0.065	0.068	0.071		
		80	0.049	0.051	0.055	0.058	0.061	0.063	0.066		

Answers to Selected Problems

CHAPTER 2

2–1 $P = 1.1453 \times 10^3$ kPa

2–3 (a) Two
     (b) Intensive P, T
         Extensive V
     (c) $Pv = RT$
     (e) $\rho = 12.634$ m³/kg

2–5 $P = 13.279$ kPa

2–7 (a) yes
     (b) no

2–9 $W = +10$ kJ

2–11 (a) $W = +7.2$ kJ
      (b) $\dot{W} = 120$ W
      (c) $W = -7.2$ kJ
      (d) yes

2–13 $W = 2.250$ kJ

2–15 (a) $W = 86.97$ kJ

2–17 $W = -0.1360$ J

CHAPTER 3

3–1 (a) Superheated vapor
     (b) Compressed liquid
     (c) Saturated liquid line
     (d) Liquid-vapor region
     (e) Compressed liquid
     (f) Superheated vapor
     (g) Saturated vapor line
     (h) Superheated vapor
     (i) Compressed liquid

3–3 (a) liquid vapor; $T = 175.38°$ C
     (b) liquid-vapor;
         $v = 6.92 \times 10^{-3}$ m³/kg
     (c) compressed liquid;
         $v = 1.012 \times 10^{-3}$ m³/kg
     (d) superheated vapor;
         $P = 0.249$ MPa
     (e) superheated vapor;
         $v = 53.62 \times 10^{-3}$ m³/kg

3–5 $T = 828.6°$ C

3–7 (a) $x = 0.00743$
     (b) % vol = 27.07%

3–9 $P = 333.4$ kPa

3–11 $v = 4.26 \times 10^{-3}$ m³/kg

3–13 $c_v = 0.8989$ kJ/kg·K

3–15 (a) $P = 100.0$ kPa, $T = 70.97$ K

(b) $P = 666.7$ kPa,
$T = 473.15$ K

(c) $P = 1246.9$ kPa,
$T = 884.9$ K

CHAPTER 4

4–1 $\dot{Q} = 0.7165$ kW

4–3 $W = -5.60$ J

4–5 $Q = -34.65$ kJ/kg

4–7 $\dot{Q} = 1.50 \times 10^6$ kW

4–9 (a) $\dot{W} = 1.26$ kW
(b) Actual power greater because finite temperature differences are required to transfer heat thus introducing irreversibilities.

4–11 (a) $W = -322.43$ kJ
(b) $I = 0.8575$ kJ/K

4–13 (a) $T = 290.5$ K
(b) $Q = -2145.8$ kJ

4–15 $\Delta h = 9.9$ kJ/kg

4–19 (b) $x = 0.923$
(c) $w = 208.3$ kJ/kg

4–20 (a) $V = 3.474$ m³
(b) $W = 348.5$ kJ

CHAPTER 5

5–1 (a) $\mathbf{V} \cdot d\mathbf{A} = -acxdy - bcdx$
(b) $\mathbf{V} \cdot (\mathbf{V}\,d\mathbf{A}) = axc\,(-axdy - bdx)\,i + bc(-axdy - bdx)\,\mathbf{j}$
(c) $\int \mathbf{V} \cdot d\mathbf{A} = 0$
(d) $\int \mathbf{V} \cdot d\mathbf{A} = -2bc$

5–3 $\mathbf{V}_2 = 61.67$ m/s

5–5 $V = 0.0476$ m/s

5–7 $\dot{m}_2 = 916.3$ kg/s

5–9 $\mathbf{R} = 838.47$ N (to the left)

5–11 $P_1 - P_2 = 2522.2$ N/m²

5–13 (a) $V_2 = 7.875$ m/s
(b) $P_2 - P_1 = 104.79$ kPa

5–15 $\eta = 44.9\%$

5–17 $\dot{W} = 603.7 \times 10^3$ kw

5–19 (a) $T_d = 316.18$ K
(b) $P_d = 1.199$ MPa
(c) $A_d/A_i = 10.54$

5–21 (a) $w_{cv} = w_{sys}$
(b) $w_{cv} = \gamma w_{sys}$

5–23 $d = 0.120$ m

5–25 $\dot{w} = -29.253$ kw/kg

5–27 $x = 0.57$

CHAPTER 6

6–1 (a) $z = 2.54$ cm
(b) $z = 2.54$ cm
(c) $z = 40.62$ cm
(d) $z = 40.62$ cm

6–3 $V = 37.30$ m/s

6–5 $P = 807.4$ kPa

6–7 $\mathbf{F} = 4.882 \times 10^3$ N

6–9 (a) $V = 711.07$ m/s
(b) $V = 553.57$ m/s
(c) $V = 2087.4$ m/s

6–11 (a) $M = 0.299$
(b) $P_o = 75.54$ kPa
(c) $T_o = 283.12$ K

6–13 (a) $V = 296.65$ m/s
(b) $\dot{m} = 0.791$ kg/s

6–15 (a) $P = 139.05$ kPa;
$T = 353.6$ K
(b) $A_d/A_i = 1.641$

CHAPTER 7

7–1 (a) $\delta = 1.574 \times 10^{-2}$ m
(b) $\tau_w = 0.5325$ N/m²

7–3 $\mu = 1.319$ N·s/m²

7–5 (a) $\delta = 22.83 \times 10^{-3}$ m
$\tau_w = 0.6677$ N/m²
(b) $D/b = 5.194$ N/m

7–9 $D_T = 118.24 \times 10^3$ N

7–11 $U_t = 18.77$ m/s

7–13 $\dot{Q}/b = 123.8$ kW/m

7–15 $\dot{q}'' = 16.59$ kW/m²

7–17 $\dot{Q} = 30.55$ W

7–19 $\dot{Q} = 999.8$ W

7–21 $\dot{Q} = 3.959$ W

CHAPTER 8

8–1 $Re = 4.537 \times 10^3$

8–3 $V/V_{max} = 2/3$

8–5 $P = 451.5$ kPa

8–7   (a) $\dot{W} = -1338.$ kW
       (b) inlet; $h_m = 2.749$m, exit;
            $h_m = 6.545$ m
       (c) $\dot{W} = 1371.$ kW

8–9 $h_x = 383.5$ kW/m²·K

8–11 $\dot{q}''_w = 16.04 \; kW/m^2$

8–13 $T = 42.35°$ C

8–15 $\bar{h} = 3.794 \times 10^3$ W/m²·K

8–17 $\dot{Q} = 4.41$ W

8–19 $\bar{h} = 19.17$ W/m²·K (concrete)
       $\bar{h} = 8.84$ W/m²·K (smooth duct)

8–21 $T = 76.08°$ C
       $A = 7.22$ m²

8–23   (a) $A = 1.199$ m²
       (b) $A = 0.822$ m²

8–25   (a) 119 tubes/pass
       (b) $A = 353.8$ m²
       (c) $T = 18.91°$ C
       (d) $L = 14.86$ m

CHAPTER 9

9–1 $T_1 = 74.84°$ C
       $\dot{Q}/L = 78.53$ W/m

9–3   (a) $U = 0.782$ W/m²·K
       (b) $\dot{Q}/L = 63.87$ W/m

9–5 $\dot{Q}/L = 385.57$ W/m

9–7   (a) $\dot{q}'' = 1.132$ kW/m²
       (b) $\dot{q}'' = 47.68$ kW/m²

9–9 $T = 150.9°$ C

9–11 $T(1) = 80.000°$ C
       $T(2) = 71.56°$ C

$T(3) = 65.42°$ C
$T(4) = 61.46°$ C
$T(5) = 59.52°$ C
$T(6) = 80.00°$ C
$T(7) = 71.27°$ C
$T(8) = 65.14°$ C
$T(9) = 61.21°$ C
$T(10) = 59.28°$ C

9–13 $t = 0.395$ s

9–15 $T = 82.4°$ C

9–17 $t = 9.84$ hr

9–19 $t = 1.90$ s

9–21   (a) $T = 37.38°$ C (lumped pa-
            rameter)
       (b) $T = 39.9°$ C (chart)

CHAPTER 10

10–1   (a) $E_b = 1.164 \times 10^4$ W/m²
       (b) Fraction = 0.1316

10–3   Ultraviolet: fraction = 0.1245
       Visible: fraction = 0.3669
       Infrared: fraction = 0.2798

10–7   (a) $F_{1–2} = 0.46$
       (b) $F_{1–2} = 0.06$

10–9   (a) $\dot{Q} = 6.536$ kW
       (b) $\dot{Q} = 4.468$ kW
       (c) $\dot{Q} = 11.00$ kW

10–11 (a) $\dot{Q} = 41.88$ kW
       (b) $\dot{Q} = 80.72$ kW

10–13 $\dot{q}'' = 3.934$ W/m²

10–15 (a) $\dot{Q} = 4.254$ W
       (b) $T = 216.2°$ C

INDEX